The Chemistry of Textile Fibres

The Chemistry of Textile Fibres

Robert R Mather
Mather Technology Solutions, Selkirk, Scotland, UK

Roger H Wardman
School of Textiles & Design, Heriot-Watt University, Netherdale, Galashiels, Scotland, UK

RSC Publishing

ISBN: 978-1-84755-867-1

A catalogue record for this book is available from the British Library

Published by The Royal Society of Chemistry,
Thomas Graham House, Science Park, Milton Road,
Cambridge CB4 0WF, UK

Registered Charity Number 207890

For further information see our web site at www.rsc.org

Preface

Textiles are ubiquitous materials that many of us take for granted in our everyday lives. We rely on our clothes to protect us from the environment, for modesty, and to make us look good. Textiles also enhance our domestic lives, in the form of soft furnishings, such as carpets and curtains, as well as towels and bed linen, *etc*. Some people may also be aware of the uses of textiles in high-performance garments, such as waterproof breathable materials for outdoor wear, or performance sportswear, such as thermoregulating materials. Fewer are aware, however, of the applications of textiles outside of these areas, such as in transport, healthcare, building, marine uses and a whole range of other industrial uses.

What lies behind the fibres and fabrics that possess such a wide variety of properties is often a complex and demanding chemistry. Fibres are polymeric materials and the production of fibres that are suitable for their intended end uses depends not only on the successful optimisation of the formulation of the polymers, but also on their physical chemistry. In turn, the manufacture of many of the synthetic fibres has presented considerable challenges to both chemical and process engineers, and many of the fibres on the market today represent the successful integration of development work by chemists and engineers alike.

The journey – from the first efforts to produce man-made fibres by chemically modifying natural fibres (principally cotton), through the marketing in the 1940s of the first nylons, to the present day breeds of fibres, some of which are many times stronger than steel whilst being a good deal lighter – is a remarkable story. It is a story of chemistry at its

The Chemistry of Textile Fibres
By Robert R Mather and Roger H Wardman
© Robert R Mather and Roger H Wardman 2011
Published by the Royal Society of Chemistry, www.rsc.org

best, and moreover, one that involves all three of the classical branches of chemistry – organic, inorganic and physical chemistry.

What we have tried to do in this book is to give an overview of the various types of textile fibres that are available today – ranging from the natural fibres, right through to the high-performance fibres that are often very technologically advanced. Whilst the focus of the book remains the chemistry of the fibres, we will also investigate the reasons why some individual fibre types are more suitable for certain application than others. The book does not go into advanced detail about the chemistry of formation of the fibres or the intricate technicalities of their production. There are many specialised authoritative textbooks on the market that serve this purpose for the researcher or specialist user. In addition, it is not always possible to give much technical detail about production methods, since these are often kept confidential by the manufacturers. Instead this book is aimed at students following 'A' level courses, or equivalent, at school or technical college, and first-year undergraduate students following textile technology courses at university. A bibliography at the end of each chapter directs the reader to further sources of information.

In Chapter 1, we start with a description of the classes of fibres and indicate the relative quantities produced annually on a global basis. The requirements of polymers in order to have fibre-forming properties are then explained, as well as the general chemical and physical properties necessary for fibres to be commercially viable.

Chapter 2 deals with the natural cellulosic fibres, the so-called 'vegetable' fibres, which include the fibres that can be obtained from the stem, leaf or seed of plants, all of which are cellulosic in nature. Later chapters in the book will explain the sophisticated developments that have led to synthetic fibres with remarkable properties, but it is important not to forget that the humble cotton fibre is the second most produced fibre in the world and as such, still has a major role to play. Chapter 3 covers the other major class of natural fibres – the protein fibres of wool, hair and silk – which are also commercially important fibres in their own way.

The first man-made fibres were produced from naturally occurring polymers, chiefly cellulosic materials. Fibres from these sources were extracted by often complex chemical processes and some of these fibre types, the so-called 'regenerated' fibres, are still important today. These fibres are the subject of Chapter 4.

Synthetic fibres, the fibres made from chemicals derived from oil, are discussed in Chapters 5 and 6. Chapter 5 deals with the methods by which man-made fibres are formed, that is, how the polymers are extruded into the fine fibre filaments required for textile manufacture. It goes on to cover the chemistry of the 'mainstream' synthetic fibres. In Chapter 6, we explain the chemical principles behind the development and manufacture of a selection of high-performance fibres. Chapter 7 covers other specialty fibres, fibres with specific functions, or what might be termed 'functional fibres'. This is an exciting area of development, one that is taking textile fibres into the areas of the so-called 'smart' or 'intelligent' textiles. Leading the way in this class are the nanofibres, but equally exciting in their own way are the electrically conducting fibres, optical fibres (for use in textiles) and electroluminescent fibres.

The point has been made above that the production of textile fibres involves all three classical disciplines of chemistry. The performance properties of textile fibres can be significantly modified by alteration of their surface characteristics and so surface chemistry has an important role to play. In Chapter 8, the techniques that are available for enhancing fibre performance by surface treatment are discussed and examples are given of the effects that can be produced.

Finally, blends of different fibre types are widely used, especially for apparel and for interior furnishings. Chapter 9 explains the reasons for blending different fibre types, typical blends used and their applications.

We are greatly aware of the need for an introductory book on textile fibre chemistry, particularly in light of the recent innovations taking place in this important sector of chemistry. We hope therefore that readers new to textiles will find the book stimulating, and will be inspired to read further. For this reason a short bibliography is given at the end of each chapter.

In the preparation of this book, we are much indebted to Mr Jim McVee, Instrument Technician, and to research students, Miss Uzma Syed and Mr Abdulalhameed Abdrabbo of Heriot-Watt University, for their assistance in preparing microscopic images of fibres. We are especially grateful to Dr Josef Schmidtbauer and Mr Jim Taylor of Lenzing Fibres GmbH for their invaluable advice on viscose and Lyocell fibres. We have also valued the advice of Prof. John Wilson of Heriot-Watt University on photonic crystal fibres and Mr Gavin Dundas on the extent of chemical knowledge that may be expected of students following courses to 'A' level and equivalent standards.

Finally, we are extremely grateful to the Royal Society of Chemistry for giving us the opportunity to share our fascination of textile chemistry with the wider community of chemists. We sincerely hope we have risen to this challenge satisfactorily.

Robert Mather and Roger Wardman

Dr Mather gratefully acknowledges the support of his wife Rosemary.
Prof. Roger Wardman dedicates the book to his late wife Rosemary.

Contents

The Chemistry of Textile Fibres
By Robert R Mather and Roger H Wardman
© Robert R Mather and Roger H Wardman 2011
Published by the Royal Society of Chemistry, www.rsc.org

CHAPTER 1

The Scope of Textile Fibres

1.1 INTRODUCTION

There is a very wide range of textile fibre types available in the market-place. They vary not only in chemical type but also in physical characteristics, reflecting the wide variety of applications they have. Many people relate textiles to apparel and to materials for domestic uses, such as carpets, bedding and soft furnishings, but in fact textiles also have many specialised industrial applications. These textile products are referred to as 'technical textiles' and are produced mainly for their functional and technical performance, rather than their aesthetic characteristics. There are no clear distinctions between apparel and technical applications either, in the sense that the 'performance' apparel market, for which garments are produced to meet specific requirements in terms for example functionality or personal protection, is a hugely important one. Indeed, performance apparel in 2009 represented some 10% of the sportswear market and is forecast to be worth over US$ 4.3 billion by 2012.

For centuries the textile industry was comprised exclusively of the natural fibres, particularly cotton, wool and silk. Indeed, in the UK the textile industry was dominated by wool, because it was not until the beginning of the eighteenth century that cotton began to be imported. Towards the end of the nineteenth century the first 'man-made' fibres were commercialised, these being the regenerated fibres, such as viscose rayon, based on cellulose. The textile industry then made considerable technological advances from the 1930s with the development of many types of commercially important synthetic polymers. In the period since

The Chemistry of Textile Fibres
By Robert R Mather and Roger H Wardman
© Robert R Mather and Roger H Wardman 2011
Published by the Royal Society of Chemistry, www.rsc.org

the emergence of the first synthetic polymers, the nylons and polyesters, considerable development of this class of fibres has taken place, with the aim of producing fibres of very high technical specifications.

1.2   CLASSIFICATION OF TEXTILE FIBRES

It is useful to classify the various types of textile fibres, and since many have similar chemical characteristics the best method of classification is according to chemical type. Before the various chemical groups are established, however, the various fibre types can be broadly classed as either natural or as man-made, as shown in Figure 1.1.

The natural fibres fall into three chemical classes:

- Cellulosics, which are the fibres obtained from various parts of plants, such as the stems (bast fibres), leaves and seeds.
- Protein (keratin) fibres, which are the fibres obtained from wool and hair and silk.
- Mineral (the only naturally occurring mineral fibre is asbestos but its use is banned in many countries because of its toxicity).

The man-made fibres (also referred to as 'manufactured fibres') can also be further sub-divided into three broad groups:

- 'Regenerated' fibres, which are fibres derived from natural sources comprising organic polymers by chemical processing to both extract the fibre-forming polymer and to impart novel characteristics to the resulting fibres.
- Synthetic fibres that are produced from non-renewable sources.
- Inorganic fibres, such as ceramic and glass fibres.

Textile fibre types are given what are called *generic* names and in Europe the organisation responsible for allocating generic names is la Bureau International pour la Standardisation des Fibres Artificielles (BISFA). IUPAC nomenclature does not meet the needs of the textile industry for naming actual fibres and BISFA established a method and published its first list of generic names in 1994. The generic fibre names are based on a common chemical group which imparts characteristic properties, such as:

–CONH–	for polyamides
–COO–	for polyesters
–(CH$_2$–CH.CN)–	for acrylics

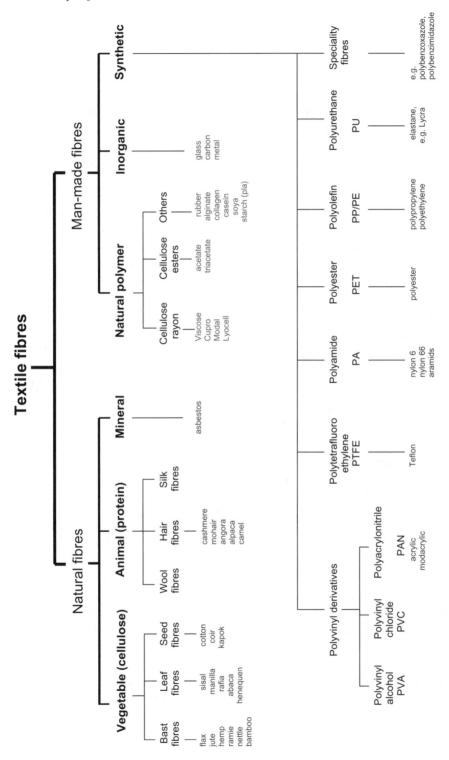

Figure 1.1 Classes of textile fibres.

It is possible that a particular fibre type has two generic names, a prime example being the names *polyamide* and *nylon*, which both cover nylon 6, nylon 46, nylon 66 and nylon 11. Generic names are used extensively in garment labelling and facilitate global trading, avoiding the need for countless chemical names and trade names. There are some instances, however, where trade names are used in garment labels. Often trademarks (symbol®) are used, which can cover a broad range of fibre types and suffix names for a very specific category of a manufacturer's fibres, so that consumers can identify particular qualities or performance characteristics with a specific manufacturer.

The European Commission, in Directive 2008/121/EC, requires (among other things) all textile products to be labelled with an indication of fibre content by reference to the recognised fibre names specified in the Directive. This Directive is up-dated from time to time as new fibres are developed and the fibre names specified in the Directive correspond to those established by BISFA. Each member state is obliged to implement this Directive and in the UK its requirements are enforced under the Textile Products (Indications of Fibre Content) Regulations.

In the USA, the Federal Trade Commission (FTC) assigns generic names and there are instances of different names being used in the USA and the EU for the same fibre types, such as: *elastane* in the EU and *spandex* in the USA, and *viscose* in the EU and *rayon* in the USA. Unfortunately such duplicity can lead to some confusion amongst consumers when buying clothes.

1.3 FIBRE PRODUCTION STATISTICS

The 2008 total worldwide demand for textile fibres amounted to approximately 76.5 million tonnes, distributed among the principal types of fibres as shown in Table 1.1. The total world production has grown steadily by some 20 million tonnes over the 10 years to 2007, as populations have risen and general standards of living have increased, with the consumption of textiles rising from 7.5 to 9.7 kg capita^{-1}. In the early 1990s, roughly equal amounts of natural and synthetic fibres were produced, but now a greater proportion of synthetic fibre is produced, mainly due to a rapid increase in the demand for polyester. However, the global recession has had an impact over the last two years and world fibre consumption has fallen back by some 6.7%, with all fibre types suffering a decline, though to different extents. The reasons for the decline are not just due to a decrease in sales of clothing, but also due to a decline in car manufacture, for which textiles are extensively used, and a decreased demand for home furnishings, as house sales have fallen back.

Table 1.1   Worldwide demand of some textile fibres (thousand metric tonnes), 2008.

Synthetic fibres		Regenerated fibres		Natural fibres	
Polyester[a]	30 650	Cellulosic[a]	2 545	Cotton[a]	24 442
Polyamide[a]	3 511	Lyocell[d]	150	Wool[a]	1 209
Acrylic[a]	1 913			Silk[a]	148
Polypropylene[a]	5 939			Jute[b]	3 250
Spandex[a]	612			Linen[b]	772
Aramid[c]	69			Ramie[b]	250
Carbon fibre[c]	38			Hemp[b]	68
				Coir[b]	954
Total	**42 732**		**2695**		**31 093**

[a]Sources: *Fiber Organon*, 2009, **80**(6), 95–112.
[b]A. G. Saurer, *A World Survey on Textile and Nonwovens Industry*, 2006, 6, (data for 2005).
[c]*Textile Month Int.*, 2009, 4, 2–5.
[d]estimated.

Table 1.2   Top ten synthetic fibre producing countries, 2008.

Country	Percentage share of global production
China	54.8
United States	7.2
Western Europe	6.7
India	5.7
Taiwan	5.0
South Korea	3.5
Indonesia	3.0
Japan	2.3
Turkey	2.2
Thailand	1.7

Source: *Fiber Organon.*, 2009, **80**(6), 95–112.

It is noteworthy that nearly all of the regions of the world are involved in fibre production, though over the last decade a considerable amount of synthetic fibre production has shifted to China (see Table 1.2). Whereas in 1994 13.5% of world synthetic fibre production was in China, by 2008 that proportion had increased to nearly 55%. Over the same period, the proportion manufactured in the USA decreased from 17.7% to 7.2%. It is in the polyester manufacturing sector where the growth rate in China has been so fast: in 1990 the Chinese share of this market was just 12%, by 2000 it had risen to 27%, but by 2008 the share was 66%.

1.4   CHARACTERISTICS OF TEXTILE FIBRES

Textile fibres are perhaps most obviously characterised by their fineness; they are long and very thin. There are numerous fibrous structures in nature, but only those that can be converted into yarns are suitable for constructing textile fabrics. Most natural fibres exist as staple fibres, ranging from 2–50 cm in length and 10–40 μm in cross-section. Staple fibres have to be converted into yarns by a spinning process, which requires that the individual fibres possess some degree of surface roughness, in order that they adhere to one another in the yarn. Spinning of staple fibres is also promoted by fineness and consistency of fibre cross-section.

Synthetic fibres, on the other hand, are produced as continuous filament yarns. So too is silk, produced from silk-worms (and spiders!). Nevertheless, continuous filament yarns are sometimes cut, to convert them to staple fibres of a required length. The yarn produced from this staple is softer and 'fuller'. Such yarn is used for knitwear and carpets, and for blending with natural fibres. Figure 1.2 schematically illustrates continuous filament and staple yarns.

The geometry and morphology of fibres are also important characteristics. Many natural fibres possess quite complex morphologies. A cotton fibre is ribbon-like but with many convolutions throughout its length (*ca.* 10 per mm). As described in Chapter 2, each cotton fibre is structurally divided into concentric zones, and there is a hollow central core. The surfaces of wool fibres consist of overlapping scales. Thus, the friction in the root-to-tip direction along the surface of each fibre is much greater than that in the opposite direction. Continuous filament yarns may possess cross-sections of quite complex shape. Whereas most filaments are produced with a circular cross-section, some are produced

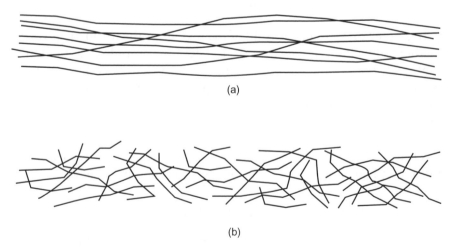

(a)

(b)

Figure 1.2   Schematic illustrations of (a) a continuous filament, (b) staple yarns.

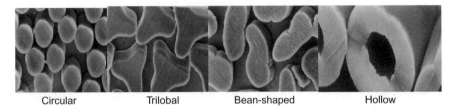

| Circular | Trilobal | Bean-shaped | Hollow |

Figure 1.3 Cross-sectional shapes of man-made fibre filaments. (Photographs courtesy of BISFA).

with cross-sections that are triangular, multilobal, bean-shaped, or even hollow, as illustrated in Figure 1.3. Nevertheless, many of the major properties that a textile requires to be successful in application are strongly influenced by the chemical architecture and properties of the polymers comprising the textile fibres. It is, therefore, necessary to start by surveying the nature of fibre-forming polymers.

1.5   REQUIREMENTS OF FIBRE-FORMING POLYMERS

Although accounts of individual polymers are given later in this book, some general observations can be usefully made here. Because textile fibres possess a very high ratio of length to cross-section, so too must the molecules that comprise them. Thus, the component polymers have to be linear. Whereas polymer chains consisting only of bifunctional repeating units are linear, the presence of trifunctional units results in branched chains. If the concentration of trifunctional units is too high, then cross-linking of the chains occurs, giving rise to a three-dimensional network.

Flexibility of the linear polymer chain is also important. It reflects the capacity of the chain to change its conformation, and arises from rotation around single bonds in the chain backbone. The ability to change from one conformer to another is determined by the potential energy barrier between them, which in turn is governed by the nature of the bonds in the chain backbone and also by the side-groups. Rigid entities in the chain, notably aromatic rings, lead to a considerable increase in chain rigidity.

Intermolecular attractions between chains affect the strength of a fibre and also its ability to bend. Although intermolecular attractions exist between polymer chains of any chemical construction by virtue of van der Waals forces, strong dipoles and groups that can form hydrogen bonds significantly increase the strength of these attractions, and thereby affect fibre strength and fibre stiffness. Thus, nylon chains are held together through extensive hydrogen bonding between $>C{=}O$ and $>N$-H groups in adjacent chains, as shown in Figure 5.7 in Chapter 5. Furthermore, ionic bonds (salt bridges) are formed between polymer chains in protein fibres. Perhaps the strongest intermolecular links amongst fibre-forming polymers are those that exist in wool, where the keratin chains are substantially held together by covalent cross-links.

Amongst synthetic fibres, chain flexibility and intermolecular attractions profoundly influence melt temperature, T_m. They also influence glass transition temperature, T_g, the temperature at which the polymer softens and becomes more rubbery.

The lengths of the constituent polymer chains are also important for the properties of the fibres they comprise. If the chains are too short, either the polymer cannot be extruded into fibres, or the extruded fibres are too weak. If the chains are too long, they tend to become more entangled. Entanglement can reduce the ability of the chains to crystallise and hence reduce the fibres' strength. However, no polymer is composed solely of chains of one particular length; there is a distribution of chain lengths and indeed this distribution of lengths can also affect some fibre properties. In fact, it is customary to characterise a polymer by at least two average molar masses, the types of average molar mass determined depending on the technique adopted. Where in effect the number of polymer chains of each size is counted, a number average molar mass, M_n, is determined. If instead the method depends on the size of the constituent polymer chains, a weight average molar mass, M_w, is determined. The ratio of the two, M_w/M_n, is called the molar mass dispersity and provides an indication of the breadth of the distribution of molar masses. Typical values of M_n and M_w for polypropylene fibres are shown in Table 1.3 and a typical molar mass distribution is illustrated in Figure 1.4. Another parameter often reported for polymers is the degree of polymerisation (*DP*), which represents the average number of repeat units in a polymeric chain. In fibres, *DP* ranges from 500 to 10 000.

Other specifications for polymers are also reported in the trade literature. Grades of polyester are often denoted by their intrinsic viscosity, determined in a suitable solvent. However, grades of polyethylene, for example, are generally distinguished by melt flow index

Table 1.3 Values of M_n and M_w for two samples of polypropylene.

M_n (number-average molar mass)	M_w (weight-average molar mass)	M_w/M_n
34 000	261 000	7.7
42 000	211 000	5.0

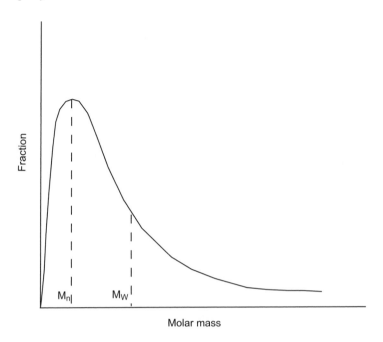

Figure 1.4 A typical distribution of molar masses for a synthetic fibre.

(*MFI*). The *MFI* is the mass of polymer in grams extruded in 10 minutes at 190 °C under an applied load of 2.16 kg using particular specified equipment. Thus, an increase in M_w generally gives rise to a reduction in *MFI*. However, there is no simple relation between *MFI* and M_w.

It is important too that the polymer chains are oriented more or less in the direction of the fibre axis. In many natural fibres, the chains exist in the form of helices whose central axis lies parallel to the fibre axis. This helical structure can exist in some synthetic fibres (polypropylene fibres are a good example), but in other synthetic fibres the chains adopt a more zig-zag configuration. In a number of high-performance synthetic fibres, the chains are almost exactly aligned in the direction of the fibre axis.

For many textile fibres, the ability of the constituent polymer to crystallise is desirable, in that crystallisation can add strength to the fibre. Forces of attraction between neighbouring polymer chains hold them together, and so bring cohesion to the fibre structure and maintain chain orientation. The ability to crystallise is considerably enhanced if there are regular repeating units in the polymer. If the repeat units of the polymer contain an asymmetric centre, then regularity of configuration is also required, as in the polymeric chains comprising polyacrylonitrile,

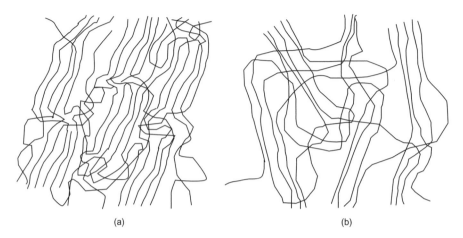

(a)   (b)

Figure 1.5   (a) Fringed micelle and (b) fringed fibril structures of fibrous polymers.

polyvinyl chloride and polypropylene fibres. In any crystalline region, only a segment of each polymer chain is usually accommodated. Indeed, different segments of a chain may lie in different crystalline regions. Figure 1.5 illustrates schematically the origins of crystallinity in a fibre. The two models shown are (a) the fringed micelle and (b) the fringed fibril structures, the latter being the currently preferred model.

Some degree of chain movement is, nevertheless, often desirable to provide sufficient extensibility to a fibre, and to allow access of moisture, dyes and any other additives required. This goal may be achieved by disrupting the crystallinity of the polymer chains through the deliberate introduction of units of different constitution. This strategy is used with acrylic fibres, for example. Copolymerisation of acrylonitrile with small amounts of other monomers serves to reduce T_g and to provide sites at a lower temperature for attracting dye molecules. Units of different constitution may, however, be unwanted, as with the presence of impurities in the precursor monomer or with side reactions, although low concentrations of these units are unlikely to suppress significantly the capability of the polymer to crystallise.

1.6   PROPERTIES OF TEXTILE FIBRES

It has already been noted (Preface) that textile products encompass a huge range of applications, and this diversity of applications requires fibres with vastly differing properties. Thus, for clothing, softness, flexibility, absorption of moisture and the ability to be dyed are key characteristics. Industrial textiles, on the other hand, are constructed for

Table 1.4 Important fibre properties.

Property	Examples
Mechanical	Tension
	Compression
	Fatigue failure
Thermal	Melt temperature
	Thermal decomposition
	Glass transition temperature
	Flammability
Electrical	Electrostatic charging
	Electrical conductivity
Optical	Refraction
	Absorption
	Reflection
Surface	Wetting
	Adhesion
	Friction
Biological	Resistance to microorganisms
	Toxicological
	Biocompatibility
	Controlled degradation *in vivo*

much more demanding applications: strength, a greater rigidity and fatigue resistance assume particular importance. Wear in brake linings or failure in a seat belt or air bag in a vehicle accident may cause fatalities. Such failure in everyday clothing is, arguably, less serious!

A range of fibre properties should be taken into account when a particular end use is considered, and the relative importance of each property will be governed by that end use. Table 1.4 lists some important fibre properties. For some end-use properties, however, there is greater influence from the way in which the fibres are assembled and from the treatments to which the end product will be subjected. In addition, the choice of fibre will be influenced by the fabrication stages necessary for the end product to be achieved. Each property listed in Table 1.4 is now considered briefly in turn.

1.6.1 Mechanical Properties

It is customary amongst textile scientists and technologists to evaluate many fibre mechanical properties from stress–strain curves. A progressively increasing stress is applied along the axis of the fibre under investigation. The stress causes elongation of the fibre. A stress–strain curve is created by plotting stress against strain, where strain is defined as the fractional elongation of the fibre. The stress–strain curve for a model fibre is shown in Figure 1.6. Different types of fibre produce

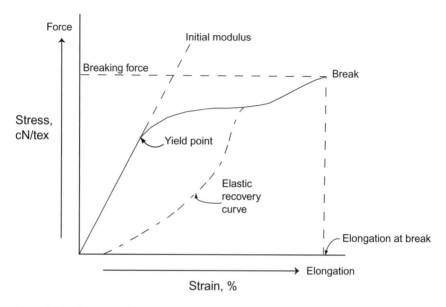

Figure 1.6 Stress–strain curve for a 'model' fibre.

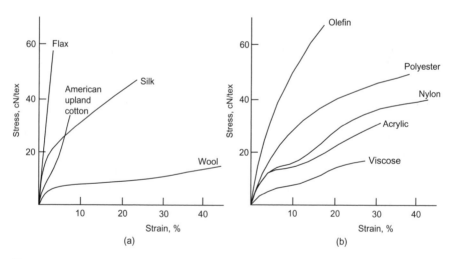

Figure 1.7 Stress–strain curves for (a) natural fibres, (b) synthetic fibres.

different stress–strain curves, as illustrated in Figure 1.7. The nature of each curve is profoundly influenced by the structure of the fibre.

The values of a number of important mechanical properties can be obtained from stress–strain curves. One of these properties is fibre tenacity, the stress applied to the fibre at its breaking point. Figure 1.7 shows that flax and polypropylene fibres have high tenacities, compared with those of wool and acrylic fibres. Differences in fibre tenacity can generally be attributed to several factors: the degree of polymerisation, the strength of bonding between adjacent polymer chains, their degree of orientation in the direction of the fibre axis and the extent of crystallinity.

Another important property is elongation to break, the strain that has been undergone by the fibre when it breaks. Fibres that possess higher crystallinity, higher chain orientation and strong inter-chain bonding will generally exhibit lower values of elongation to break. Thus, wool fibres possess considerably greater extensibility than flax fibres (see Figure 1.7a). When fibres are stretched, it is desirable that they recover their original length. For small amounts of stretch most fibres will re-cover by almost 100%, but over longer extensions it is quite common for them to not recover fully. The consequence of this is that garments can become baggy around joints such as elbows or knees, where repeated stretching often occurs. Thus the parameter of elastic recovery is important. Often there is a hysteresis pattern to the stretching – relaxation modes of measurement, as illustrated in Figure 1.6.

Initial modulus, which is a convenient measure of a fibre's stiffness, is the slope of the stress–strain curve at its origin. It provides a useful indication of the ease with which a fibre extends under a particular stress. Thus, nylon is well suited for women's hosiery, as it stretches considerably even under low stress. By contrast, in a rope, for example, there must be low stretch even under very high stress, so fibres with a high initial modulus would be essential in a rope.

Work to rupture, or toughness, is another important measure, this being the total energy input up to the breaking point of the fibre. Work to rupture is determined from the area under the stress–strain curve. It is used as a measure of a fibre's ability to withstand high amounts of energy, such as would be required when a parachute opens or when a seat belt comes into operation.

It should be noted that stress–strain curves are normally measured at moderate rates of fibre extension. However, useful though these measurements are, they may not always be reliable indicators of mechanical performance, for example, performance under impact (short durations) or under continuous loading (long durations). In addition, some properties are time-dependent, such as creep, which is a progressive increase in strain when the fibre is subject to a continuous stress.

1.6.2 Thermal Properties

The thermal properties of a textile also have a practical impact. When the temperature is high enough, textile fibres either melt or begin to degrade. This aspect is particularly important in textile processing, as in the melt extrusion of synthetic fibres (discussed in Chapter 5) and in end use, as in the behaviour of a fabric on ironing or of performance fabrics for fire safety clothing. Natural fibres are generally decomposed by heat. Thus, cotton turns yellow after several hours at 120 °C and is markedly degraded in air at 150 °C. Wool turns yellow at 130 °C, and silk is decomposed at *ca.* 170 °C. A few common synthetic fibres, such as acrylic fibres, are also degraded by heat. Many, however, are produced by melt extrusion, so the ability to exist in the melt phase without significant decomposition is a key requirement. Most high-performance fibres have been developed to withstand high temperatures with little reduction in mechanical performance.

Another significant property of synthetic fibres is the glass transition temperature (T_g.). When a fibre is heated above its T_g, its internal structure is altered from glass-like to rubber-like. Above T_g, the distance between adjacent polymer chains in the non-crystalline regions of the fibre is now large enough to render the interchain bonding forces ineffective. Thus, the strength of interchain bonding plays a key role in governing the value of T_g. The value of T_g can have important practical implications. Above T_g, many synthetic fibres can be readily dyed, because they become more accessible to incoming dye molecules. On cooling below T_g, after the dyeing process, the dye becomes restrained within the fibres.

The flammabilities of textiles hold the key to their application as well, and many flame-retardant materials have been developed that considerably reduce flammability in textiles. Ignition of textile fibres is a complex process, and no one laboratory test can properly determine the ignition hazard of a particular textile. A common method of comparing flammabilities is through determination of the limiting oxygen index (LOI). LOI can be represented as the minimum fraction of oxygen in an oxygen/nitrogen atmosphere required to sustain combustion of the fibres after ignition. The higher the index, the more resistant to ignition the fibres are expected to be. However, LOI relates to 'candle-like' burning, which far from always gives a realistic indication of the burning properties of a particular textile fabric.

1.6.3 Electrical Properties

Electrical properties of fibres are particularly important in relation to static electrification, arising from their low electrical conductivity. The accumulation of static electricity can be problematic in fibre processing, for example in the repulsion of warp yarns prior to weaving and the adherence of yarns to processing machinery. Static can also cause problems in the behaviour of end products, as in the soiling of garments and the unpleasant – and even hazardous – effects arising from electrical discharges.

The severity of the problem can, in practice, be linked to the extent of moisture in a fibre. The greater the moisture content, the greater is the electrical conductivity. It is customary to assess a fibre's capacity to absorb moisture by determining its moisture regain, which is defined as:

$$(\text{Mass of absorbed water in fibre}/\text{mass of dry fibre}) \times 100\%$$

at 20 °C and 65% relative humidity. Many natural fibres possess high moisture regain values, as shown in Table 1.5, and so in these fibres, static electrification presents little problem. By contrast, in many synthetic fibres, whose moisture regain is much smaller, the problem of static electrification is particularly severe. Even in nylon, which possesses one of the highest moisture regains amongst common synthetic fibres, the electrical conductivity is still too low to suppress static electrification in clothing.

The practical approach normally adopted to dissipate static charges is to develop means of increasing electrical conductivity in fibres and fibre assemblies. In fibre processing, charge dissipation is achieved by coating the fibres with a suitably formulated spin finish. Spin finishes are normally applied as emulsions and, in addition to antistatic agents, contain lubricants and biocides. For the end products, a number of approaches are available and these are normally adopted during fabric construction. An everyday example is the blending of polyester and cotton fibres for shirt fabrics. Alternatively, suitable finishing treatments can eliminate static charges and examples are given in Section 8.3.3.5. In the lifetime of the end product, fabric conditioners also serve to reduce static charges. Many fabric conditioners are based on quaternary ammonium salts

Table 1.5 Moisture regain values at 20 °C and 65% relative humidity.

Fibre	Moisture regain, %
Cotton	7–8
Wool	14–18
Silk	10–11
Viscose	12–14
Polyester	0.4
Nylon 6,6	4.1
Acrylic	1–2
Polypropylene	0

containing long aliphatic hydrocarbon chains. They are designed to impart a thin hydrophobic layer to the surface of a fibre, but also one that is electrically conductive by virtue of their cationic nature (see Section 8.3.3.2).

Some types of fibres exist with high electrical conductivities, and there are several ways of producing them. They can be processed from electrically conducting polymers, such as polyaniline, polypyrrole and polythiophene, and can achieve conductivities in the order of up to $100 \, S \, cm^{-1}$. (For comparison, the conductivities of copper and graphite are *ca.* 6×10^5 and $700 \, S \, cm^{-1}$ respectively). Other approaches include coating conventional fibres with metals or conductive polymers. Electrically conducting fibres are discussed in detail in Section 7.2.

1.6.4 Optical Properties

The appearance of a textile is, of course, essential for outlets such as clothing and household furnishings. It is also becoming increasingly important for more technical applications of textiles. To enhance their appearance, colour is imparted to textile fibres. Coloration of textiles is outside the scope of this book and is the subject of a companion volume.

The appearance of a textile is determined by the interaction of the textile fibres with visible light. Light that falls onto a fibre may be partly refracted, absorbed or reflected. Each of these aspects plays a part in determining the appearance of a fibre, although the way in which the fibres are assembled to form a fabric also has a significant role. It is difficult to relate the appearance of a fibre to fundamental optical properties, and it is even more complicated for fibre assemblies. Nevertheless, it is important to identify those features of fibres that may contribute to appearance and which, therefore, need to be carefully controlled during production and processing.

The reflection of light from a fibre will depend on the angle of incidence of the light and the refractive index of the fibre. The refractive index will in turn depend on the organisation and chemical constitution of the polymer chains comprising the bulk of the fibre. However, the refractive indices of most fibres fall within quite a narrow interval, 1.45–1.60, though polyester has the unusually high value of 1.72. It is aspects of fibre geometry, therefore, (*e.g.* surface roughness and cross-sectional size and shape) which, by virtue of their effects on the distribution of angles of incidence, assume the dominant role in reflectivity.

Light entering the bulk of a fibre is refracted. However, because of the highly anisotropic nature of textile fibres, the refractive index in the direction of the fibre axis, $n_{//}$, is different from that across the fibre, $n_\perp$.

Optical birefringence, $(n_\parallel - n_\perp)$, is widely used to evaluate the overall orientation of the constituent polymer chains. The degree of orientation, f, can be determined from the ratio of the birefringence of the fibre to that of an equivalent idealised fibre, whose polymer chains are all oriented along the fibre axis. Thus:

$$f = (n_\parallel - n_\perp)/(n'_\parallel - n'_\perp),$$

where $n'_\parallel$ and $n'_\perp$ correspond to the idealised fibre. It follows, therefore, that for an ideally oriented fibre, $f = 1$, and that for a completely isotropic fibre exhibiting no birefringence, $f = 0$.

An important aesthetic property of textile fabrics is their lustre. Light may be reflected from the constituent fibres in a *specular* manner, in which the angle of reflection equals the angle of incidence, or *diffusely* with varying intensity over a wide range of angles, or through a combination of both. The visual appearance of the fabric arising from these reflections determines its lustre.

Optical fibres have assumed considerable prominence in the transmission of data over long distances. Many optical fibres are made of glass though some, for transmission of data over short distances, are composed of polymers (see Section 7.3). For communication, polymer optical fibres are much cheaper to produce and install than glass fibres.

For textile fibres that are to be used in certain environments, their long-term stability depends very much on their resistance to the action of light. Their photostability is influenced not just by the light itself (its intensity and wavelength distribution) but also by the combined actions of oxygen, moisture and heat. Photostability is an important consideration for fibres in a wide range of applications, from outdoor uses such as awnings and tarpaulins, to fabrics used in car interiors and to home textiles such as curtains, carpets and other soft furnishings.

Cellulosic fibres are relatively stable to near-UV radiation and visible light. Intense radiation will reduce the *DP* and increase the number of carbonyl and carboxyl groups, with a consequent loss in tensile strength. Wool fibres are much less stable to light, with a very noticeable yellowing occurring on exposure to near-UV radiation. Wool keratin is a complex polymer formed from 18 different α-amino acids (see Chapter 3) and these various amino acid residues decompose in different ways, and according to the wavelength of irradiation.

Of the synthetic fibres, nylon is fairly susceptible to photodegradation by sunlight which limits its use for outdoor applications. It is likely that scission of the amide bond occurs in an initiation stage, followed by reaction of the carbonyl and amine radicals with the methylene groups, in reactions possibly involving oxygen. The result of the degradation is a loss in tensile strength and significant yellowing. Nylon fibre intended for outdoor uses contains light stabilisers, which are added at the polymerisation stage of manufacture.

Polyester undergoes rapid photodegradation when exposed to direct sunlight because of its absorption in the range 290–310 nm. Behind glass, which filters out most of these wavelengths, polyester is much more resistant, and so it is suitable for applications in car interiors for example. In general, acrylic fibres show excellent resistance to photo-degradation and these fibres are widely used for outdoor structures such as awnings, tarpaulins, car covers, *etc.*

The chemical processes that occur during photodegradation are complex and the presence in the fibres of dyes and other additives such as delustrants can also make fibres more sensitive to light. It is common practice for manufacturers to add a delustrant to increase the opacity of their fibres. Typically the white pigment titanium dioxide is used, either as the rutile or the anastase form, though anastase is pre-ferred because it is less abrasive. However when applied at the 2% level required to give what are termed 'dull' fibres, it has a serious effect on the breaking strength of both nylon and polyester, as illustrated in Table 1.6.

Some dyes increase the sensitivity of fibres to photodegradation, whereas others have a photoprotective effect. Cotton and other cellu-losic fibres dyed with vat dyes can be highly sensitised, leading to the breakdown of the cellulose chain through the action of light, moisture and heat, which is called 'phototendering'. In outline, the photo-activated state of the dye is quenched by oxygen and the 'activated' oxygen then attacks the cotton, either directly, or through the action of hydrogen peroxide formed by the reaction of the activated oxygen with moisture. The overall mechanism can be represented by the scheme:

$$\text{Dye} + \text{h}\nu \rightarrow \text{Dye}^*$$
$$\text{Dye}^* + \text{O}_2 \rightarrow \text{Dye} + \text{O}_2^*$$

Table 1.6 Effect of TiO_2 on the breaking strength of nylon 6,6 and polyester fabric after exposure to sunlight behind glass, expressed as a per-centage of the value for the unexposed fabric.

	Nylon 6,6	*Polyester*
0% TiO_2	75	88
2% TiO_2	10	27

Then:

(a) Under dry conditions $\quad$ $O_2^* + \text{cell-OH} \rightarrow$ oxidation products
(b) Under moist conditions $\quad$ $O_2^* + 2H_2O \rightarrow 2H_2O_2$
$\qquad\qquad\qquad\qquad\qquad$ $H_2O_2 + \text{cell-OH} \rightarrow$ oxidation products

1.6.5 Surface Properties

The surface properties of a fibre have particular technological import-ance. They govern the way in which yarns interact with one another in a fabric, and they play a major role in terms of resistance to abrasion. They also affect fibre friction, a key factor both in the production and behaviour of fabrics and in the processing of staple fibres into yarn. The diffusion of liquids through fibres, the transport of liquids through fibre assemblies, and the soiling and cleaning of fibres are all strongly influ-enced by the surface energy of the fibres. Surface properties are also relevant to the adhesion of coatings to fibres and to the bonding of fibres to a matrix in a composite material, and they can influence the bio-compatibility of a fabric.

All these aspects are influenced by the chemistry and morphology of the individual textile fibres. In addition, it is often the case that the composition of the surface of a fibre differs markedly from its bulk composition. Fibre surface properties are often deliberately modified by temporary or permanent coatings, in order to confer durability to the fibre during processing or end use. Some natural fibres inherently pos-sess distinctly different surface and bulk compositions: cotton is one such example, as shown in Table 2.1. Moreover, the physical structure of the surface of a synthetic fibre is often different from that of its bulk: this is the so-called 'skin-core' effect. Surface properties of textiles and their modification are discussed in Chapter 8.

1.6.6 Biological Properties

Biological properties comprise a number of aspects: resistance to attack by microorganisms and insects, biocompatibility and even controlled degradation *in vivo*. Attack by microorganisms and insects results in fibre discoloration and degradation. The microorganisms in question include bacteria, mildew, fungi – and even algae. Warmth and high humidity accelerate their action. In general, cellulosic fibres are the most prone to attack by microorganisms. Protein fibres are less prone, and synthetic fibres are almost completely resistant, unless they contain impurities or additives that are susceptible. Amongst insects, clothes moths and several types of beetle can degrade wool fibres, whilst silverfish attack cellulosic fibres. Synthetic fibres are resistant to insects.

Textile fibres are being used increasingly in medical and healthcare applications. Table 1.7 categorises these applications and Table 1.8 provides examples of the types of fibres utilised. One important feature of any textile to be selected for a biomedical application is that it should be biocompatible. The textile must not invoke any deterioration or degradation of living cells with which it comes into contact, nor should it trigger any adverse immune reaction. The surface features of the textile fibres are particularly important in the way cells respond to them.

Resorbable fibres are becoming increasingly important in medical device technology. They can be specifically designed and developed to retain their mechanical properties *in vivo* for just a predetermined

Table 1.7   Types of biomedical textile.

Product	Examples
Non-implantable	Bandages, plasters, wound dressings, pressure garments, prosthetic socks
Implantable	Artificial tendons and ligaments, artificial skin, hernia patches, sutures, vascular grafts
Extracorporeal devices	Artificial kidneys, lungs and livers
Protective & healthcare	Cloths, wipes, bedding, surgical gowns
Hygiene products	Sanitary wear, nappies, incontinence pads, dental floss/tape

Table 1.8   Textile fibres used in medicine.

Textile fibre	Medical applications
Synthetic	
Polyamide	Wound dressings, compression bandages, sutures, surgical hosiery
Polyester	Orthopaedic bandages, artificial kidneys, sutures, artificial tendons and ligaments, cardiovascular implants
Polypropylene	Orthopaedic bandages, sutures, mechanical lungs
Polyethylene	Wound dressings, artificial tendons and ligaments, orthopaedic implants
Natural	
Cotton	Wound dressings, bandages
Silk	Wound dressings, sutures, artificial tendons
Regenerated	
Viscose rayon	Wound dressings, bandages, artificial kidneys and livers
Carbon fibre	Hernia patches
Resorbable fibres	Sutures, controlled drug delivery

period. The most widely used resorbable fibres are polylactic acid and polyglycolic acid and their copolymers. *In vivo* they are progressively hydrolysed. Other resorbable fibres include polycaprolactone co-polymers and polydioxanone and its copolymers.

Resorbable textiles are used as scaffolds after surgery, for example, to enable tissue regeneration and replacement. In many cases, the scaffold eventually becomes redundant and would have to be removed by further surgery. Scaffolds constructed from biodegradable fibres, however, progressively degrade, and surgery for removal of the scaffold is rendered unnecessary. In some instances, permanent textile scaffolds are required after surgery: these may be constructed from synthetic fibres. Scaffolds consisting of both permanent and temporary components have also been constructed. Resorbable fibres are discussed further in Section 7.5.

SUGGESTED FURTHER READING

1. The European Commission website gives an overview of the European textile industry, including information on trade statistics, policy on environmental issues as they relate to the textile industry, the impact of REACH legislation, together with links to relevant EU publications: http://ec.europa.eu/enterprise/textile/index_en.htm
2. The website of Bureau International pour la Standardisation des Fibres Artificielles (BISFA), the international association of man-made fibre producers which establishes the terminology of man-made fibres, is a useful source of information: www.bisfa.org
3. S. B. Warner, *Fiber Science*, Prentice Hall, New York, 1995.
4. W. E. Morton and J. W. S. Hearle, *Physical Properties of Textile Fibres*, Woodhead Publishing, Cambridge, UK, 2008.

CHAPTER 2

Cellulosic Fibres

2.1 INTRODUCTION

Cellulosic fibres are derived from plants and can be obtained from the
three principal component parts of any plant – the seed, the stem (the
so-called 'bast' fibres) and the leaf. The range of plants from which
cellulosic fibres are produced is extensive, from nettles to shrubs to full-
grown trees. Whilst the chemistry of cellulose was not well understood
until the 1930s, chemical processes for cellulose were developed in the
late nineteenth century and early twentieth century that led to variants
such as viscose and the acetate fibres. As a consequence, there is a vast
array of cellulosic fibres on the market today of widely ranging characteristics and properties. These cellulosic man-made fibres are covered in
Chapter 4. In this chapter, the natural cellulosic fibres, the seed, bast and
leaf fibres, are discussed.

2.2 SEED FIBRES

The seed fibres are cotton, coir and kapok. Coir is the seed hair obtained
from the outer shell of the coconut. They are brown fibres, very coarse
and resistant to abrasion. Consequently coir is used mainly for brushes,
matting and cordage. Kapok is obtained from the fruit of the kapok tree
and is yellowish-brown and slightly lustrous in appearance. However it
is a very brittle fibre and has little practical textile use. The rest of this

The Chemistry of Textile Fibres
By Robert R Mather and Roger H Wardman
© Robert R Mather and Roger H Wardman 2011
Published by the Royal Society of Chemistry, www.rsc.org

chapter is devoted to the cotton fibre, by far the most commercially important seed fibre.

2.2.1 Growth and Morphology of Cotton

Cotton is a fibre obtained from the seed of a species of plant of the genus *Gossypium*, which belongs to the *Malvaceae* order and grows wild in many parts of the world. The fibres have been used to make textiles for many centuries, indeed millennia, most notably in China and India, though it was not until the eighteenth century that cotton began to be imported into the UK.

The shrub is now cultivated in many countries, the main producers in terms of land area being India (9.4 million hectares), China (6.0 million hectares), USA (3.1 million hectares), Pakistan (2.9 million hectares) and Brazil (0.9 million hectares). Other producers include Russia, Mexico, Sudan, Egypt, Turkey and Australia. Cotton is quite sensitive about the conditions it requires for growth, favouring warm, humid climates and land that comprises a fairly sandy soil. The growing conditions have a huge influence on the quality of the fibres obtained, the quality being governed by the fineness and the length of the fibre filaments. The highest quality fibres are the Sea Island and Egyptian cottons, with fibre lengths varying between 25–65 mm. This quality is difficult to grow and is therefore more expensive. What might be termed the 'standard' cottons, those with fibre lengths of between 13–33 mm, are typically from the Americas ('American Upland' cotton). The varieties produced in the Asian countries generally have shorter fibre lengths of between 10–25 mm.

The way in which the cotton seed grows leads to fibres which have unusual morphological features. During growth, after the blossom falls off, dark green triangular pods called bolls form and within these bolls there are about 20 seeds. The seed hairs (the cotton fibre) gradually develop and on maturity the boll bursts open (see Figure 2.1), liberating the seeds covered in white cotton fibres. The cotton must be picked as soon as the bolls open otherwise they can become spoiled by weather.

As the seed hairs dry out, the shrinking that takes places results in a bean-shaped cross-section, indicating that the density of packing of the cellulose chains within the fibre is not uniform. The morphology of the fibres is complex in that there are a number of layers, so perhaps it is not surprising that the packing of the cellulose molecules within the fibres is inconsistent. There are four main parts to the fibre structure: the cuticle, the primary wall, the secondary wall and the lumen.

Figure 2.1   Cotton boll after opening.

The cuticle is a waxy outer layer which serves as a protective covering. It is water repellent and also provides resistance to attack from aqueous solutions. It is necessary to remove this waxy layer using detergents before cotton can be dyed, otherwise dyes cannot diffuse into the fibres. The cuticle is a very thin layer (just a few molecules thick) and immediately underneath it is the primary layer which is about 200 nm thick. The primary layer consists of fibrils of cellulose, each about 20 nm thick, which are arranged in a spiralling network along the fibre length. The secondary wall makes up the bulk of the cotton fibre and consists of several layers of cellulose fibrils, each about 20 nm thick, that spiral along the fibre axis. There are essentially three layers of the secondary wall, labelled S_1, S_2 and S_3 (see Figure 2.2).

In these layers, the angle at which the fibrils spiral varies, from about 20° in S_1 near to the primary wall, to around 45° in S_3 near to the lumen. Unlike the fibrils in the primary wall which have a consistent direction, the fibrils in the secondary layers show a reversal of twist from the S to the Z direction. It is this spiralling of the fibrils along the fibre axis that gives the cotton fibres their inherent high strength. Prior to opening, the cotton fibres develop in a tubular form within the boll, with a canal running down their centre, called the lumen. The lumen was the main pore through which a sap, made up of a dilute solution of sugars, proteins and minerals could pass, during the growth stage. When the boll bursts open, the sap in the seed hairs dries out, leaving residues of mainly proteins and minerals, and the lumen decreases in size. The fibres collapse from a circular cross-sectional shape to a bean shape with a hollow centre and the fibres become flatter and convoluted along their length (see Figures 2.3a and 2.3b, respectively).

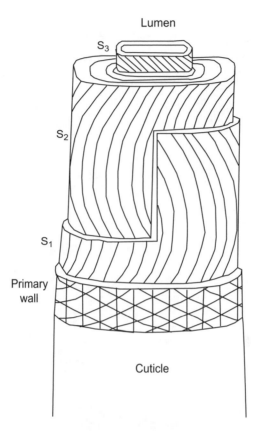

Figure 2.2 Schematic representation of cotton morphology.

As a naturally occurring fibre, cotton is subject to attack by pests and disease, any of which may prevent the cotton from growing to its full maturity. These fibres are termed 'dead cotton' and they are characterised by having virtually no secondary wall structure. They have a very thin ribbon form and easily entangle into knots which are called 'neps'. It is difficult to remove these neps from mature cotton and because they tend to reflect light off their top surface they appear as white spots on dyed cotton fabrics. Another problem is that, in addition to the normal length cotton fibres, cellulosic fibres of much shorter length called cotton 'linters' also grow in the bolls. These shorter fibres are unsuitable for spinning so they are removed from the normal cotton fibres. However cotton linters are a useful additional source of cellulose and are used typically for the production of paper, viscose and acetate rayon fibres (see Chapter 4).

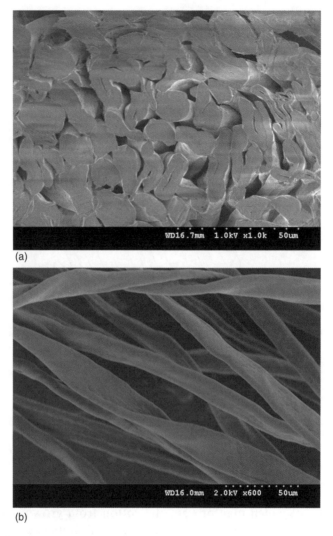

(a)

(b)

Figure 2.3 (a) Cross-sectional and (b) longitudinal views of cotton fibres. (Photographs courtesy of U. Syed, Heriot-Watt University).

2.2.2 Chemical Composition of Cotton

The main component of cotton is cellulose, though the precise proportion varies with the source of the cotton and the growing conditions. Also, for a given cotton fibre, the composition differs between the fibre surface and the interior of the fibre. An average composition by percentage of the (dry) cotton fibre is shown in Table 2.1.

Table 2.1 Total and surface chemical compositions of cotton fibres.

	Percentage of substance	
Substance	*Total fibre*	*Fibre surface*
Cellulose	88–96	52
Pectins	0.7–1.2	12
Wax	0.4–1.0	7
Proteins	1.1–1.9	12
Other organic matter	0.5–1.0	14
Ash	0.7–1.6	3

Cellulose is a polysaccharide with the empirical formula of $(C_6H_{10}O_5)_x$, but it is in fact a more complex molecule than this apparently simple formula would suggest. Cellulose is formed by the condensation polymerisation of β-D-glucopyranose, with the six-membered rings joined through 1,4-glycosidic bonds. The structure shows the convention for numbering the carbon atoms in the ring:

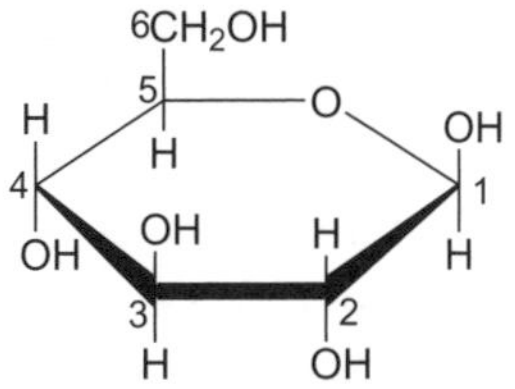

The repeating unit of cellulose is formed from cellobiose, which is made up of two glucopyranose units:

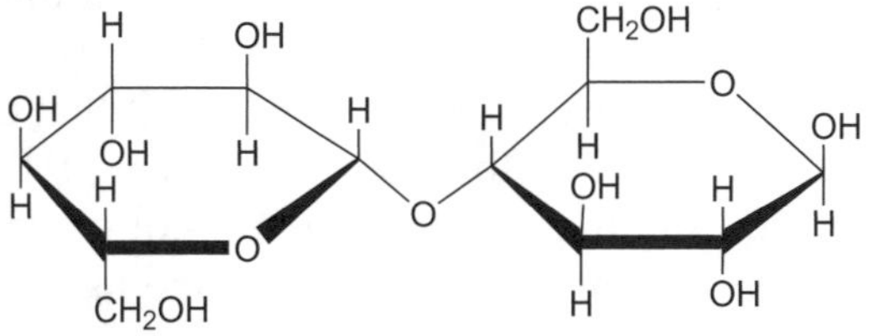

It is interesting to note that starch, which has very different properties from cotton, is formed from α-D-glucopyranose, again with the six-membered rings joined through 1,4-glycosidic bonds.

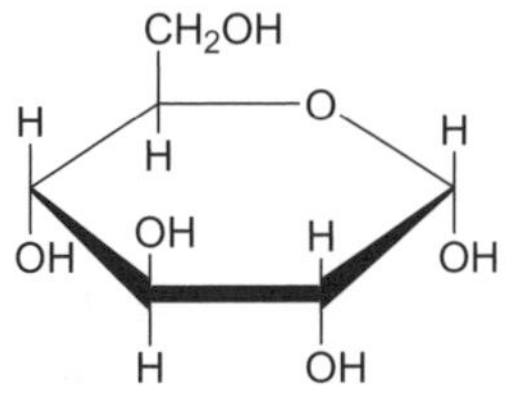

The chemical structure of cotton cellulose can be shown as follows:

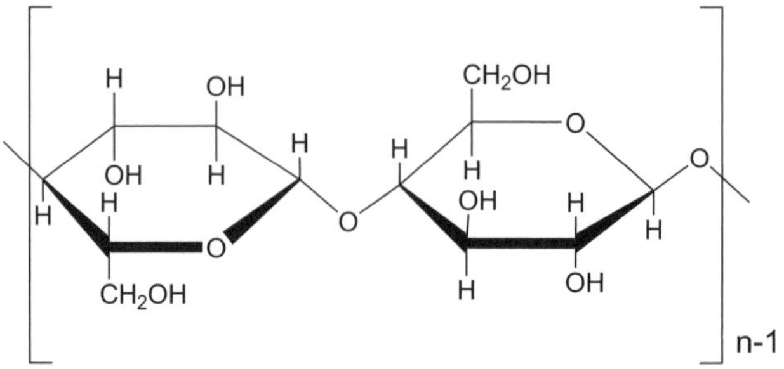

It is very difficult to give a precise value for the degree of polymerisation (*DP*) of cotton cellulose (the number of glucosidic residues in a polymer chain), for three reasons: firstly, there is considerable variation according to the source of the cotton; secondly, even for a particular sample, there is a distribution of values; and thirdly, the value depends to a certain extent on the method used for its determination. Values quoted generally for *DP* are of the order of 3000, though values in excess of 10 000 have been quoted for native cotton cellulose. Even so, a *DP* of 3000 corresponds to a molar mass of over 400 000.

The representation of the structure of cellulose by the projection formula is misleading, however, because it suggests that the pyranose rings are planar, whilst they actually have a 'chair structure' (as can be seen in the following structural diagram). The projection formula also suggests the –CH$_2$OH groups and the –OH groups are perpendicular to the plane of the ring, which is not the case. These groups lie in the same plane as the ring, so the molecule is fairly flat in shape. As a result, the molecular chains are able to align quite closely, with the result that cotton cellulose is a highly ordered, crystalline substance. The three –OH groups play an important part in contributing to the ability of cotton cellulose to form highly crystalline regions, through hydrogen bonding.

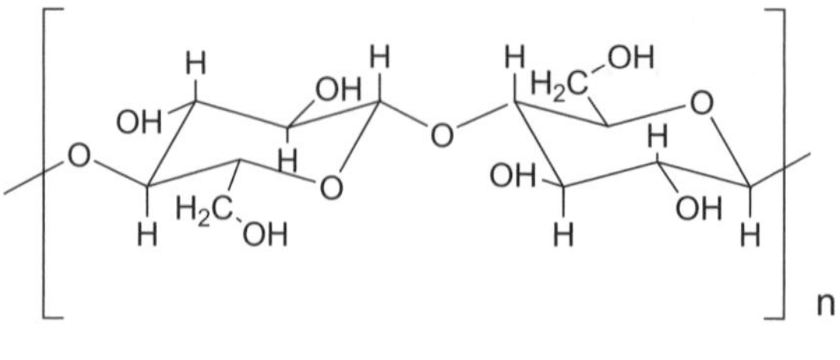

Most of the non-cellulosic substances are present in the cuticle, although some can be found in the lumen also, remaining from the growing period of the fibre. Pectin is a complex mixture of poly-saccharides and occurs often in the cell walls of plants, though its precise structure varies according to the plant species. Pectins are typically represented as a chain of α-D-galacturonic acid units, though most of the carboxylic acid groups are methylated:

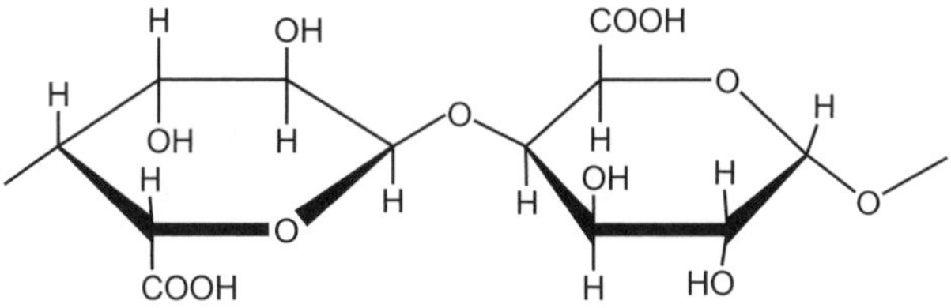

The nitrogenous matter is usually protein and simple amino acids. The wax is a complex mixture, principally comprising high molar mass alcohols, esters and acids. The esters are formed from fatty acids, R–COOH, where typically $R = C_{13}-C_{21}$, and monohydric alcohols of high molar mass ($R = C_{24}$ and higher). Also present in tiny amounts, but sufficient to have a deleterious effect on appearance, are naturally occurring pigments which are mainly chlorophyll, xanthophylls and carotene.

2.2.3 Crystalline Structure of Cotton

Four different crystalline forms of cellulose, called cellulose I, cellulose II, cellulose III and cellulose IV are known to exist, though it is the cellulose I and II forms that are the most significant in textile fibres. Cellulose I is the form found in native, untreated cotton. The cellulose II structure is formed in 'regenerated' cellulose, when cotton cellulose has been dissolved during processing, for example in the production of viscose or Lyocell fibres (see Chapter 4), or during the process of mer-cerisation, which involves treatment of cotton with a solution of sodium hydroxide of very high concentration (see Section 2.2.5.2).

In cellulose I, the polymer chains are aligned in a fully extended form parallel to one another as shown in Figure 2.4a. The repeat distance along the direction of the polymer chain is the length of the cellobiose unit. Polymer chains are configured in sheets in the ac plane. The b axis is at right angles to the ac plane (the view along the ab plane would be looking along the fibre axis). Sheets of layers are interposed over each other, as illustrated by the darker shaded chain in Figure 2.4. The stability and rigidity to the crystalline structure brought about by extensive hydrogen bonding, and both intra- and inter-molecular bonds

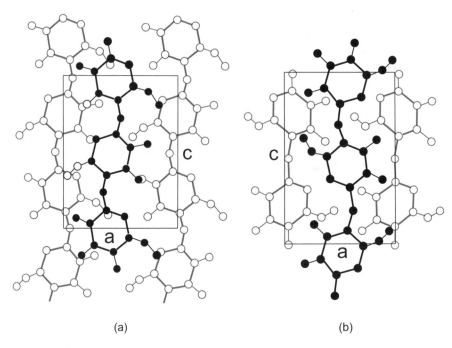

(a)   (b)

Figure 2.4 Alignment of polymer chains in cellulose: (a) parallel-chain model of
cellulose I, (b) antiparallel-chain model of cellulose II.

are important. The hydrogen bonding which has been proposed to exist
between adjacent polymer chains in a sheet is shown in Figure 2.5.

In cellulose II, the planes of the glucose do not lie perfectly flat along
the ac plane (see Figure 2.4b), and the centre chains lie antiparallel to the
corner chains. The different conformation of the hydroxymethyl group
on C-6 means that only one intra-molecular hydrogen bond can form.
Although both cellulose I and II forms have a monoclinic unit cell
structure, their cell dimensions differ.

The cellulose III crystalline structure forms when cellulose is precipitated from solution in liquid ammonia. Its structure is closely related
to that of cellulose II. The cellulose IV lattice is formed when regenerated cellulose is treated with certain liquids at high temperature under
tension. Its structure is very similar to that of cellulose I.

Different models exist for the way in which the crystalline regions co-
exist with the more amorphous regions of the polymeric structure of
fibres, of which two, the fringed micelle theory and the fringed fibril
theory are the most valid. In the fringed micelle theory, the individual
polymer chains are proposed to travel through both crystalline and

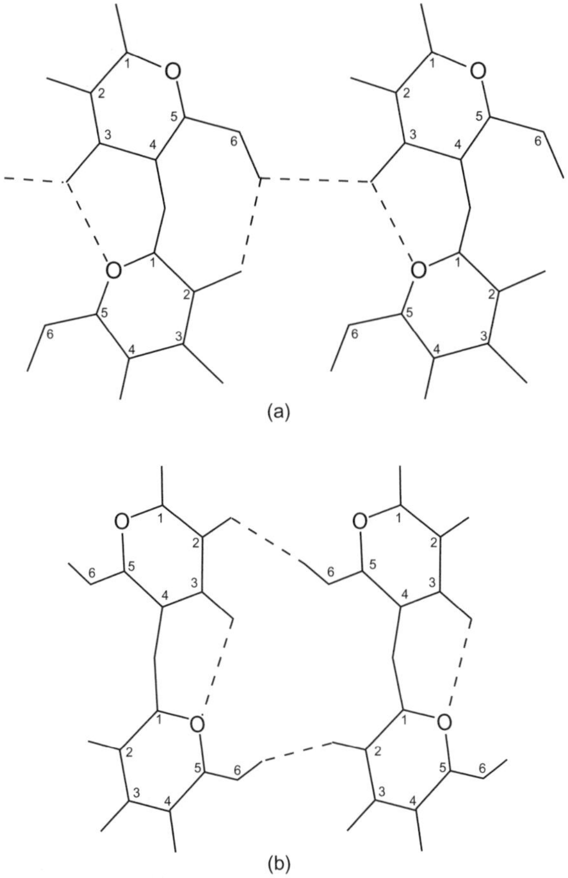

(a)

(b)

Figure 2.5 Hydrogen bonding in (a) cellulose I, (b) cellulose II.

amorphous regions and it just happens that in the crystalline regions the molecules align in either the cellulose I or cellulose II conformation. The molecules transfer from the ends of the crystalline micelles into random configurations that typify the amorphous regions. In the fringed fibril model, fringe polymer molecules can leave crystalline fibrils at any point and assume an unstructured, amorphous configuration. The two models are represented in Figure 1.5a and Figure 1.5b, respectively. For many years it was considered that cotton possessed the fringed micelle structure, but this is thought to be a rather simplistic model and more recently the description of the structure in terms of fringed fibrils has predominated.

The degree of crystallinity of polymer fibres can be determined by various methods, such as infrared spectroscopy, X-ray diffraction, calorimetry and density measurements. Values are difficult to determine with accuracy, however, which is not helped by the fact that in the natural fibres especially, the components of complex morphological regions have different crystalline structures. For cotton, X-ray studies seem to indicate a value of about 50% crystallinity.

2.2.4 Chemical Processing of Cotton

It can be seen in Table 2.1 that cotton is approximately 90% cellulose, the remaining 10% being impurities of various types. Whilst these impurities are present in relatively small amount, they have a significant impact on the appearance and properties of the fibres. The waxes in particular impart a hydrophobic, protective coating that if not removed acts as a barrier to water, preventing penetration by dyes. The pigment imparts a dull yellowy appearance to the fabrics. In addition, formulations called 'sizes' are added to the warp yarns prior to weaving cotton fabrics to reduce friction and prevent fraying. The size formulations usually contain an adhesive such as starch, polyvinyl alcohol, polyacrylic acid or carboxymethyl cellulose and a lubricant such as water-soluble gums, tallow or waxes. Up to 20% (on the weight of the yarn) of size can be added to the yarns, or more if tightly woven fabrics are being produced. After weaving, and prior to dyeing, it is necessary to remove all of the impurities from the fibres to realise their desirable characteristics for use in apparel and home furnishings, *etc.*

The first process applied to cotton fabric is that of 'desizing'. This process removes much of the water-soluble compounds present in the size formulations, but removal of the starch component is more problematic in that it is insoluble and is best removed by enzymes. The next process is 'scouring', which involves the removal of fats, waxes, tannins, pectins, protein matter, and mechanically adhering dirt or residues of seed husks. This process involves treatment of the cotton yarns or fabrics in caustic soda solution at the boil for approximately one hour. During the process the fats and waxes are converted to soluble soaps, and the protein matter, tannins and pectins are converted to water-soluble degradation products.

For the efficient processing of woven fabric, the desizing and scouring processes can be carried out in the same machine. The scouring process removes all of the impurities present in cotton fibres, with the exception of the small amount of naturally occurring pigment. The presence of this pigment is not a problem if the fabric is to be dyed to a heavy depth of shade, but if the fabric is to be dyed to only a pale depth, or the fabric is to be sold as a white, then the pigment has to be removed. Removal of the pigment is achieved by a bleaching process, the preferred bleaching agent being hydrogen peroxide.

2.2.5 Chemical Reactions of Cellulose

Cotton fibres are subject to attack by various agents which cause degradation. Even if the severity of attack is only slight, there can be a disproportionate influence on the textile properties of the fibres, especially on tensile strength. Common agents that can attack cotton are acids, alkalis, oxidising agents and heat.

2.2.5.1 Acid. Acids are the most destructive agents for cellulose, attacking the glycosidic linkages by an acid-catalysed hydrolysis reaction:

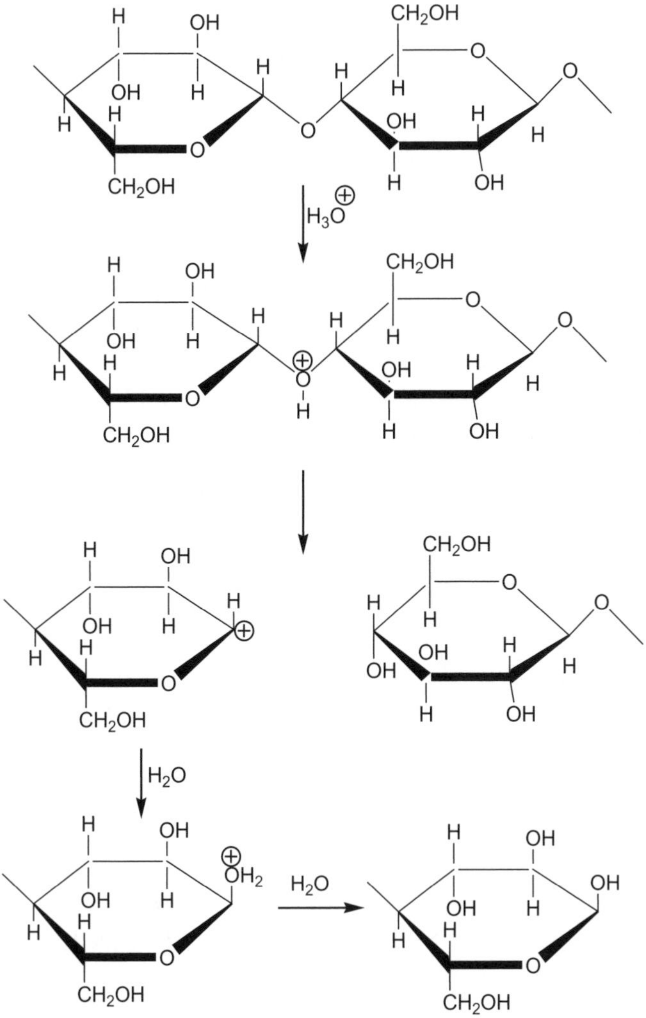

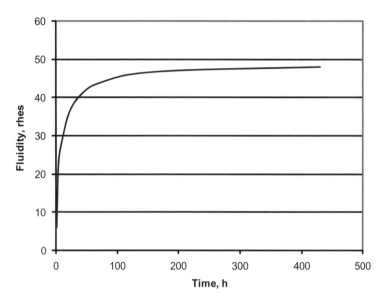

Figure 2.6   Influence of acid hydrolysis on fluidity of cotton.

The immediate effect of the hydrolysis is to split the long polymer chains
of the cellulose, so that the average degree of polymerisation (DP) is
reduced, with consequent reduction of tensile strength. The loss in DP
can be followed by measuring the fluidity of the treated cotton dissolved
in cuprammonium hydroxide, as shown in Figure 2.6. The initial rapid
increase in fluidity arises from the ease of attack of the acid in the more
amorphous, readily accessible regions of the fibre. The rate of reaction
slows down as the reaction then occurs at the chain ends of the more
inaccessible regions.

The acid hydrolysis reaction causes the formation of both reducing
and non-reducing groups at the ends of the polymer chains. The number
of reducing end groups can be monitored by measuring the amount of
metallic silver deposited from ammoniacal silver nitrate. A standard test
which provides a useful quantitative value for the degree of degradation
of cotton involves the determination of the copper number, derived from
the amount of cuprous oxide deposited from Fehling's solution.

2.2.5.2   Alkali.   The situation with alkali is very different from that
with acids, and whilst cotton fibres are readily degraded by acids, they
are much more stable in alkali. However, that is not to say the alkali has

no effect on cotton fibres and, at high temperatures especially, alkali can also cause hydrolysis of the cellulose. In this case, the reaction occurs in a stepwise manner by the removal of anhydroglucose units from the reducing ends of the cellulosic chains. Whilst the rate of reaction is slow, there is nevertheless an important consequence for the processing of cotton materials, because in scouring of cotton to remove impurities, such as fats and waxes from the fibres, alkaline conditions (soap and soda) at the boil are used. A loss in weight of the cotton, in excess of that expected if just the impurities are removed, usually results, and this can only be explained by the loss of glucosidic entities from the cellulosic chains.

The most important action of alkali on cotton is that of the swelling of the fibres. This behaviour was first noticed in 1844 by John Mercer, who found that by treating cotton yarns or fabric in solutions of 25–30% caustic soda, the fibres swell in diameter and shrink longitudinally. The process, still used today, is referred to as 'mercerisation', and is much used for the manufacture of cotton sewing threads. The fibres become denser, but the most important benefit is the increase in dyeability. Indeed the visual depth of shade of dyed mercerised fabric is higher than for unmercerised fabric, even when the same amount of dye is present. Furthermore, if the treatment is carried out with the yarn or fabric under tension to reduce the longitudinal shrinking, the fibres became more lustrous, smoother and stronger. However, not all the benefits can be fully realised together, so whilst mercerisation under tension considerably enhances lustre, the improvement in dyeability is only modest.

2.2.5.3 Oxidation. The glucosidic repeating unit structure of cellulose contains three hydroxyl groups: one primary and two secondary groups. As with all alcohols, these groups can be oxidised: the primary alcohols are oxidised firstly to an aldehyde and then to a carboxylic acid, and the secondary alcohols are oxidised to the corresponding ketone. In addition, it is possible that cleavage of the glucosidic ring can occur, both between the C-2 and C-3 carbon atoms and between the C-1 and the oxygen atom of the ring. The products of oxidation are called 'oxycelluloses', though they can vary in their chemical nature quite considerably. The types of products that can form are shown in the following reaction scheme, but it should be borne in mind that along a polymer chain, the glucose rings will have oxidised to different extents and no one single product will exist.

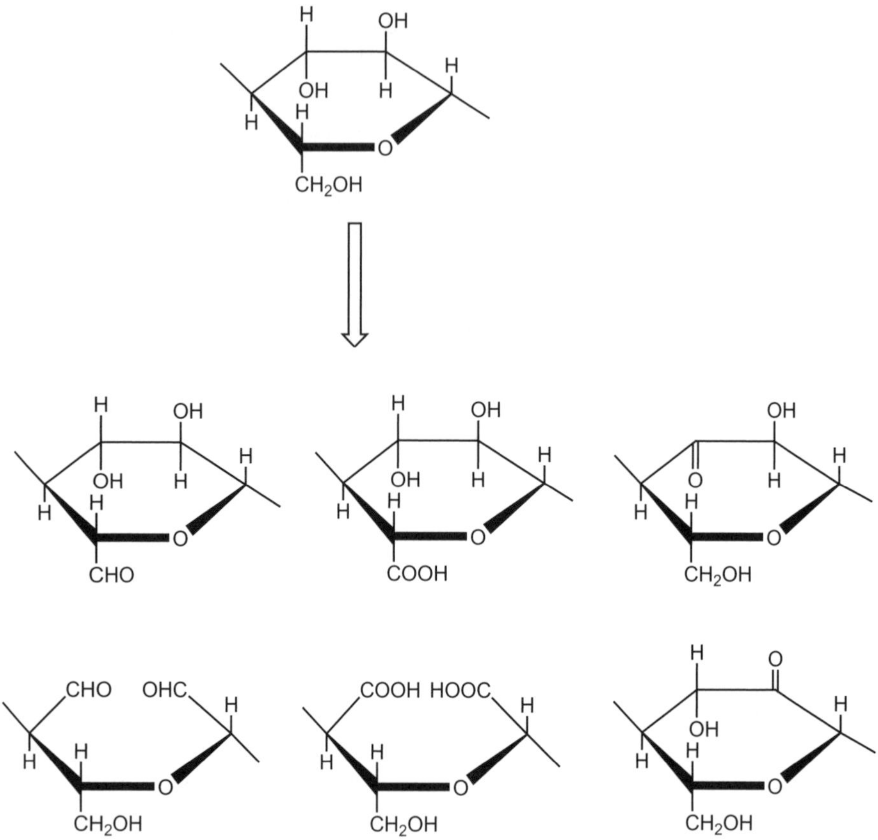

A wide variety of oxidising agents can cause such degradation, with hydrogen peroxide, sodium chlorite and sodium hypochlorite being of main interest due to their use in bleaching cotton to improve its whiteness. In order to meet environmental legislation, chlorine-based agents are no longer used (at least in the West) and hydrogen peroxide is the preferred bleaching agent. The optimum conditions required for peroxide bleaching are treatment for about two hours at a temperature of 95 °C in a sodium hydroxide solution, containing sodium silicate at pH 10.3 in the presence of stabilisers to prevent decomposition due to metal ions such as Mn^{2+} or Fe^{2+}. Nevertheless, over-bleaching can still occur, with consequent reduction in the tensile properties of the treated cotton, so accurate control of the process conditions (pH, temperature, stabilisers and time) is vital. The action of other oxidising agents, such as periodates, nitrogen dioxide and potassium permanganate have also been studied extensively. Some of these agents can be quite specific in their action, giving reactions that are not typical of those of the traditional bleaching agents.

In the bleaching process it is usually the most accessible regions of the fibre, the amorphous regions, that are attacked first and these are also the regions that are most susceptible to over-bleaching. So typically the precise nature and degree of degradation is not even throughout the fibre. Whilst the normal bleaching process is designed merely to destroy the naturally occurring pigment present, if the oxidation is carried out to excess, the fibrous structure of the cotton can be destroyed leaving a friable powder.

It is clear that a wide variety of degradation products from the oxidation reactions can be formed. Collectively these are known as oxycelluloses and they can be classed as 'reducing oxycelluloses' or 'acidic oxycelluloses'. Some oxidative reactions, generally those conducted in alkaline media, tend to produce a greater quantity of acid groups but low numbers of aldehyde (reducing) groups. Conversely, oxidation in acidic media tends to produce oxycelluloses that have lower numbers of acid groups but higher numbers of reducing groups. The measures that are used to assess the degree of damage to the cellulose are methylene blue adsorption and the copper number. The adsorption of the basic dye, methylene blue, which as a cationic dye is attracted to the negative charges on the acid groups, gives a measure of the number of acidic oxycellulose groups. Determination of the copper number can be used to obtain a quantitative value for the number of reducing groups formed on degradation. These long-established traditional methods are still widely used because most common instrumental analytical techniques, such as infrared spectroscopy, do not possess sufficient sensitivity to accurately quantify the acidic oxycelluloses or the reducing end groups. However other methods have been developed. A method based on the complexometric titration for the quantitative determination of carboxyl groups has been proposed. A method to determine carbonyl groups has also been developed in which the carbonyl groups are reacted with a specific fluorescence-labelling compound and the fluorescence intensity measured quantitatively.

2.2.5.4 Action of Heat on Cotton. Cotton fibres are stable to temperatures of up to 150 °C, but above this temperature the tensile strength reduces, as the DP decreases. Chemical degradation, through the combined effects of heat, moisture and atmospheric oxygen cleaving the glucosidic linkages, leads to the formation of hydrocelluloses, when aldehyde and carboxylic acid groups are formed. The cotton turns a yellow-brown colour, which gradually becomes darker brown as the temperature rises and oxidation of the cotton occurs. Above 200 °C the fibres completely lose their tensile strength as thermal degradation intensifies.

In a flame, cotton fibres ignite readily and burn quickly, in a manner similar to burning paper, and like paper the fibres will continue to burn if the flame source is removed. If the flame of the burning cotton fibres is extinguished, the fibres will continue to smoulder and emit smoke, with a smell of burnt paper.

2.2.5.5 Other Important Reactions of Cellulose. The presence of the three hydroxyl groups in the glucosidic ring has an important bearing on the processing of cotton. The hydrophilic nature of these groups gives cotton an inherent ability to attract water and, with a moisture regain value of $\sim 11\%$, confers comfort in wear through its absorbency. The hydroxyl groups are reactive towards a variety of chemicals and this fact can be made use of to advantage in enhancing the properties of cotton by dyeing and finishing. Whilst many types of dyes can be attracted to cotton by hydrogen bonding with the –OH groups, the dyes of one particular application class of dye, called 'reactive' dyes, actually react with the –OH groups to form a stable covalent bond with the fibre.

In general, the reactivity of the primary –OH group is higher than that of the two secondary groups, but their relative reactivities depend to a large extent on the particular compound with which the cotton reacts. It is generally accepted that in the application of reactive dyes, it is the primary alcohol group that reacts preferentially with the dyes. There are many types of reactive groups which can be incorporated into dye structures to confer fibre reactivity, but all react with the –OH groups of the cellulose by either a nucleophilic substitution reaction or a nucleophilic addition reaction. In each case, the reaction requires alkaline conditions so the dyeing process is normally carried out in the pH range of 9–11. The alkaline medium causes partial dissociation of the –OH groups of the cellulose, the resulting –O$^-$ ions providing the nucleophilic sites for the reactive group of the dye.

Typical of the dyes which react by a nucleophilic substitution mechanism are those with a chlorotriazinyl reactive group, this type of dye being the first of the reactive dyes for cellulosic fibres to be developed. Their reaction with the –O$^-$ sites in the cellulose rings occurs by the mechanism in Scheme A, p. 39.

The reactive group in the dyes which react by the nucleophilic addition mechanism is usually a vinyl sulfone system. This type of dye is marketed in the more stable form of the β-sulfatoethylsulfone derivative, which under the alkaline conditions of the dye bath loses sulfuric acid and generates the vinylsulfone group. The sulfone part of the group is highly electrophilic, imparting a strong positive character on the carbon–carbon double bond, thereby facilitating reaction with the strongly

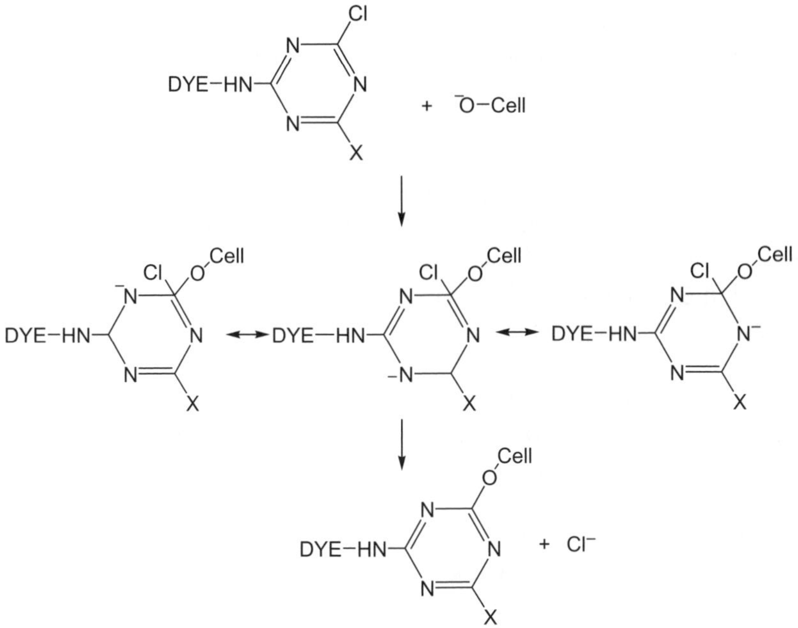

Scheme A

nucleophilic glucose rings of the cellulose (Scheme B).

$$Dye-SO_2-CH_2-CH_2-OSO_3Na$$

NaOH

$$Dye-SO_2-CH=CH_2 \quad + \quad Na_2SO_4 \quad + \quad H_2O$$

$$Dye-SO_2 \overset{\delta- \ \delta+}{-CH=CH_2} \quad + \quad {}^{\ominus}O-Cell$$

$$Dye-SO_2-CH_2-CH_2-O-Cell$$

Scheme B

In each case the covalent bond formed between the dye and the cellulose is very strong and resistant to most of the chemical environments that the dyed fabrics are likely to be subjected to during their lifetime.

The washfastness of reactive dyed cottons is especially good, so they are ideal for products which are often dyed to heavy depths of shade and are likely to be frequently washed, such as towelling, underwear or babywear.

Another use made of the –OH groups in the glucosidic rings is to react them with cross-linking agents, which are applied to improve the crease resistance of cotton fabrics. During washing, swelling of the fibres occurs as they absorb water which places stresses on them. The polymer chains in the amorphous regions move to relieve the stresses and on drying after washing, new hydrogen bonds form between them in their new arrangement, leading to creases in the fabric. Cross-linking agents reduce swelling and limit movement of the polymer chains. Many years ago, formaldehyde was used as a cross-linking agent, but the health issues associated with residual formaldehyde present in the treated cotton brought an end to its use. Typical of the cross-linking agents used nowadays is dimethyloldihydroxyethylene urea (DMDHEU), which whilst not entirely free of formaldehyde contains only very low quantities. DMDHEU reacts according to the following mechanism:

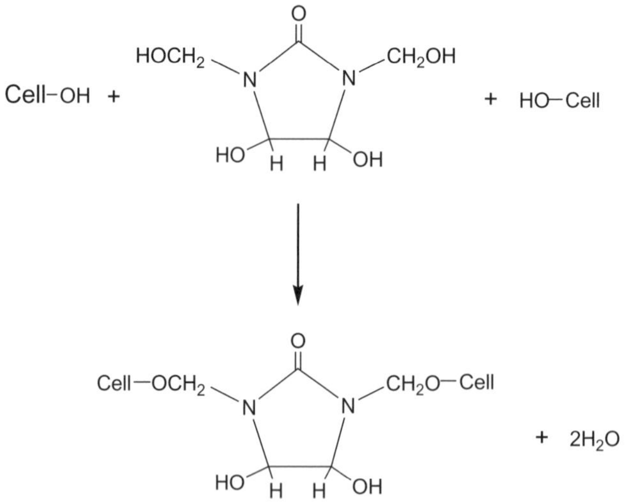

Cell–OH + [structure: HOCH₂–N, CH₂OH–N, ring with C=O, HO, H, H, OH] + HO–Cell

↓

Cell–OCH₂–N, CH₂O–Cell–N, ring with C=O, HO, H, H, OH + 2H₂O

Products (typically shirts) made of the treated cotton fabric are marketed as 'easy-care', or even 'non-iron'. Since the cellulose molecules of treated fabrics are more restricted in movement due to this cross-linking, they have lower tensile strength and tear strength than untreated fabric. In untreated fabrics, the tearing forces are distributed more widely over the molecular chains.

2.2.6 Fibre Properties

The main properties of cotton fibres are shown in Table 2.2. In general terms, cotton fibres are fairly strong, but what is more, they are 10–20% stronger when wet. Their tenacity increases with moisture content and this enhanced strength in moist conditions is one of the reasons why the processing of cotton dominated in Lancashire, rather than in Yorkshire where the climate generally is drier, because this enhanced strength facilitated the spinning and weaving operations.

The popularity of cotton results from its softness and absorbency, together with its availability at relative low cost. It is used extensively for underwear, jeans, shirting, dresses, knitwear, sportswear, leisure-wear, work wear and children's wear. It is also used for towelling and interiors, such as curtains and bedding. In addition, cotton is often blended with other fibres, such as polyester, nylon, viscose or modal. An interesting example is the trend towards the manufacture of warp-knitted terry towels with a synthetic ground structure of nylon or polyester, which adds stability and strength. This type of towelling is useful for hotels and hospitals which use specialised industrial laundries and have a high frequency of laundering. There is only about 5% of synthetic fibre present in these towels but it is enough to provide a substantial increase in strength, without detracting from their absorbency.

Table 2.2 Properties of cotton fibres.

Fibre length	High quality cottons: 25–60 mm. American Upland cottons: 13–33 mm. Indian and Asiatic cottons: 9–25 mm.
Fineness	10–20 μm.
Specific gravity	1.54 (one of the heavier fibre types).
Tenacity	25–40 cN tex^{-1}, but up to 20% stronger when wet.
Elongation at break	5–10%.
Elastic recovery	Fairly inelastic. Only 45% recovery from a 5% stretch.
Resilience	Low, but abrasion resistance is good.
Moisture regain	8.5%.
Reaction to heat	Cotton has no melting point, it is relatively heat resistant (iron up to 200 °C), but will yellow with a *very* hot iron. It burns very readily when it gives a smell like burnt paper.
Launderability	Cotton garments can be washed, boil washed, dry cleaned and tumble dried. They dry slowly and crease easily. They often shrink during the first few washes – so the fabrics are often pre-shrunk prior to garment making.

2.2.7 'Green Cotton'

Finally, it is worth saying a few words about the ecological aspects related to the growing of cotton. As has already been mentioned, such is
the world-wide demand, just about everywhere in the world that is
capable of growing cotton of satisfactory quality is doing so. However,
it comes at an ecological price. The cotton shrub is susceptible to attack
by many types of insects and by various diseases and, in order to prevent
what can be substantial losses, the plants are sprayed with pesticides.
On a world-wide basis significant quantities of such chemicals are used.
One estimate indicates that 16% of the world pesticide production is
used on cotton.

A study at the University of Utrecht has shown that the production of
cotton compares badly against the regenerated cellulosic fibres (see
Chapter 4), such as viscose and Lyocell fibres. In terms of land use,
almost 1.1 ha are required for the manufacture of each tonne of cotton
fibre, compared with 0.7 ha for viscose and only 0.25 ha for Lyocell.
There is a much more striking contrast in terms of water usage for the
production of fibres, as shown in Figure 2.7.

In the case of cotton, considerable quantities of water are required for
the irrigation of plants, whilst in the case of the man-made fibres, water

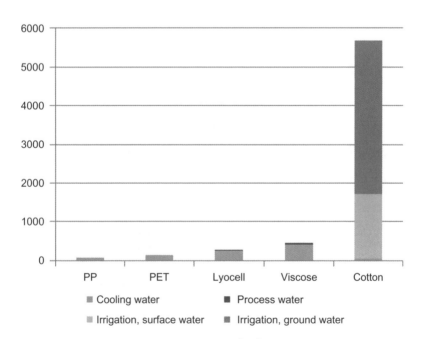

Figure 2.7 Water use, natural water origin (m^3 t^{-1} fibre).

is only required for processing. The values shown in Figure 2.7 refer only to the production of the fibres. In terms of energy and water consumption in domestic use, cotton again compares badly since it absorbs much more water in laundering and cotton fabrics are much more difficult to dry than hydrophobic synthetic fibres, such as polyester or nylon.

In response to this situation, there has been a move to the production of so-called 'organic' cotton, which in this context means cotton that is grown without the use of pesticides. Organic cotton has to be cultivated according to the EEC Regulation No. 2092/91 in the EU and the National Organic Program standards in the USA. The main criteria for growing organic cotton are:

- GMO (genetically modified organism) seeds are not allowed.
- Seeds cannot be treated with fungicides or insecticides.
- Instead of using synthetic fertilisers, strengthening of soil is based on crop rotation, which also requires less irrigation.
- Herbicides for weed control are avoided.
- Pest control is achieved by means of a balance between pests and their natural predators through healthy soil and by the use of beneficial insects and better biological and culture practices.
- Chemicals for defoliation are avoided.

As might be expected, yields are lower than those produced using pesticides and prices are therefore higher but demand for 'organic' cotton, in an increasingly environmentally aware domestic market, has caused a number of producers to now retail goods made of 100% 'organic' cotton. According to one source, sales of organic cotton textiles in the UK reached £100 m in 2008 (ten-fold more than in 2002), representing the fastest growing sector of the £30 bn UK clothing and textiles market. It is predicted that sales of organic cotton in 2011 will be double that of 2008. In 2008, the UK represented over 8% of the total organic cotton market. Naturally coloured organic cottons are produced, but not in significant quantities and with only a very limited range of colours, mainly browns and greens. Whilst their fastness properties are very good, the actual fibre properties are inferior to those of conventionally grown cottons.

Organic cotton garments have become widely available in most large retail stores in the UK, as demand has increased. It is important for consumers, however, that products have an independent certification that the cotton is organic and that the garment has been produced within a minimum set of standards. The most demanding label for organic textiles today is the Global Organic Textile Standard (GOTS), the development of which commenced in 2002 by an international group, represented in the UK by the Soil Association. The first draft was issued in May 2005 and the second in June 2008. This standard is concerned more with the processing of the organic cotton fibres into garments than with the actual growing of the cotton, and imposes a number of quite severe restrictions on the different textile processes:

- At all stages through processing, organic fibre products have to be kept separate from conventional fibre products and must be clearly identified.
- All chemical inputs (process chemicals, dyes and auxiliaries) are to be assessed and must meet basic requirements on toxicity and biodegradability; there is a list of approved chemicals to help the textile companies to fulfil this requirement.
- Exclusion of 'critical' inputs such as toxic heavy metals, formaldehyde and GMO substances.
- Restrictions for accessories (*e.g.* no PVC, nickel or chrome permitted, no plastic appliqué or inlays).
- The effluent of all wet processing units must be purified in a functional waste water treatment plant.
- Meeting social minimum criteria (based on ILO key norms) is compulsory for all processing stages.

The certification is required for the entire textile processing chain, right from post harvest handling through to final packaging and labelling, and annual inspections are carried out. Despite these tough demands and the price premium that the customer eventually has to pay, application of GOTS is increasing rapidly with well over 1000 manufacturers and traders of textile products participating in the inspection and certification system.

2.3 BAST FIBRES

Bast fibres are obtained from the stem of a plant. Bast fibres can have very long lengths, for example flax fibres can be over 1 m long, but they actually comprise a number of much shorter fibres (called the ultimate fibres) that are bound together by the gums and resins that occur naturally in the plant stems. As the stem has to hold a plant upright during growth, it is naturally stiff, due to its complex chemical nature. The cellulosic fibres are therefore commonly mixed with pectins, lignins, hemicelluloses and waxes, amongst other substances and the processing required to abstract the cellulosic fibres from the stem has to be quite rigorous.

2.3.1 Flax

Flax is the oldest textile fibre known, with evidence for its use as long ago as 8000 BC. It was a highly valued fibre, especially by Egyptian, Babylonian and Phoenician rulers, and linen fabrics were used to wrap bodies of the nobility in burial chambers. Much later, the Romans valued flax because of its durability and white colour and it became a symbol of divine purity. During the Renaissance it was used by artists, such as Leonardo da Vinci, for canvasses for their paintings, and in Flanders for making tapestries. Due to its high strength and durability, flax was also used to make sail cloths for sailing ships. By the nineteenth century, flax was widely used throughout Europe, but with the introduction of artificial rayons towards the end of the century, and especially with the introduction of synthetic fibres in the twentieth century, the demand for linen declined. Nowadays, flax is considered a luxury fibre.

Flax fibres are obtained from the stem of the plant (of the species *Linum usitatissimum*). Flax fibre is termed 'linen' once it is spun into a yarn, then woven or knitted into a fabric. There are other species of the flax plant, and some are grown for their seed, from which linseed oil can be obtained. The flax plant requires a temperate climate and can be grown in many parts of Europe. The best quality flax fibres are obtained from the coastal regions of northern France (72%), Belgium (15%) and the Netherlands (4%). Smaller quantities are produced in Poland (2.6%) and the Czech Republic (2.5%). At one time the cultivation of flax and linen manufacture was very strong in Northern Ireland, though it ceased there many years ago. Figure 2.8a shows flax growing in Northern Ireland during the last century. The bulk of flax production is in China, although the quality of the fibre is not good enough for the production of fine linen fabrics. A significant quantity of flax is also produced in Russia.

Flax plants grow to a height of about 1 m, with stems of about 2–3 mm in diameter. The flax fibres lie at the surface of the stem and run for its whole length, so, in contrast to cotton, flax fibres can be very long indeed. There are approximately 1000 flax fibres in each stem.

The process for obtaining the flax fibre from the plant stems is fairly long, though in chemical terms, not especially complex. The flax fibres are surrounded by cellular tissue, waxes and pectins, which bind them to the woody centre that gives the stem its mechanical strength. This structure has to be broken down so that the flax fibres can be easily removed by physical means. The process by which this is achieved is called 'retting' and there are different methods by which this softening of the sheath structure is brought about, each producing slightly different qualities of flax.

(a)

(b)

Figure 2.8   (a) A field of flax and (b) dam retting. (Photographs courtesy of the Irish Linen Centre and Lisburn Museum).

(a) *Dew retting* is the most widely used method, and quite simply involves spreading the stems on the ground for some 3–12 weeks, during which time the enzymatic action of bacteria dissolves the cellular tissues. The process is aided by the moisture from dew falling on the stems during the night and the heat of the sun during the day.

(b) *Dam retting* involves laying the stems in a dam, pond or bog. Figure 2.8b shows workers pulling retted flax from a dam.

(c) In *Stream retting* the stems are laid in a slow moving stream.

(d) *Chemical retting* involves treating the stems with agents, such as dilute alkali, soaps or even dilute acids. The use of chemicals adds to the cost, however, and the quality of the linen produced does not justify the extra expense.

After retting, the stems are dried, and then the woody parts are removed by the mechanical processes of breaking and 'scutching'. In the breaking process, the dried stems (essentially straw) are passed between fluted rollers where the woody parts (called 'shives') are broken along their length. The flax fibres remain undamaged and are separated from the shives by beating with blunt blades in a scutching machine. The shives are used for the manufacture of fibreboards. The mechanical separation leaves the flax as bundles of fibres which are stuck to each other by a gummy substance. In the next stage, the bundles of fibres are combed (or 'hackled') which involves passing them through sets of pins, each set being finer than the previous set so that the fibre bundles become progressively finer. The longest fibres are called 'scutched flax', or 'lines', and the shorter ones 'tow'. The 'tow' consists of short length flax fibres, just a few centimetres long, but they can be spun into yarns and woven into linen for casual wear or furnishing fabrics. The 'scutched flax' has much longer staple length and is also the finest in terms of diameter, and is used to produce the highest quality linens. The fibre lengths are 40–60 cm, but after hackling the average length is reduced to about 30–40 cm. The resulting flax fibres are still in bundles, held by a gummy substance. So, prior to spinning the fibres into a yarn, they are degummed in an alkaline bath at 65–90 °C. If desired the fibres can also be bleached, to remove the natural creamy colour.

Flax fibres are not as pure as cotton in terms of cellulose content; indeed they contain only about 60% cellulose. In addition, they contain other substances such as hemicelluloses and lignin, as well as proteins, waxes, pectins and natural colouring matters (see Table 2.3).

Hemicelluloses are polysaccharides, somewhat similar to cellulose but composed of various types of sugar units and possessing a high degree of chain branching. The main sugar residue present in hemicelluloses is D-xylose (shown below), but D-mannose is also very common.

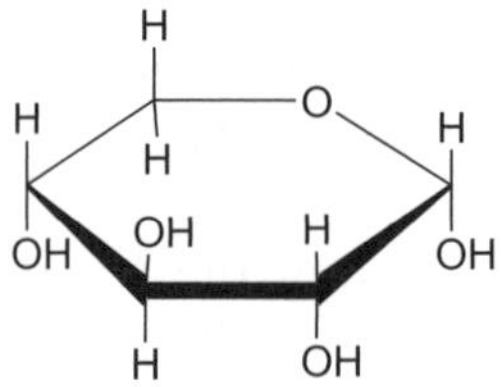

The exact composition of hemicelluloses depends on the plant source. They occur in the cell walls of plants, often with lignin. In contrast to cellulose, the molecules of which are highly crystalline, hemicelluloses are amorphous structures with low *DP* values (around 150), and are easily solubilised by dilute alkali or acid which break them down into simple sugars.

Lignins are amorphous cross-linked polymeric materials of molar mass $\sim 10\,000$ and are associated with the woody stem of plants and trees. They are fairly resistant to retting, which is why they are present in the final flax fibres after processing. They are aromatic in character and the main building blocks appear to have the following structures:

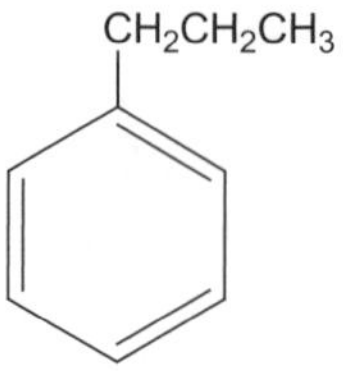

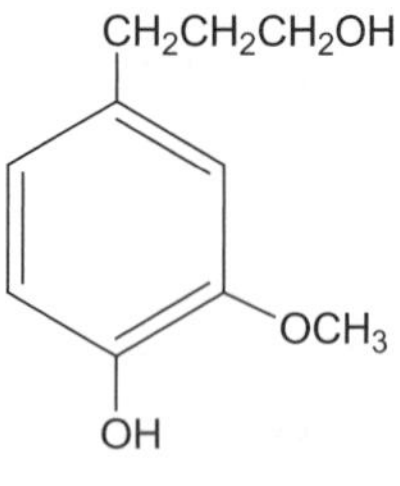

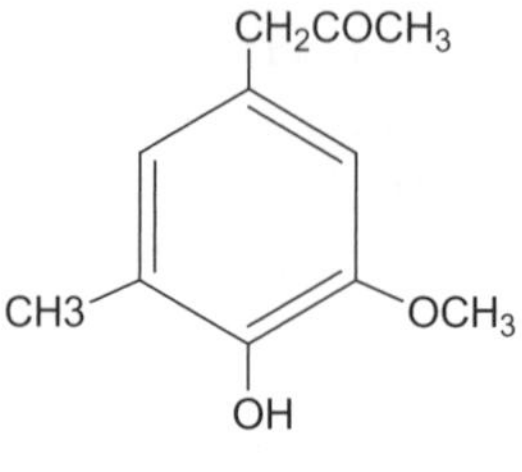

Table 2.3 Percentage chemical composition of flax fibres.

Substance	Unretted flax	Retted flax
Cellulose	62.8	71.3
Hemicelluloses	17.1	18.5
Pectins	4.2	2.0
Lignin	2.8	2.2
Waxes	1.4	1.7
Water soluble materials	11.9	4.4
Natural pigment	trace	trace

Table 2.4 Properties of flax fibres.

Fibre length	Varies between 6–65 mm, but on average is about 20 mm.
Fineness	$\sim$20 µm.
Specific gravity	1.54.
Tenacity	Stronger than cotton at $\sim$55 cN tex^{-1}, but about 20% stronger when wet.
Elongation at break	Flax is not very extensible and stretches only 1.8%.
Elastic recovery	It will recover almost completely from what stretch can be given to it.
Resilience	Good.
Moisture regain	12% – more absorbent than cotton.
Reaction to heat	Like cotton flax has no melting point and it is very heat resistant but will yellow with a *very* hot iron. It burns very readily when it gives a smell like burnt paper.
Sunlight	Flax gradually loses strength on exposure to sunlight.
Launderability	Like cotton linen can be washed, boil washed, dry cleaned and tumble dried. It dries slowly and creases very easily.

It is likely that these compounds are linked through C–C bonds between the propyl group or *via* ether linkages.

The physical characteristics of flax fibres are shown in Table 2.4. Flax fibres have a soft handle and are fairly lustrous in appearance. The final length of the fibres ranges between 10–60 mm and they have a diameter of about 20 µm. The fibres are not as convoluted as cotton fibres, but as with cotton fibres they have a lumen (or central canal) down the centre.

Flax fibres are amongst the strongest of the naturally occurring fibres, but they do not stretch much at all and even the elongation at break is only 1.8%. In terms of comfort of wear of garments made from linen, it is regarded as a summer fibre, readily absorbing moisture and having a cool, pleasant handle. Flax is a good conductor of heat and so linen sheets feel cool in the summer. It is a dense fibre but this helps linen to drape well and not become slack or fluffy. One drawback, however, is the ease with which linen creases. The ability of flax to absorb water makes linen useful for tea-towels and cloths for cleaning glass, since it removes water effectively, but without leaving any stray fibres behind.

2.3.2 Hemp

During the eighteenth century hemp was produced extensively in Britain. Its main use was in shipping, when there was a big demand for ropes, sails and sacking, and hemp fibres, being very long and strong, were ideal for this purpose. At one time canvas was made from hemp, the word 'canvas' being derived from cannabis. During the nineteenth century, however, production declined as costs of jute and cotton decreased and served these markets. In more recent times, whilst the versatility of hemp has been recognised, it has had a chequered image because of its relation to the hallucinogenic drug marijuana. Hemp and marijuana are produced from the same genus (*Cannabis sativa*) but from different cultivars, and the hemp cultivar contains only trace amounts of the chemical, Δ^9-tetrahydrocannabinol (THC), which induces hallucinations. Other cultivars contain much higher amounts of THC.

Due to the association between the two cultivars, the growing of hemp was made illegal in both the USA and the UK. In the UK, the restriction has been lifted, however, and since 1993 hemp has been allowed to be grown, but only under a licence issued by the Home Office. The Home Office also offers guidance on siting commercial hemp plantations and also has a 'Fibre Processing Aid Scheme', where aid is payable according to the tonnage of fibre produced by the authorised primary processor. Cultivation is still illegal in the USA, though it is legal to import hemp fibre into the country. The main producing countries of hemp fibre are China, Spain, Korea and Russia, and the plant is very tolerant of varying climatic conditions. Hemp does require, however, soil with a pH just on the alkaline side of neutral that is rich in humus. It is a fast growing plant, reaching maturity in about 3–4 months and it is claimed that in terms of land usage, hemp produces about 2–3 times as much fibre as the same acreage of cotton. Although the plants do not normally require spraying with pesticides or insecticides, some types of insects can feed on the plant and destroy the stem.

As with flax, hemp is a bast fibre and is obtained from the stem of the plant. The fibres are extracted using the same retting process as that for flax. The resulting fibres are much lower in cellulose content than cotton, but unlike cotton, contain hemicelluloses and lignin as shown in Table 2.5. These two compounds often occur together in the cell walls of plants and give them strength, an important requirement for the stems of plants.

Table 2.5 Composition of hemp fibres.

Substance	% by mass
Cellulose	77.9
Water	6.6
Hemicellulose	6.1
Solubles	4.3
Protein	2.7
Lignin	1.7
Waxes	1.5
Pectins	1.4
Ash	1.2

Table 2.6 Properties of hemp fibres.

Fibre length	On average is about 15 mm.
Fineness	$\sim 20\,\mu m$.
Specific gravity	1.54.
Tenacity	53–62 cN tex^{-1}.
Elongation at break	Hemp is not very extensible and stretches only 1.5%.
Elastic recovery	Poor.
Resilience	Good.
Moisture regain	12%.
Sunlight	Stable, but prolonged exposure gradually weakens it.

The presence of lignin causes hemp fibres to have a harsher handle than cotton and to be stiffer and more brittle. The physical and chemical properties are similar to those of flax. Some hemp fibres are relatively bright and lustrous, but others are darker which detracts from their appearance. Most hemp is sold naturally coloured, since dyeing is difficult and the inclusion of a dyeing step is contrary to its otherwise eco-friendly credentials. Hemp is used in the manufacture of clothing; it is strong, durable and absorbent, and has a good resistance to UV radiation. It is often blended with cotton, but it also has many other uses, such as composites and insulation materials, and in construction for making bricks. It is also used for making specialised papers, though for this purpose hemp pulp is very expensive, when compared with wood pulp. The properties of hemp fibres are shown in Table 2.6.

2.3.3 Jute

Although the use of jute in Europe has declined substantially over the last 30 years, it is still a very important fibre worldwide and after cotton is the most widely used of the natural fibres. Like flax and hemp, it is a bast fibre, being obtained from the stem of the jute plant. Unlike these fibres, however, the jute plant requires warm, humid conditions for growth, typically temperatures in the range 25–40 °C and a humidity of 70–90%. Additionally considerable rainfall is required, so not unexpectedly, the main producing countries are India, Bangladesh and China, with lesser quantities being produced in Myanmar and Thailand.

There are two main species of the jute plant: *Corchorus capsularis* (white jute) and *Corchorus olitorius* (Tossa jute), which grow to heights of between 1.5–4.3 m, with stem diameters of up to 20 mm. Since good yields are obtained without the need for fertilizers or pesticides, jute fibres are relatively cheap. Cultivation is labour intensive because the plants are harvested by manual cutting, though as is typical in the Asian countries, labour pay rates are low. After harvesting, the plants are retted by the same methods used for flax, after which the jute strands are stripped from the stems by hand, washed then dried on hangers for 2–3 days. The resulting jute strands are up to 3 m long, but they are actually composed of many very short fibres (between 0.5–6.0 mm long) held together by lignocelluloses. The fibres contain between 61–71% cellulose, with large amounts of hemicelluloses (14–20%) and lignin (12–13%) present, together with pectin (0.2%). The properties of jute fibres are shown in Table 2.7.

Jute has a reasonable strength, though it is weaker than flax or hemp. Not surprisingly, bearing in mind the level of impurities present, jute fibres are at best an off-white colour (the Tossa jute variety being more golden in colour). It is also slightly lustrous. Where the final end use and added cost justifies it, jute can be bleached and fibres of a much better quality of white are obtained. Jute can be dyed, though this is not the norm. Jute is mainly used for the production of low-value products, such as ropes and sacking. Hessian is the fabric obtained by weaving jute. It is also used for carpet manufacture; jute is the most common carpet

Table 2.7   Properties of jute fibres.

Fibre length	Only 0.5–6.0 mm.
Fineness	26–30 μm.
Specific gravity	1.3–1.5.
Tenacity	30–45 cN tex^{-1}.
Elongation at break	1.7%.
Elastic recovery	Low.
Resilience	Good, but it deteriorates fairly quickly in the presence of moisture.
Moisture regain	12.6%, but it can absorb up to 23% of water under conditions of high humidity.
Sunlight	Stable, but turns brown on prolonged exposure.

backing material. It is also used for home interiors, for example, in furnishing fabrics and wall hangings.

For some of its uses jute has been replaced by synthetic fibres, especially by polypropylene for sacking and also for carpet backings. Polypropylene is a relatively cheap fibre which has a greater resistance to water and microbial attack, and it also retains its strength when wet. At one time there was a huge jute industry based in Dundee in Scotland, but this has now totally disappeared as the demand for jute has fallen away with the introduction of synthetic fibres. Nevertheless, jute is likely to be an important fibre in the future, bearing in mind its low cost, the fact that it is produced from renewable resources and that it is fully biodegradable.

2.3.4 Ramie

Ramie is one of the oldest textile fibres and it has been cultivated in China for many centuries. Indeed Egyptian mummies were wrapped in ramie fabrics during the period 5000–3000 BC. It is a relatively little used fibre outside the East Asian countries, but it is increasingly finding its way into apparel for the European market, not so much in the form of 100% fabric but in blends, typically ramie/cotton (55 : 45) where the ramie increases the lustre, or in blends with wool where it reduces shrinkage.

The plant is native to East Asia and is commonly known as Chinese grass, white ramie, green ramie and rhea. There are two types of plant: *Boehmeria nivea,* which is native to China, and *Boehmeria tenacissima,* which is cultivated in the more tropical countries of South America, Indonesia and the Philippines. Nowadays, Brazil is also an important producer. The plant, a member of the nettle family (but without the stinging hairs), is a hardy perennial and whilst a crop is not possible in the first year of planting, in subsequent years, two to four, sometimes up to six, crops per year can be harvested. The plants have a life of between 6–20 years. The best conditions for the plant are high temperatures and high humidity, together with high rainfall, but it is important that the rainfall is consistent all the year round. With such high productivity the plants make heavy demands on the soil, so large amounts of fertilisers are required to replenish nutrients to maintain the production of fibres of adequate quality.

Ramie is a bast fibre, so the fibres have to be extracted from the stem of the plant. However this is not as easy as with flax, hemp or jute. The ramie fibres exist as fibre bundles, in which the ultimate fibres (the individual fibres) are cemented together by a gum. The gum is made up of waxes, hemicelluloses, lignin and pectins that are difficult to remove, and a simple retting process is not effective. Firstly, the fibres are extracted from the stems by a mechanical process called 'decortication', which is a beating process that crushes the stalks to remove the bark, so that the fibre bundles can be pulled out. The decorticated fibres are washed then dried in the sun for 2–3 days. The resulting fibres contain between 20–30% gum which must be removed to obtain spinnable fibres.

The gum is removed either by microbial degumming or chemical degumming. The latter process is more effective and involves treatment of the fibres in alkaline solutions, usually caustic soda, though other alkalis such as sodium carbonate, sodium tripolyphosphate or sodium silicate are also used, which break down the pectins. However the precise alkalis used and their concentrations tend to be kept secret by the mills. One method involves boiling the decorticated ramie fibres in 1% sodium hydroxide solution for two hours, then washing and drying the fibres. This reduces the gum content to about 5%. A combined microbial and chemical treatment is also very effective and economical. The resulting ramie fibres have a cellulose content of 91–93%, with about 2.5% hemicelluloses, 0.63% pectin and 0.6–0.7% of lignin. The degummed fibres have an improved fineness, lustre and tenacity. They have a pale creamy colour but can be bleached if necessary using hydrogen peroxide or sodium hypochlorite, to give a whiter colour and a softer handle. The harvesting of the stems and the consequent extraction of the fibres are both very labour-intensive operations.

The properties of ramie fibres are given in Table 2.8. Ramie is the strongest of the bast fibres and, as with most of the natural cellulosic fibres, its strength increases when it is wet. The ultimate fibres have the longest ultimate fibre length of all the bast fibres, and are the most durable, having good resistance to bacteria, mildew and insect attack. The fibres have a smooth, lustrous appearance and can be dyed reasonably easily. The main disadvantage of ramie is its low elasticity,

Table 2.8 Properties of ramie fibres.

Fibre length	120–150 mm, can be over 200 mm.
Fineness	40–60 μm.
Specific gravity	1.51–1.55.
Tenacity	45–88 cN tex^{-1}, increases by 25% when wet.
Elongation at break	3–7%.
Elastic recovery	Very low.
Resilience	Poor.
Moisture regain	12%.
Sunlight	Does not change colour with exposure to sunlight.

which means it is stiff and brittle. This causes difficulty in producing yarns and the hairy surface of the yarns causes a lack of cohesion between them when weaving fabrics. The main uses are in apparel, for dresses, suits, skirts jackets, skirts, though wrinkle resistance is poor, as is abrasion resistance. Other uses are in interior textiles, such as curtains, upholstery bedspreads and table linens.

2.3.5 Bamboo

Bamboo is the fastest growing of all woody plants. Depending on the particular species, the plants can grow to heights of 30 m, some at a rate of 0.3 m per day. Whilst bamboo can grow successfully in many climates most of that used for fibre production is grown in China, where its cultivation is strictly controlled to ensure ecologically acceptable methods that do not involve the use of any fertilisers or pesticides. The plants can be re-harvested, so there is no need for re-planting and because they can grow densely, very high yields (up to 60 tonnes per hectare) can be obtained which are vastly in excess of the values of 20 tonnes per hectare for trees (the source of cellulose for viscose and Lyocell fibres) and of 1–2 tonnes per hectare for cotton. The plants also require very little irrigation.

Much of the bamboo fibre is produced in a process similar to that for making viscose (see Section 4.1), so it is, in effect, a regenerated cellulosic fibre. The hard woody stalks are steamed then mechanically crushed to extract the pith. Sodium hydroxide solution is then added, when a process of what is described as 'hydrolysis–alkalization' takes place resulting in a starchy pulp that can be extruded through a spinneret into a bath of dilute acid, when fine bamboo fibre filaments are formed. So, although the fibres are derived from a renewable resource, they have to be extracted by a chemically aggressive process with significant implications for environmental impact. In countries outside Europe and the USA, the policing of what legislation does exist to control emissions to the environment is not always rigorous.

An alternative process is to crush the woody stalks mechanically then subject them to a retting process, in which the natural enzymes break down the cell walls and the bamboo fibres can be separated. The resulting fibres, which are less soft, are called litrax bamboo, or, because it is produced in a similar way to flax, 'bamboo linen'. Bamboo fibres contain about 61% cellulose and 32% lignin. The properties of bamboo fibres are given in Table 2.9. The characteristic features of bamboo are its absorbency and coolness, together with a very soft handle. It is therefore ideal for garments worn in hot, humid climates.

Table 2.9   Properties of bamboo fibres.

Fibre length	38–76 mm.
Specific gravity	0.8.
Tenacity	23 cN tex^{-1}, but decreases to 13.7 cN tex^{-1} when wet.
Elongation at break	23.8%.
Moisture regain	13%.
Sunlight	Does not change colour with exposure to sunlight

These attributes also make it very suitable for bathrobes and towels. The fibres are smooth and round and are ideal for skin that is sensitive to allergies or irritable materials. Bamboo garments are also claimed to give a score of 50 on the UV protection scale, which is equivalent to a reduction of 98% in UV energy reaching the skin.

In the advertising material of some manufacturers and distributors, it is claimed that bamboo fibres are naturally anti-bacterial, due to the presence of an agent called 'bamboo-kun', which also provide the fibres protection from pests during growth, thus avoiding the need for pesticides. It is claimed that even after 50 washes bamboo fibres still retain their natural anti-bacterial, bacteriostatic and deodorisation qualities. However, independent tests of bamboo fibres from a wide range of suppliers have shown such claims to be entirely false.

2.3.6   Nettles

A common perception of nettles is that they are plants of no real use to anyone, growing as weeds. Nettles have, however, been used as a source of textile fibres for many centuries, though never in huge quantities and after World War II they all but disappeared as other cheaper fibres became more available. This situation is beginning to change, however, with the increasing demand for textiles made from plants that have a low environmental impact, that are tolerant of climatic and soil conditions, and from which it is not too difficult to extract the fibres. In these respects nettles show promise. In recent years, considerable effort has gone into developing cultivars that have higher fibre content than that of wild nettles.

After the stems have been harvested, they are retted to remove the bark from the core. The bark is then boiled, which releases the fibres and the fibres are then combed. The fibres typically contain between 79–83% cellulose, 7.2–12.5 hemicellulose and 3.5–4.4% lignin, depending on the part of the stalk from which the fibres come. Fibre lengths, diameters and tenacities may also vary by large amounts, depending on whether the fibres come from the bottom, middle or top of the stem. Although

Table 2.10 Properties of nettle fibres.

Fibre length	48–52 mm.
Fibre diameter	30–35 μm.
Specific gravity	0.72.
Tenacity	24–62 cN tex^{-1}.
Elongation at break	2.3–2.6%.

the fibres are fairly coarse, they can be very strong. The fibres are a creamy white colour and have a soft handle. The properties of the fibres are shown in Table 2.10. Traditionally nettle fibres have been used to make ropes, twines and sailcloth, but increasingly applications in clothing are being investigated. The main issue, however, is the productivity of the plant and the weight of fibre that can be produced per hectare of land. Initial experiments have shown the value to be about 1.7 tonnes per hectare which, although it compares favourably with yields for cotton, is low in comparison with bamboo for example.

2.3.7 Uses of Bast Fibres

In general, the uses of bast fibres depend on their quality, ranging from the very fine, white flax fibres used to make high quality linens, to fibres such as jute which are rougher and used for applications where strength and resilience are more important, such as cordage and ropes. However, of emerging interest, especially in developing countries, is their use as geotextiles in applications such as soil reinforcement in land engineering projects. Fibres such as flax, hemp and jute have very high tensile strengths, higher than standard grade polyester fibres for example, and indeed their strength increases when they are wet. They stretch very little and have low elasticity, so they possess all of the mechanical properties required for this application. Whilst they are fairly resilient, even in 'hostile' environments, they are biodegradable so will eventually decompose through the action of microorganisms. This takes some time, however, and during this period the natural geotextile material will maintain the structure of an embankment, for example, giving the soil adequate opportunity to consolidate and for the root systems of vegetation to become established and provide further structure to the ground. Synthetic fibres have been engineered for this purpose but, whilst very effective, they are permanent and much more expensive. For many ground engineering projects, it is not necessary for the geotextile to remain permanently in the ground and the biodegradable property of natural fibres is actually useful.

Bast fibres are also used to manufacture nonwoven products for automotive uses. In particular they are used for moulded shapes for noise insulation in flooring and for trim components. The basic raw material is cotton which is hardened by adding thermoset binding agents, and jute, kenaf or bast fibres are added to increase rigidity.

2.4 LEAF FIBRES

Leaf fibres have limited commercial value, mainly because they are coarser than the bast fibres and the uses to which they can be put are limited. The most important fibres of this class are sisal, henequen and abaca, and they are used for the production of cordage and ropes. The fibres are usually obtained from the leaves by mechanically scraping away the non-fibrous material. They are then washed and dried, sometimes in the sun where a degree of bleaching also takes place.

2.5 OTHER POLYSACCHARIDE FIBRES

In recent years, some other naturally occurring polymers, the poly-saccharides, such as alginate, chitin and chitosan have become useful for the production of wound dressings. The polymers, extracted from various sources of plant and animal cells, have long been known for their ability to accelerate healing. Much development work has been carried out on these polymers to produce fibres so that woven, knitted or non-woven fabrics can be made.

2.5.1 Alginate Fibres

Alginate is a substance that occurs in brown seaweed. It is a block co-polymer of α-L-guluronic acid (G) and β-D-mannuronic acid (M), the GG and MM blocks occurring in various lengths and proportions. Blocks of MG also occur in the polymer chains:

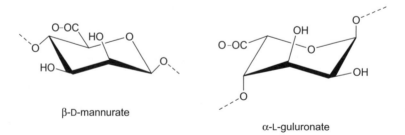

β-D-mannurate

α-L-guluronate

The polymer chain has the ability to interchange cations (notably sodium and calcium), and the particular cation bound determines the properties of the polymer. Thus sodium alginate is soluble in water, but calcium alginate is insoluble, and the polymer containing both types of cation shows intermediate behaviour. In contact with a wound exudate, ion exchange occurs when calcium ions are transferred from the polymer to the wound and replaced by sodium ions. The presence of sodium ions causes the fibre in the dressing to absorb water and swell to form a soft, protective gel. When the wound has healed, this gel can be removed easily by washing in warm saline solution.

Alginate fibres usually contain a high proportion of either the G or the M monomers. Calcium ions are more firmly bound in the high G alginate fibres than the high M types, with the result that the high G types swell only slowly due to the slow rate of ion exchange. Wound dressings are commercially available of both high G and high M alginate forms, so that the different needs in wound treatment can be met.

Alginate fibres are produced by wet spinning technology (see Section 5.1.3), in which a solution of sodium alginate is extruded through a spinneret into a bath of calcium chloride solution. The fibres precipitate as the calcium alginate form. The fibres are most easily produced for wound dressings as a non-woven web. They are of fairly low tenacity (only 14–18 cN tex^{-1}) and have quite low elongation (about 2–6%), so they are difficult to process by conventional textile machinery into yarns for weaving or knitting. Specially developed processes are required for making yarns from these fibres.

2.5.2 Chitin and Chitosan Fibres

Chitin is the main component of the cell walls of fungi and the exoskeletons of crustaceans such as crabs, lobsters and shrimps. It is a very abundant naturally occurring polymer and is produced commercially from the shell wastes of crustaceans. Chitin is a polymer of *N*-acetylglucosamine and, apart from the fact it contains nitrogen, it is structurally similar to cellulose:

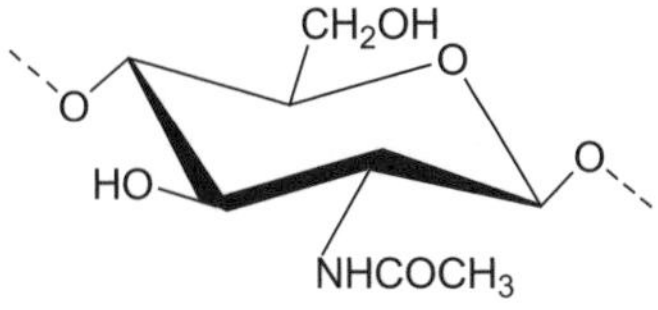

Chitin 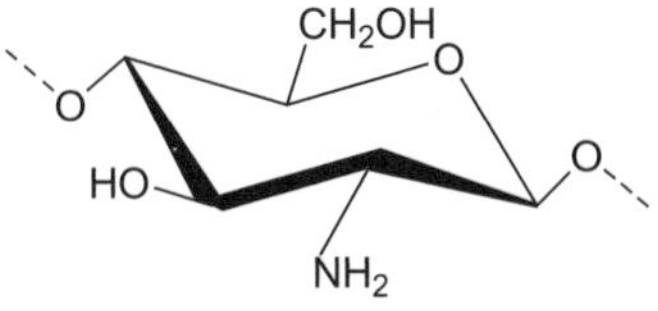   Chitosan

Chitosan is the deacetylated version of chitin. The wound healing abilities of both chitin and chitosan have long been known. Both fibre types are reasonably strong (tenacities 15–25 cN tex^{-1}) and elastic (elongation 8–20%), and in addition to wound dressings they can also be used as sutures. The fibres are highly hydrophilic, biocompatible and non-toxic.

A process for the production of chitin fibres has been difficult to develop. Chitin is not readily soluble and even with the more aggressive solvents, such as concentrated mineral acids, trichloroacetic acid or formic acid, dissolution is difficult. In the 1980s, a process was developed that involves firstly treating the chitin with *p*-toluene sulfonic acid in isopropanol, after which it dissolves easily in dimethyl acetamide containing a small amount of lithium chloride. The solution can then be extruded as fibres by wet spinning into a coagulation bath containing either water or methanol solution.

In contrast, chitosan dissolves readily in aqueous solutions of most organic and inorganic acids, due to the presence of the basic primary amine group. The usual method of fibre production is by wet spinning a dilute acid solution of the chitosan into a bath containing dilute alkali solution.

SUGGESTED FURTHER READING

1. P. J. Wakelyn, N. R. Bertoniere, A. D. French, D. P. Thibodeaux, B. A. Triplett, M. A. Rousselle, W. R. Goynes, J. V. Edwards, L. Hunter, D. D. McAlister and G. R. Gamble, in *Handbook of Fiber Chemistry*, ed. M. Lewin, Taylor & Francis Group, Boca Raton, USA, 2007, ch. 9.
2. T. P. Nevell and J. A. Zeronian, *Cellulose Chemistry and its Applications*, Wiley, New York, 1985.
3. D. Klemm, B. Philipp, T. Heinze and H. Wagenknecht, *Comprehensive Cellulose Chemistry*, Wiley-VCH, Weinheim, Germany, 1998.

CHAPTER 3
Protein Fibres

3.1 INTRODUCTION

Protein fibres are derived mostly from animal hair. Although wool from sheep is commercially by far the most important type of fibre derived from hair, there are a variety of other hair fibres, as shown in Figure 1.1. The principal protein constituent of hair fibres is keratin, although (as discussed in Section 3.2.6) there are many different types of keratin. A notable feature of keratin is the high content of the sulfur-containing amino acid, cystine, compared with other natural proteins. Silk, however, is primarily processed from secretions of the silk-worm, *Bombyx mori*. The predominant protein in silk, fibroin, contains no cystine. There is also an increasing interest in silk derived from spiders. Spider-silk possesses outstanding mechanical properties, and it is thought that successful copying of the secretion mechanism that a spider employs could lead to industrial production of silk fibres of high mechanical performance. This chapter provides a discussion on wool fibres, followed by an overview of other hair fibres used in textiles, and finally a review on silk fibres.

3.2 WOOL

3.2.1 Origins and Morphology

There is enormous variation in the structure of wool fibres. Different qualities of fibre are obtained from different breeds of sheep, and even

The Chemistry of Textile Fibres
By Robert R Mather and Roger H Wardman
© Robert R Mather and Roger H Wardman 2011
Published by the Royal Society of Chemistry, www.rsc.org

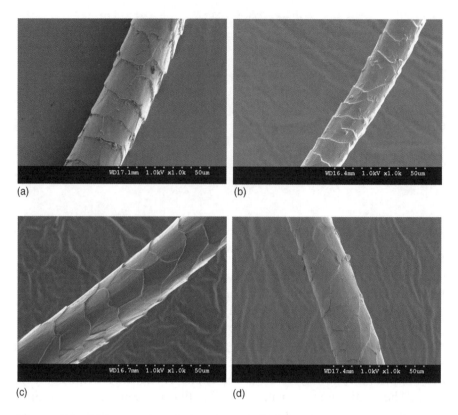

Figure 3.1 Electron micrographs of wool fibres: (a) Merino, (b) lambswool, (c) Shetland wool, (d) carpet wool. (Photographs courtesy of J. McVee, Heriot-Watt University.)

from a single fleece. Some examples are shown in Figure 3.1. The diet of the sheep and the conditions under which it has been grazing are also influential. Generally, the sources of variation can be identified as genetic, physiological and nutritional.

Wool is classified according to the average length and diameter of the individual fibres. Table 3.1 shows some examples. Coarser wool fibres with the largest lengths and diameters are harder wearing and are used extensively for rugs and carpets. Such fibres are derived from breeds such as Lincoln, Leicester and Romney sheep. The finest wool is derived from Merino sheep, a breed developed originally in Spain, but which is now associated particularly with Australia. Many wool types of intermediate fineness, typically 'lambswool', have also been developed.

Table 3.1 Average lengths and diameters of wool types.

Breed	Range of fibre diameters (μm)	Range of fibre lengths (mm)
Merino	17–25	60–100
Cheviot	28–33	75–100
Lincoln	39–41	175–250

Raw wool contains a wide variety of impurities, which can account for between 30% and 70% of the total mass. The impurities consist of wool grease, secreted from the sebaceous glands in the skin, suint, produced from the sweat glands, and also dirt and sand. Wool grease consists chiefly of esters, formed from the combination of sterols and aliphatic alcohols with fatty acids. Suint consists primarily of the potassium salts of organic acids. The nature of the dirt and the amount of sand reflect the conditions under which the sheep has been reared. Raw wool may also contain up to 5%, or even more, of vegetable matter.

The morphology of wool fibres is complex. A schematic diagram showing the morphology of fine wool is shown in Figure 3.2. There are essentially two main categories of cell: cortical cells, which account for *ca.* 90% of each fibre, and cuticular cells, which are present at the fibre surface. The cuticle is separated from the underlying cortex by a cell membrane complex, which also holds adjacent cortical cells together. The cell membrane complex is, therefore, the only continuous phase in a wool fibre. Coarser wool fibres often contain a medulla as well. The medulla comprises a central core of cells, either continuous or intermittent, arranged along the fibre axis between the cortical cells.

3.2.2 Cuticle

The structure of the cuticle is responsible for the surface properties of wool (discussed in Chapter 1). Each cuticular cell approximates to a rectangular sheet of length *ca.* 30 μm, breadth *ca.* 20 μm and thickness *ca.* 0.5 μm. As can be seen from Figure 3.2, there is considerable overlap of adjacent cuticular cells, giving rise to a scaly type of fibre surface. One consequence of this arrangement is the so-called directional frictional effect: the coefficient of friction from the root to the tip is considerably less than that from the tip to the root. The directional frictional effect is responsible for the innate ability of wool to felt, a

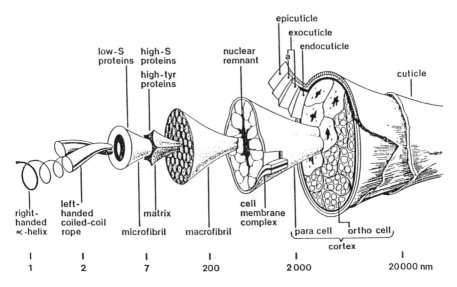

Figure 3.2 Schematic diagram of the morphology of a fine wool fibre.

unique property amongst textile fibres. Felting involves the progressive entanglement of wool fibres when they are subjected to mechanical action (particularly, agitation in water), the fabrics becoming thicker and bulkier. The process is utilised for the production of woollen felts, such as those used for overcoats and the playing surfaces of snooker tables. However, felting is a major factor causing shrinkage of woollen garments during washing.

The cuticle accounts for *ca.* 10% of the fibre. It consists of three layers: the epicuticle, the exocuticle and the endocuticle (see Figure 3.3). The epicuticle is identified as a thin surface membrane, containing predominantly keratin chains with a small proportion of lipid. Although it is traditionally considered as an individual component of the cuticle, it is now realised to be part of the resistant membrane that surrounds all cuticular and cortical cells. The exocuticle, *ca.* 0.3 μm thick, comprises 60% or so of each cuticular cell. There are two subcomponents of the exocuticle, although neither is well defined. The A-layer, which lies adjacent to the epicuticle, appears to possess a much higher cystine content than the B-layer, which forms the rest of the exocuticle. The endocuticle is a well defined layer, which lies below the exocuticle. In fine wool, the endocuticle is *ca.* 0.2 μm thick and comprises *ca.* 40% of the cuticle. The level of cystine in the endocuticle is very low; its protein content is non-keratinous. Mechanically, the endocuticle is a weak part of the wool fibre. It has been noted for example that, as carpets become

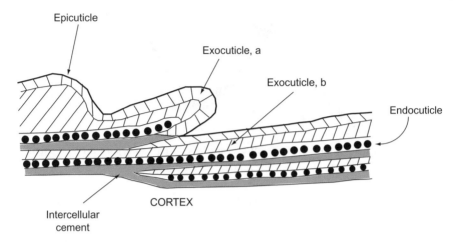

Epicuticle

Exocuticle, a

Exocuticle, b

Endocuticle

CORTEX

Intercellular cement

Figure 3.3 Schematic diagram of the cuticle of a wool fibre.

more worn through use, there can be preferential fracture in the endocuticle.

3.2.3 Cortex

The cortex comprises over 85% of a wool fibre, and is the main component governing its mechanical properties. It is extremely complex structurally, as Figure 3.2 indicates. It consists of spindle-shaped cells, generally *ca.* 100 µm long and 3–6 µm wide, aligned along the fibre axis. There are two principal types of cortical cell: the orthocortex and the paracortex. They are distinguished from each other by the distribution of non-keratinous material within each cell. In paracortical cells, the non-keratinous material is concentrated in particular regions called 'nuclear remnants'. In fine wools, such as Merino wool, the orthocortex and paracortex lie alongside one another throughout each fibre, as illustrated in Figure 3.2. The proportion of orthocortex is *ca.* 66% and that of paracortex is *ca.* 30%. The bilateral segmentation of the two types of cortical cell is considered to give rise to the natural crimp of fine wool fibres. The orthocortex is oriented towards the outside of the curl of the crimp (Figure 3.4). In coarser wools, as in Lincoln wool for example, the orthocortex is surrounded by the paracortex. In some cases, mesocortical cells are present at the boundary between the orthocortex and paracortex, but where present they normally comprise <4% of each fibre.

Each cortical cell consists of macrofibrils oriented along the fibre axis. The macrofibrils are cylindrical, and the diameter of each

Figure 3.4   Relationship between fibre crimp and *ortho-* and *para*-cortical segmentation.

cylindrical unit is *ca.* 0.3 µm. The length of a macrofibril can vary from
10 µm up to 100 µm, *i.e.* the length of an entire cortical cell. Each macrofibril is a cluster of hundreds of microfibrils embedded in a matrix.
Microfibrils are *ca.* 7 nm in diameter, and are formed from keratin
chains, comprising four sections of α-helix separated by three non-helical segments. The matrix consists of amorphous keratinous material.
Paracortical cells contain a higher proportion of matrix than the
orthocortical cells.

3.2.4   Cell Membrane Complex

The cell membrane complex is a continuous network throughout each
fibre and provides adhesion between the cells. It accounts for 3–3.5% of
the fibre. The complex consists of three principal components: an
intercellular cement (consisting of a non-keratinous protein), a lipid
component and a chemically resistant membrane. The membrane consists of keratinous material and forms the boundary between each cell
and the rest of the cell membrane complex.

   The cell membrane complex is relatively weak, probably due to its
component lipid material. Despite its low concentration in each fibre, it
therefore has a surprisingly large influence on fibre mechanical properties. Thus, the abrasion of a fabric during use can give rise to fracture
in the spaces between adjacent cortical cells.

3.2.5   Chemical Composition of Wool

Clean wool fibres consist overwhelmingly of proteins, with only <1%
containing non-protein matter. About 80% of the proteins are keratins,
which differ from other proteins in possessing high contents of sulfur.
The wide variation in fibre structure, described in Section 3.2.1, implies

several different types of keratin in different proportions from one type of wool to another.

Keratins, like all proteins, are condensates of naturally occurring amino acids, whose general formula can usually be represented as follows:

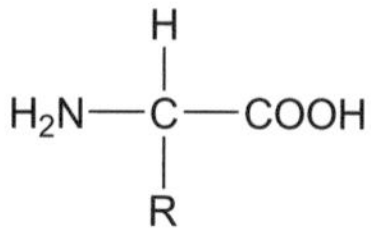

$$H_2N-\underset{R}{\overset{H}{\underset{|}{\overset{|}{C}}}}-COOH$$

The amino acids are joined by peptide (–CO–NH–) linkages. Thus, the repeat unit in each keratin chain can be represented as shown below:

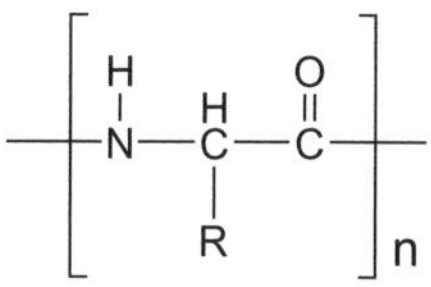

$$\left[-\underset{}{\overset{H}{\underset{|}{N}}}-\underset{R}{\overset{H}{\underset{|}{C}}}-\overset{O}{\overset{||}{C}}-\right]_n$$

Moreover, at one end of the chain there is a primary amine group (–NH₂) and at the other end a carboxyl group (–COOH). The nature of the side-chain, R, (or residue, as it is often called) distinguishes one amino acid from another. With the exception of glycine, for which R is hydrogen, the amino acids possess asymmetric carbon atoms, which render them optically active and, as in other natural proteins, laevo-rotatory. They are, therefore, L-α-amino acids. Two exceptions to the general formula are cystine, which plays a key role in wool chemistry, and proline.

Hydrolysis of wool keratins with strong acids (*e.g.* 6 M hydrochloric acid solution for 24 h) yields, with the exception of tryptophan, the amino acids listed in Table 3.2. Tryptophan is destroyed, but its content can be determined after hydrolysis with a mixture of toluene-4-sulfonic acid and tryptamine:

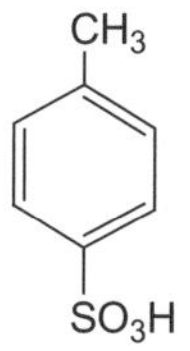

Toluene-4-sulfonic acid 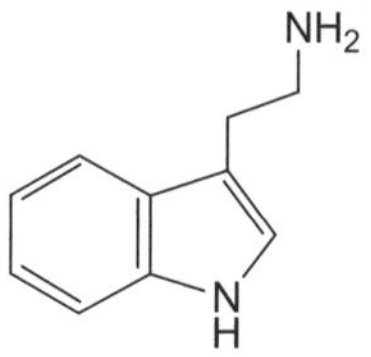   Tryptamine

Table 3.2 Amino acids present in wool.

No.	Amino acid	Structure of side chain (R)	Mol %
1	Glycine	H–	8.4
2	Alanine	CH_3–	5.4
3	Phenylalanine	C_6H_5–CH_2–	2.9
4	Valine	$(H_3C)_2CH$–	5.6
5	Leucine	$(H_3C)_2CH$–CH_2—	7.7
6	Isoleucine	H_3C-H_2C, H_3C CH–	3.1
7	Serine	HO–CH_2–	10.4
8	Threonine	H_3C, HO CH—	6.4
9	Tyrosine	HO–C_6H_4–CH_2–	3.8
10	Proline (complete formula)	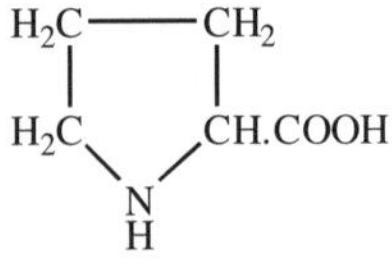 H_2C—CH_2 / H_2C CH.COOH \ N H	6.6
11	Methionine	CH_3–S–CH_2–CH_2–	0.5
12	$\frac{1}{2}$-Cystine	–CH_2–S–S–CH_2–	10.3
13	Arginine	H_2N—C—N—$CH_2CH_2CH_2$— (NH, H)	6.9
14	Lysine	H_2N–$CH_2CH_2CH_2CH_2$–	2.9
15	Tryptophan	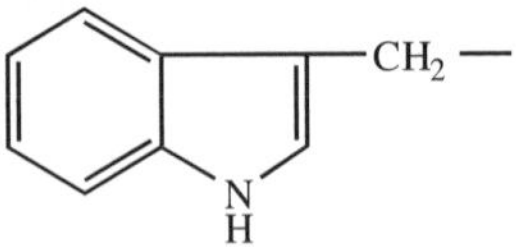 indole–CH_2—	0.5
16	Histidine	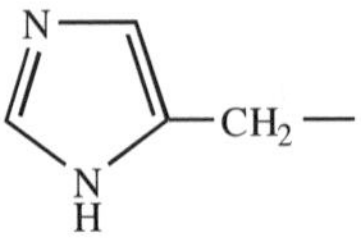 imidazole–CH_2—	0.9
17	Aspartic acid	HOOC–CH_2–	6.5
18	Glutamic acid	HOOC–CH_2CH_2–	11.9

The content of tryptophan obtained by this method is *ca.* 0.5 mol%. The residues, R, in Table 3.2 vary extensively in size and chemical nature. The R-groups of the top six amino acids are hydrocarbons and hence non-polar. The next three contain hydroxyl groups. Of the remaining amino acids, lysine, arginine and histidine possess basic groups (the content of histidine is quite small), and aspartic acid and glutamic acid possess carboxyl groups. It should be noted, however, that keratins also contain asparagine and glutamine residues:

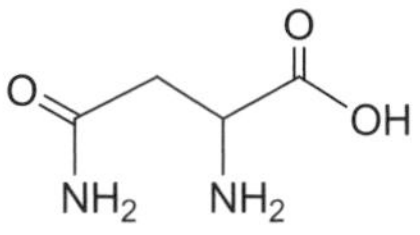

Asparagine 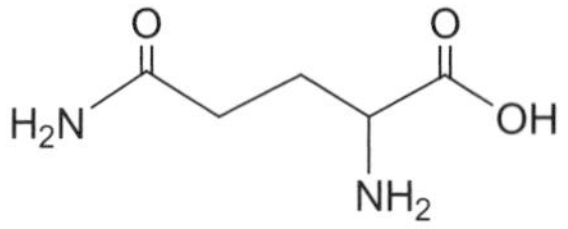   Glutamine

During acid hydrolysis, these residues are converted to aspartic acid and glutamic acid, respectively. From studies of the enzymatic digestion of wool, it has been found that the converted asparagine accounts for *ca.* 60% of the aspartic acid content in Table 3.2 and that glutamine accounts for *ca.* 45% of the glutamic acid. Thus, the true content of aspartic acid in keratin is *ca.* 2.6 mol% and of glutamic acid *ca.* 6.5 mol%. It is significant too that there is then an approximate balance between the content of basic and acidic amino acids in keratin. Where a unit of proline (which is strictly an imino acid!) occurs in a keratin chain, the adjacent peptide bonds are almost at right angles to each other, as a result of the orientation of the imino and carboxyl groups. Proline units, therefore, exert a marked influence on the conformation of the keratin chain, and tend to impede the formation of regular α-helices.

A particularly significant amino acid in wool chemistry and technology is cystine. The disulfide bonds in cystine can link adjacent keratin chains through covalent cross-links:

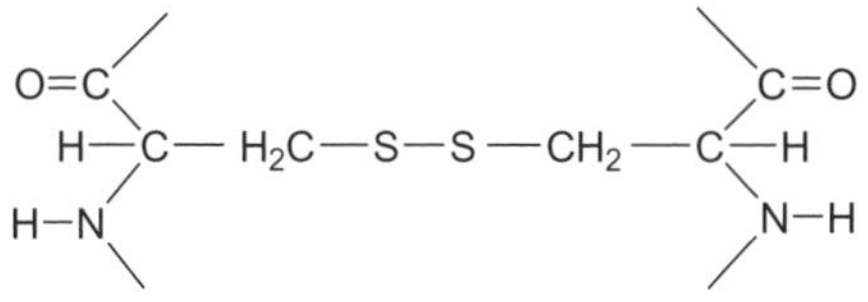

Hence, the content of cystine is given in Table 3.2 in terms of 1/2-cystine. Disulfide bonds can also link different parts of the same keratin chain.

Moreover, the sulfur content of keratins is almost completely derived from cystine; methionine, the other main naturally occurring amino acid containing sulfur, is almost completely absent. The cleavage and subsequent rearrangement of disulfide bridges between keratin chains is the basis for important technological processes, such as shrinkproofing and setting.

Although disulfide bridges account for most of the cross-links, a number of other types of cross-links are also present, as shown in the structural diagrams on p. 71. Isopeptide cross-links occur to a small extent: these join the ε-amino group of a lysine residue to a γ-carbonyl group of a glutamic acid residue (or, more rarely, to the β-carboxyl group of an aspartic acid residue). Non-covalent links, such as hydrogen bonds and salt bridges, are also present. An amino or hydroxyl group in one keratin chain can be linked through hydrogen bonding to a carboxyl group in an adjacent keratin chain. Furthermore, there is extensive hydrogen bonding within the α-helical structure. Salt bridges are ionic bonds, linking amino and carboxyl groups. These groups account for the amphoteric nature of wool. In environments of low pH, for example, the amino groups in keratin are predominantly protonated, with the result that wool carries an overall positive charge, whilst at neutral pH there is a slight overall negative charge. The isoelectric point, the pH at which wool is neutral, is 4.5–5.0, depending on the nature of the fibres.

The values of the amounts of the individual amino acids listed in Table 3.2 are subject to appreciable variation, but nevertheless they show that some amino acids are much more prevalent (such as cystine and serine) than others (such as histidine and methionine). As indicated in Section 3.2.1, genetic, physiological and nutritional factors all influence the content of each amino acid. Moreover, wool contains *ca.* 170 different types of protein altogether, so the variation observed in the overall amino acid content of wool is perhaps not surprising.

3.2.6   Isolation and Location of Keratins

Different groups of keratins can be isolated using a reduction–carboxylation method. The cystine residues linking the keratin chains are first reduced to cysteine residues:

$$-S-S- \rightarrow -SH + HS-$$

a reaction which breaks the disulfide bridges and produces thiol (–SH) groups. The thiol groups produced are immediately treated with iodoacetate:

$$-SH + ICH_2COO^- \rightarrow -S-CH_2COO^-$$

Reoxidation of the thiol groups is prevented, and the treated proteins are solubilised by virtue of their anionic groups. The resulting material is known as *S*-carboxymethylkeratin (SCMK). Non-keratinous proteins, whose cystine content is much lower, are not solubilised by this technique, and they are separated out as a solid residue.

Three protein fractions can be isolated from SCMK, for example by gel electrophoresis: low-sulfur proteins, high-sulfur proteins and high-glycine/high tyrosine proteins. Some characteristics of these proteins are given in Table 3.3. Each of these fractions is composed of several different families, and each family comprises a number of closely related members. The amino acid sequences of many of the constituent proteins are now known. Low-sulfur proteins have partially α-helical conformations, and are rich in lysine, arginine, aspartic acid, glutamic acid and leucine. These amino acids favour the formation of α-helices. Low-sulfur proteins are predominantly located in the microfibrils. In high-sulfur proteins, cystine, proline, serine and threonine constitute >50% of the total content of amino acid. The high level of cystine confers a high

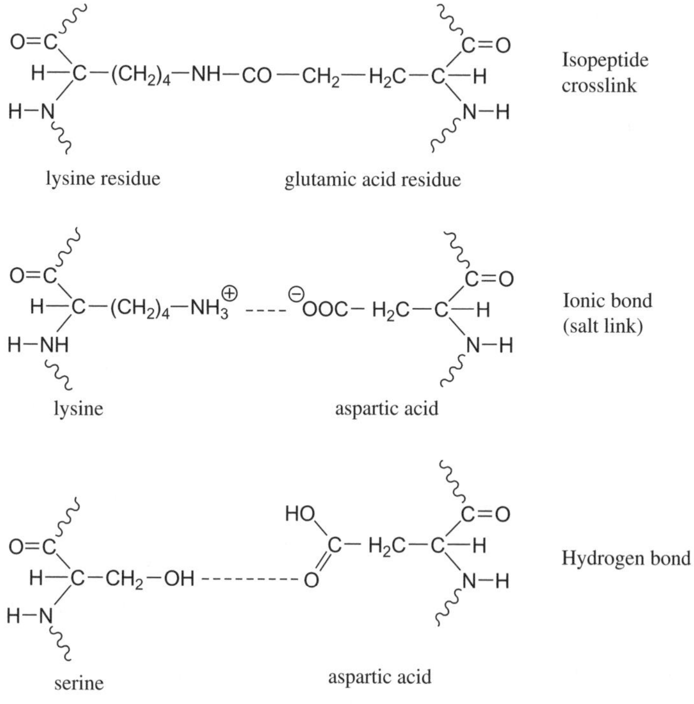

Isopeptide crosslink

lysine residue   glutamic acid residue

Ionic bond (salt link)

lysine   aspartic acid

Hydrogen bond

serine   aspartic acid

Table 3.3 Characteristics of proteins in wool.

Protein fraction	Amount, %	Sulfur content, %	Molar mass[a]
Low-sulfur	58	1.5–2.0	45 000–50 000
			45 000–60 000
High-sulfur	18	4–6	14 000–28 000
			11 000–23 000
Ultrahigh-sulfur	8	8	28 000
			37 000
High-Gly/Tyr	6	0.5–2.0	9 000–13 000
			11 000–12 000

[a]Values vary according to source.

density of disulfide bridges between adjacent keratin chains. In turn, the high density of disulfide bridges, together with the high content of proline, suppresses the formation of α-helices in high-sulfur proteins. The high glycine/high tyrosine proteins also possess a high content of serine. High sulfur and high glycine/tyrosine proteins are located chiefly in the matrix.

The overall content of cystine in the wool fibre cuticle is higher than in wool as a whole. It appears that the cuticle contains some proteins not found in the cortex. The cuticle is rich in proline, glycine, serine and valine – and also in cysteic acid, an oxidation product of cystine (see Section 3.2.8.5). The formation of α-helices is largely precluded. The keratin content is largely concentrated in the exocuticle, especially the A-layer. The proteins in the endocuticle are non-keratinous.

3.2.7 Processing of Wool Fleeces

The impurities described in Section 3.2.1 need to be removed from a wool fleece before it is ready to be processed into yarn, and then cloth. The principal process involved is scouring, in order to clean the fleece, but subsequent bleaching is also sometimes required, in order to remove yellowness, especially if the wool is to be dyed to a pale shade. Carbonising, using dilute sulfuric acid to remove excess vegetable matter (which is primarily cellulosic), may be required too. The action of acids on cellulose has already been discussed in Section 2.2.5.1.

Wool grease and suint can be removed by emulsification in a solution containing alkali and non-ionic detergent. Although this approach is capable of removing the high level of impurities present in the raw wool, the principal difficulty is that wool is readily degraded by alkaline solutions, particularly at higher temperatures, as discussed in

Section 3.2.8.2. Strong alkalis, such as sodium hydroxide, are therefore never used in practice, and even weak alkalis, such as sodium carbonate, are sometimes avoided. To preserve the wools' properties, scouring is best conducted under conditions as close to neutral pH as possible.

The extent of yellowness in wool depends on the breed of sheep and the conditions of rearing. Moreover, fibre tips are often quite heavily tinted by weathering. Bleaching can be achieved either by oxidation or reduction. For oxidative bleaching, hydrogen peroxide is preferred. Its bleaching action is especially effective in alkaline solution. Alkali releases the perhydroxy anion (OOH^-) which reacts further with hydrogen peroxide to yield the superoxide radical anion ($\cdot O_2^-$), believed to be the principal species responsible for bleaching. As already noted, however, alkali degrades wool fibres, and indeed bleaching with hydrogen peroxide is generally conducted at pH 8–9, obtained using ammonia, for 1–2 h at 50–60 °C. Stabilisers are present too, to slow down the decomposition of hydrogen peroxide into water and oxygen. Traces of heavy metals present in the processing water can bring about decomposition of hydrogen peroxide. Tetrasodium pyrophosphate and sodium tripolyphosphate are both used as stabilisers, although there is now increasing concern about the presence of phosphates in effluents. Silicates and oxalates have also been used as stabilisers.

To avoid the risk to fibre damage that alkalis promote, wool is sometimes bleached with hydrogen peroxide under weakly acidic conditions, *e.g.* at pH 5 for 1 h at 80 °C. The degree of whiteness achieved is poorer, but still quite often adequate for subsequent processing, for example, if the wool is to be dyed to a heavy depth of shade.

Wool can be bleached by a reductive process. The reducing agent most commonly utilised is sodium dithionite ($Na_2S_2O_4$) at pH 5.5–6.0 for approximately 1 h at 50–60 °C. Sodium hydroxymethanesulfinate ($Na^+HOCH_2SO_2^-$) and zinc hydroxymethanesulfinate ($Zn^{2+}(HOCH_2SO_2^-)_2$) are less reactive than sodium dithionite, and so are used at a higher temperature: 90 °C at pH 3 for 30 min. However, all three reducing agents tend to cause fibre damage. Fibre damage can, however, be lessened if reduction is effected in the presence of cross-linking agents, such as 1,2-dibromoethane (CH_2Br-CH_2Br) or ethanedial (glyoxal; $O{=}CH-CH{=}O$). The action of cross-linking agents is discussed later in Section 3.2.8.7.

An alternative reducing agent, which causes less fibre damage, is thiourea dioxide, applied at pH 7 for 1 h at 80 °C. The active bleaching species is sulfinic acid:

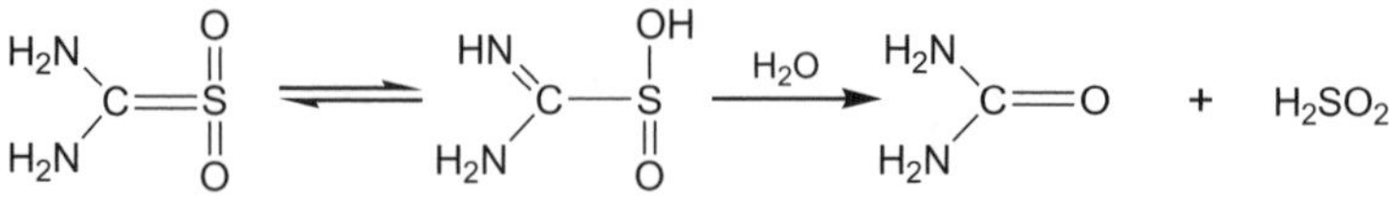

$$H_2N \diagdown C=S \diagup ^{O}_{O} \quad \rightleftharpoons \quad HN \diagdown C-S \diagup ^{OH} _{O} \quad \xrightarrow{H_2O} \quad H_2N \diagdown C=O \quad + \quad H_2SO_2$$

3.2.8 Chemical Reactions of Wool

3.2.8.1 Introduction. Wool fibres can be degraded by a variety of chemical agents. Even mild chemical degradation can exert a marked influence on the fibres' physical and mechanical properties. Wool chemistry is dominated by the reactions of the cystine residues, because of the structural integrity conferred on keratins by the disulfide bridges. Indeed, some of these reactions are the basis of several important technological modifications to wool, especially setting and shrinkproofing. Nevertheless, some of the other constituent amino acids can also be chemically modified.

3.2.8.2 Alkali. As noted in Section 3.2.7, wool is particularly prone to damage by alkali. Indeed, it can be dissolved completely in 5% sodium hydroxide solution at 100 °C. Alkali attacks the disulfide bridges and can hydrolyse the peptide linkages along each keratin chain. Thus, the use of alkali in wool processing technology has to be severely restricted.

The chief products formed from attack by alkali are lanthionine and lysinoalanine residues. Minor products, such a β-aminoalanine and ornithinoalanine residues, have also been isolated:

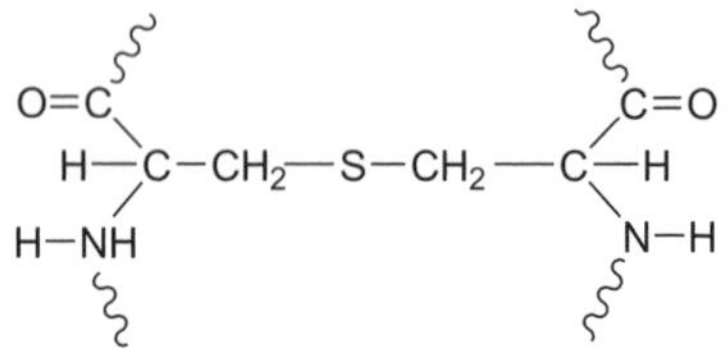

lanthionine residue

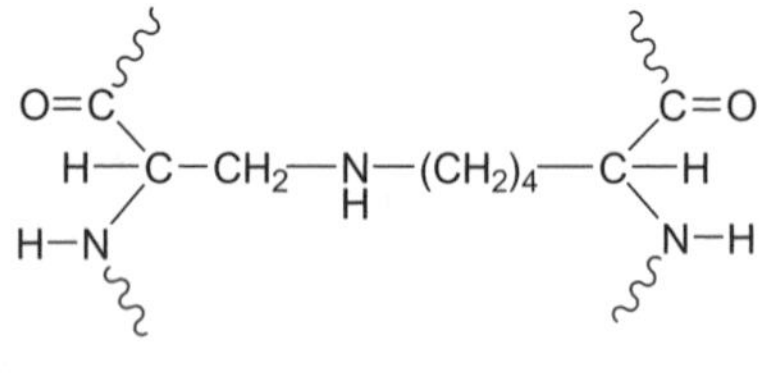

lysinoalanine residue

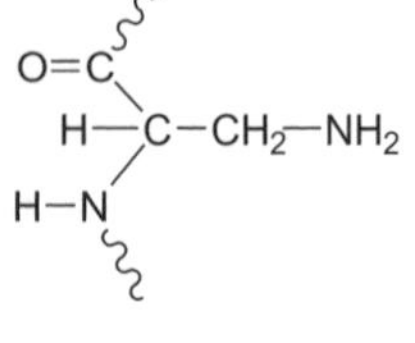

β-aminoalanine residue

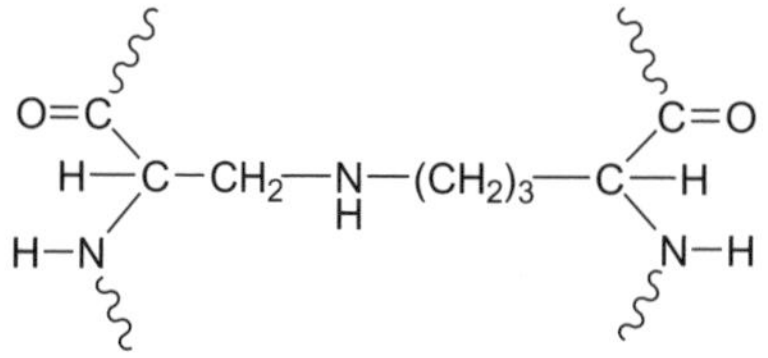

ornithinoalanine residue

All these products arise from attack on the disulfide bridges. The underlying mechanism is complex and probably involves several pathways. The initial steps appear to involve deprotonation of β-carbon atoms in disulfide bridges and the formation of dehydroalanine residues. Cysteine residues and sulfur are also produced. Reaction Schemes 3.1 and 3.2 can account for these observations.

In the first reaction scheme, attack by alkali on one of the β-carbon atoms yields a dehydroalanine residue and perthiocysteine residue. The latter then decomposes to sulfur and a cysteine residue. In the second reaction scheme, there is attack by alkali on both β-carbon atoms in a disulfide bridge. Lanthionine and lysinoalanine residues are then formed through addition of a cysteine or lysine residue, respectively, to the dehydroalanine residue as follows:

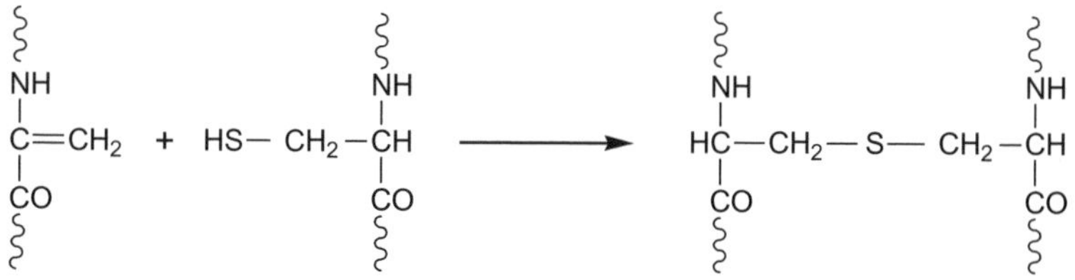

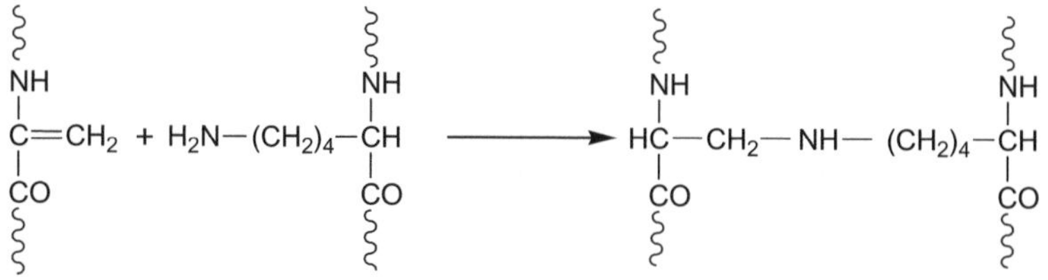

The changes to keratin structure affected by alkali form the basis of an alkali solubility test for wool damage, which proved useful before more modern analytical techniques became routinely available. The test involves the extraction of wool by 0.1 M sodium hydroxide at 65 °C for

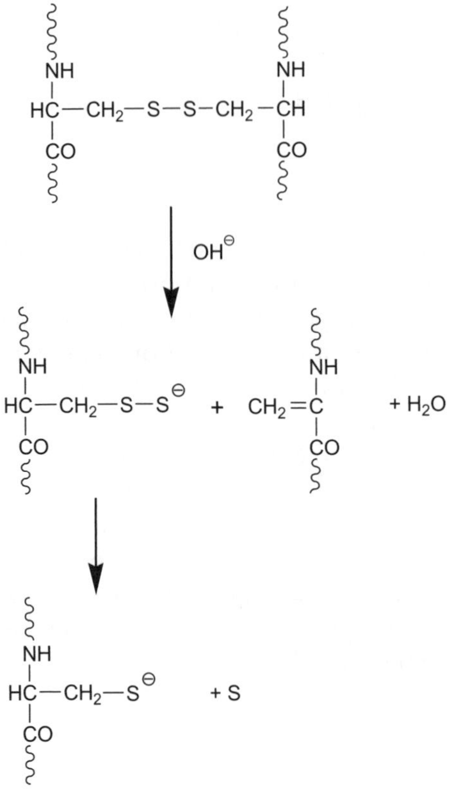

Scheme 3.1

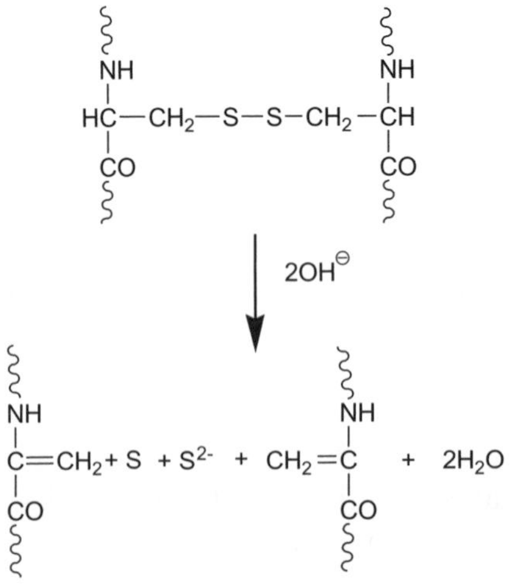

Scheme 3.2

1 h. Thus, comparisons can be made of the damage caused to wool fibres by different treatments, such as bleaching and carbonisation.

It is noteworthy that primary aliphatic amines have a similar effect to alkali on wool fibres. Dehydroalanine residues are again formed as a result of deprotonation of β-carbon atoms in the disulfide bridges. These residues then react with the primary amine, to yield β-(*N*-alkylamino) alanine residues as shown:

$$
\begin{array}{c}
| \\
NH \\
| \\
C{=}CH_2 \\
| \\
CO \\
|
\end{array}
\;+\; H_2N{-}R \;\longrightarrow\;
\begin{array}{c}
| \\
NH \\
| \\
HC{-}CH_2{-}HN{-}R \\
| \\
CO \\
|
\end{array}
$$

In addition, lanthionine can also be formed from the action of cyanide ions on wool:

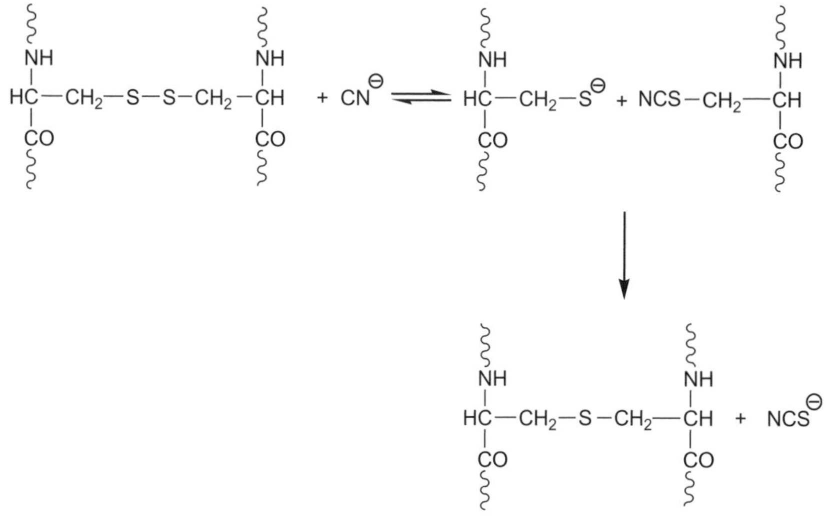

3.2.8.3 Acid. Wool fibres are more resistant to acids. Indeed, they are often dyed under acidic conditions, with some types of dye being applied at a pH as low as 2. As already explained in Section 3.2.5, the amino groups present at the ends of the keratin chains and in some of the amino acid residues along the chains become protonated. Dyes applied to wool are anionic and therefore at lower pHs there is extensive ionic attraction between dye and fibres.

Under much harsher conditions, acids hydrolyse the peptide linkages in keratins to their constituent amino acids, with the exception of tryptophan, which is normally destroyed. It has been noted above (see

Section 3.2.5) that hydrolysis forms the basis of the quantitative analysis of amino acids in proteins. The mechanism can be summarised as follows:

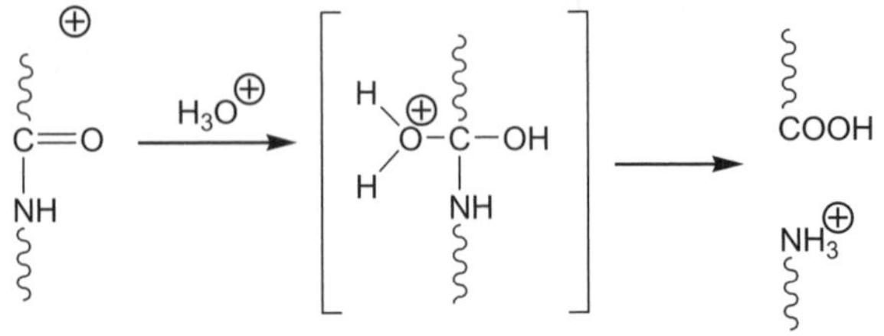

$$
\begin{array}{c}
\overset{\oplus}{}\\
C{=}O \xrightarrow{H_3O^{\oplus}} \left[\;\;\overset{H}{\underset{H}{\diagdown}}O{-}\overset{\oplus}{C}{-}OH \;\;\right] \longrightarrow COOH \\
| \qquad\qquad\qquad\qquad\qquad | \\
NH \qquad\qquad\qquad\qquad\quad NH \qquad\qquad NH_3^{\oplus}
\end{array}
$$

Acid also converts asparagine and glutamine residues to aspartic acid and glutamic acid residues, respectively. These reactions are considerably faster than the hydrolysis of the peptide linkages.

Acidic conditions also prompt an $N \rightarrow O$ rearrangement of serine and threonine residues, both of which possess hydroxyl groups. This reaction is illustrated for serine:

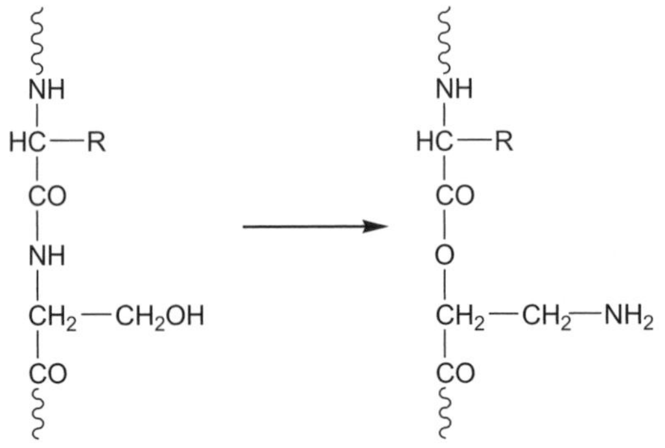

$$
\begin{array}{ccc}
NH & & NH \\
| & & | \\
HC{-}R & & HC{-}R \\
| & & | \\
CO & & CO \\
| & \longrightarrow & | \\
NH & & O \\
| & & | \\
CH_2{-}CH_2OH & & CH_2{-}CH_2{-}NH_2 \\
| & & | \\
CO & & CO
\end{array}
$$

The ester bonds formed as a result of this rearrangement are much more readily hydrolysed than the original peptide bonds.

3.2.8.4 Reduction. Reduction of wool is important technologically, as it forms the basis of the setting of creases and pleats in wool fabrics (see Section 3.2.9). The reagents most commonly used to reduce wool are thiols and phosphines. Amongst all the amino acid residues, only the cystine residues are affected.

Reduction by thiols occurs through reversible displacement reactions:

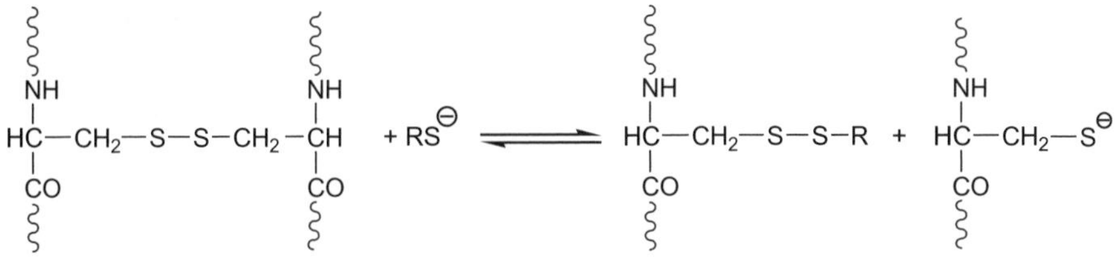

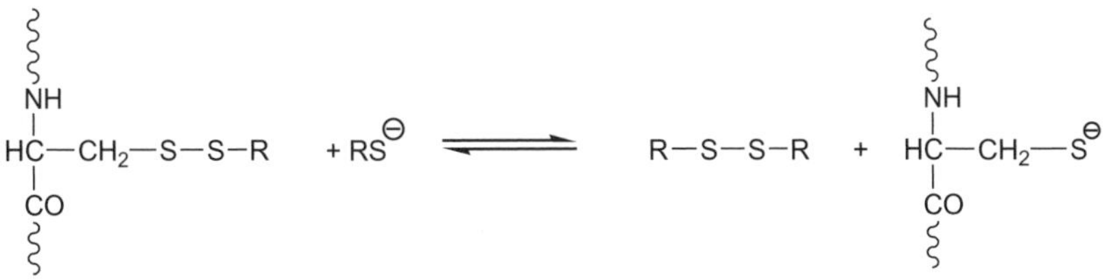

For complete reduction, therefore, a large excess of thiol is needed. Among the thiols that have been applied are: thioglycolic acid (HS–CH$_2$–COOH), its ammonium salt and toluene-ω-thiol (C$_6$H$_5$–CH$_2$–SH). In particular, the thiol must be ionised, so alkaline conditions are often necessary. As discussed in Section 3.2.8.2, however, alkali attacks wool, so the use of thiols at high pHs is best avoided. Some thiols, nevertheless, can be successfully used at pH 5–6, such as 2-hydroxyethanethiol (HO–CH$_2$–CH$_2$–SH) and 1,4-dithiothreitol (HS–CH$_2$–CH(OH)–CH(OH)–CH$_2$–SH).

The thiol groups formed in the reduced wool are, however, rapidly converted back to disulfide groups by oxidation in air. Their oxidation can be prevented by subsequent treatment with alkylating agents. As described in Section 3.2.6, this approach, using iodoacetate as the alkylating agent, is used for isolating keratin fractions.

Reduction by phosphines can be represented as shown in the following reaction mechanism:

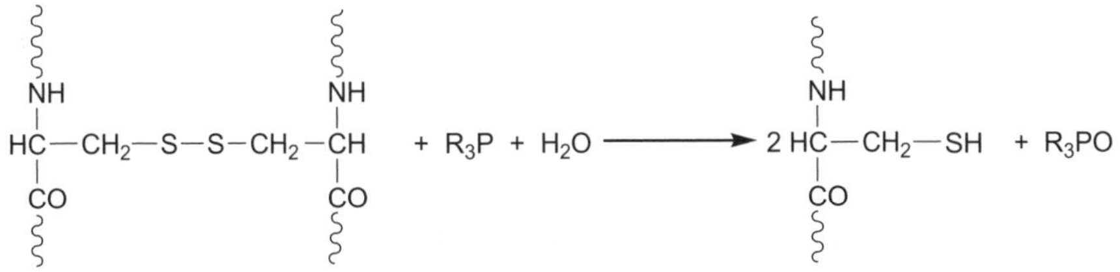

One particularly effective phosphine is tri-*n*-butylphosphine. The disulfide bridges in wool keratins are reduced in almost quantitative yield in ambient temperatures at pH 1–8, without any damage to the fibres. Another effective phosphine is tetrakis(hydroxymethyl) phosphonium chloride (THPC). Strictly, this is the precursor of the active reducing agent, tris(hydroxymethyl) phosphine, to which THPC is converted in aqueous solution:

$$
HOH_2C \underset{\underset{CH_2OH}{|}}{\overset{\overset{CH_2OH}{|}}{\overset{\oplus}{P}}} CH_2OH \quad Cl^{\ominus} \longrightarrow \underset{\underset{CH_2OH}{|}}{\overset{\overset{CH_2OH}{|}}{P}} {-} CH_2OH \quad + \; HCHO \; + \; HCl
$$

However, it can be seen that the reaction also yields methanal (formaldehyde), so the use of THPC is in practice limited. Moreover, some unstable derivatives of cysteine and tyrosine can occur when THPC is applied. Phosphines also possess the extra advantage that they react only slowly with alkylating agents. Reduction of disulfide bridges and subsequent alkylation can, therefore, be achieved in the same solution.

Reduction of wool can also be achieved by sulfitolysis, using for example sodium sulfite. A number of reactions with wool can occur, however, because the dissolution of sodium sulfite brings about an equilibrium between sulphurous acid (H_2SO_3), bisulfite and sulfite ions. Even metabisulfite ($S_2O_5^-$) ions may be present as well. The proportions of each species depend on the pH. A thiol and a sulfosulfonate anion are produced by sulfitolysis:

$$
\underset{\underset{CO}{|}}{\overset{\overset{NH}{|}}{HC}}{-}CH_2{-}S{-}S{-}CH_2{-}\underset{\underset{CO}{|}}{\overset{\overset{NH}{|}}{CH}} + SO_3^{2-} \rightleftharpoons \underset{\underset{CO}{|}}{\overset{\overset{NH}{|}}{HC}}{-}CH_2{-}S^{\ominus} + \underset{\underset{CO}{|}}{\overset{\overset{NH}{|}}{HC}}{-}CH_2{-}S{-}SO_3^{\ominus}
$$

$$
\underset{\underset{CO}{|}}{\overset{\overset{NH}{|}}{HC}}{-}CH_2{-}S{-}S{-}CH_2{-}\underset{\underset{CO}{|}}{\overset{\overset{NH}{|}}{CH}} + HSO_3^{\ominus} \rightleftharpoons \underset{\underset{CO}{|}}{\overset{\overset{NH}{|}}{HC}}{-}CH_2{-}SH + \underset{\underset{CO}{|}}{\overset{\overset{NH}{|}}{HC}}{-}CH_2{-}S{-}SO_3^{\ominus}
$$

The extent of sulfitolysis is, in practice, influenced by several factors. Although pH is the most significant, the concentration of sodium sulfite and the nature of the buffer ions also play a part. It has been reported that, in the absence of buffer, maximum sulfitolysis is achieved at pH 3.4, but in citrate buffer at pH 4.5 and in ethanoate (acetate) buffer at pH 5.0–5.5. In all cases, however, the maximum extent occurs at acidic pHs. At these pHs, the bisulfite ion is by far the predominant species, and so it has been proposed that it is therefore the reactive species in sulfitolysis. However, the sulfite ion appears to react about 1000 times more quickly than the bisulfite ion with the amino acid, cystine, so even in weakly acidic conditions, reaction with sulfite ions may also be significant.

Sulfitolysis of wool is important technologically, in that it is the process underlying the setting of creases and pleats in wool fabrics (see Section 3.2.9). Moreover, if sulfitolysis takes place in the presence of urea, which breaks the interchain hydrogen bonds, all the disulfide bridges are broken; the reaction, therefore, is very useful for the determination of the cystine content of wool samples. Indeed, a urea/bisulfite solubility test has been developed to assess damage to wool fibres during processing. In general, damage by acid increases solubility due to hydrolysis of peptide linkages in the keratin chains. By contrast, damage by alkali tends to reduce solubility, because of the replacement of disulfide bonds by lanthionine and lysinoalanine cross-links, which are incapable of reduction.

3.2.8.5 Oxidation. Cystine, cysteine, methionine and tryptophan are the amino acid residues that are most susceptible to oxidation. However, as the content of cystine is by far the greatest amongst these amino acids, the major effect of treating wool with oxidising agents is oxidation of the cystine residues. Oxidation appears to proceed *via* a series of products, with complete oxidation yielding cysteic acid residues, through cleavage of sulfur–sulfur bonds:

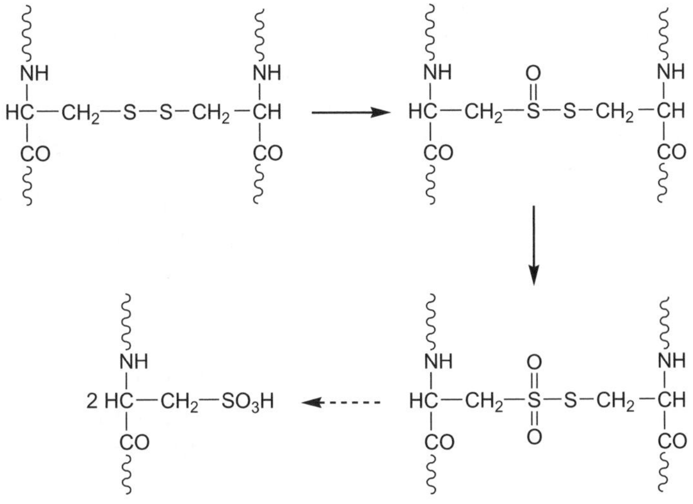

Indeed, complete oxidation forms the basis of a method for isolating different protein fractions from wool, and this method was originally used extensively, before the reduction/carboxylation method became widespread.

To isolate wool protein fractions using an oxidation approach, peracetic acid and performic acid have been used as oxidising agents. Performic acid is arguably the better reagent to use, in that it oxidises cystine residues quantitatively and does not cleave the peptide bonds in the protein chains. Unlike peracetic acid, however, performic acid is unstable in aqueous solution and has to be applied from formic acid solution. Three types of oxidised protein, known as keratoses, can be extracted: α-keratose, which corresponds to the low-sulfur proteins described in Section 3.2.6, γ-keratose, which corresponds to high-sulfur proteins, and β-keratose, believed to be derived from the cell membrane complex.

Another important oxidising agent is hydrogen peroxide, and its use as a bleaching agent has already been noted in Section 3.2.7. Under alkaline conditions, hydrogen peroxide oxidises cystine residues to cysteic acid residues, although the alkali present will itself react with cystine residues as well. Under acidic conditions, considerable amounts of the intermediate oxidation products can be detected. Hydrogen peroxide also cleaves peptide bonds.

3.2.8.6 Chlorination. Chlorine has traditionally been utilised extensively in the shrinkproofing of wool, although increasingly there are toxicological concerns about its use. Chlorination is normally carried out in acidic or neutral conditions. However, the nature of the oxidising species is dependent on pH. At very low pHs (<2), chlorine itself predominates. At pH 3–6, hypochlorous acid (HOCl) is the principal species, and hypochlorite (OCl^-) at pH >8. Thus, the nature and extent of chlorination are strongly influenced by pH. For example, hypochlorite reacts more slowly than chlorine or hypochlorous acid.

Chlorination oxidises cystine residues, to form cysteic acid residues. In acidic solutions, oxidation occurs through the series of intermediate products. In addition, there is extensive cleavage of peptide bonds at points in the keratin chains where tyrosine residues are present.

It is noteworthy that when wool is immersed in chlorine water, sacs appear almost immediately on the fibre surfaces. The sacs are enclosed by a thin membrane of the epicuticle. This phenomenon is known as the Allwörden reaction. The sacs, however, collapse on exposure to concentrated salt solutions, so it is evident that they are formed by osmotic swelling through absorption of water. It is thought that disulfide bridges and maybe peptide bonds are broken just below the semi-permeable surfaces of the cuticular cells. The consequent formation of soluble polypeptides triggers diffusion of water into the sacs by osmosis. Bromine water produces a similar effect.

3.2.8.7 Cross-Linking. Amongst textile fibres, wool is unusual in containing covalent cross-links between polymer chains. However, it is clear that these disulfide bonds succumb to a variety of treatments, and so there can be value in applying more resistant cross-linkages.

Methanal (formaldehyde) was for a long time the most established cross-linking agent for wool though, as observed above, the use of methanal is nowadays generally discouraged. Its mechanism of action is not clearly understood, but it seems that it can form hydroxymethyl ($-CH_2OH$) groups with several amino acid residues, principally lysine, with the possibility of the subsequent formation of methylene ($-CH_2-$) cross-links. Oxymethylene ($-CH_2-(OCH_2)_n-$) links may also be formed.

Several dialdehydes have also been investigated as alternatives to methanal. Glutaraldehyde is arguably the most effective, and it too reacts mainly with lysine residues. Indeed, the use of a bifunctional reagent that reacts with wool would seem to be an obvious approach to inducing further cross-linking. However, reaction conditions must be such that each functional group in the reagent reacts with a residue on a different keratin chain.

One class of cross-linking reagents comprises diepoxides, which can introduce cross-links between the amino groups of neighbouring keratin chains. The general formula of many diepoxides is:

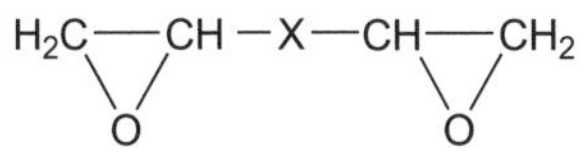

The cross-link formed between two lysine residues can be illustrated as follows:

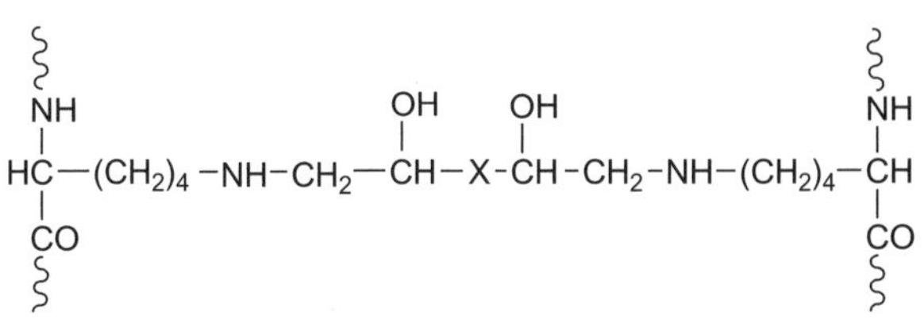

Another interesting cross-linking agent is 1,4-benzoquinone:

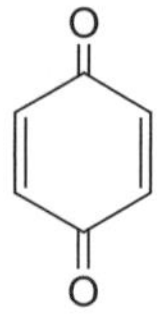

1,4-benzoquinone can react with primary amines as shown in the following reaction:

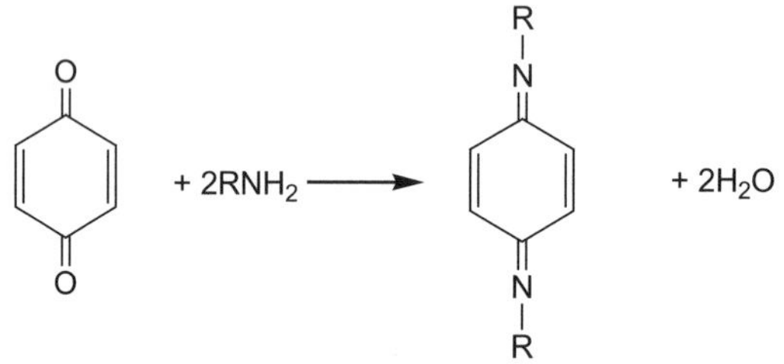

$$\text{benzoquinone} + 2RNH_2 \longrightarrow \text{diimine product} + 2H_2O$$

On this basis, it is considered that the main sites for reaction on keratin chains are the amino groups of the lysine residues. Other effective cross-linking agents include aryl and alkyl dihalides, and bifunctional acid chlorides and isocyanates.

3.2.8.8 Reactive Dyes. Reactive dyes can be applied to wool. Indeed, wool fibres possess several types of functional group with which these dyes are capable of reacting: –OH, –NH$_2$ and –SH. Reaction may be through nucleophilic substitution or addition, in an analogous manner to reaction with cellulosic fibres, as discussed in Section 2.2.5.5. An unusual reaction is with α-bromoacrylamido dyes, which react with primary amine groups in wool to yield aziridine derivatives (where D indicates the rest of the dye molecule):

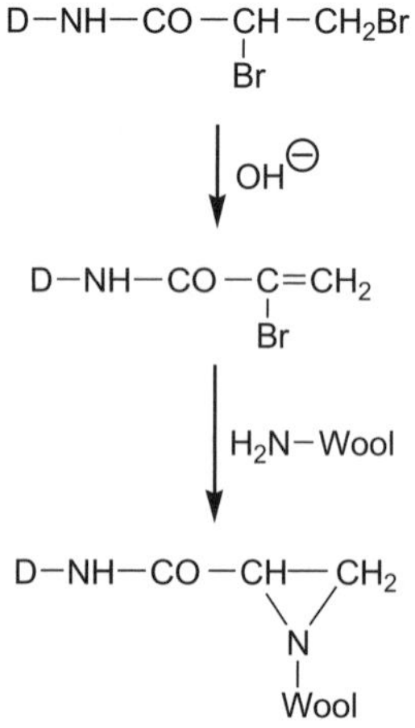

$$\text{D}-\text{NH}-\text{CO}-\underset{\underset{\text{Br}}{|}}{\text{CH}}-\text{CH}_2\text{Br}$$

$$\downarrow \text{OH}^{\ominus}$$

$$\text{D}-\text{NH}-\text{CO}-\underset{\underset{\text{Br}}{|}}{\text{C}}=\text{CH}_2$$

$$\downarrow \text{H}_2\text{N}-\text{Wool}$$

$$\text{D}-\text{NH}-\text{CO}-\text{CH}-\text{CH}_2$$
$$\underset{\underset{\text{Wool}}{|}}{\text{N}}$$

The aziridine derivatives may react further with primary amine groups in wool to yield the following product:

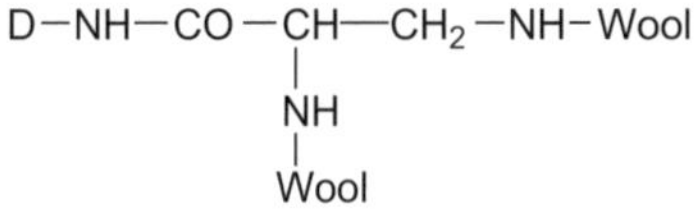

$$D-NH-CO-CH-CH_2-NH-Wool$$
$$|$$
$$NH$$
$$|$$
$$Wool$$

3.2.8.9 Action of Heat. Wool fibres are required to withstand steam or boiling water during a number of processing stages, such as scouring and dyeing. During use, wool fabric is also required to withstand hydrothermal conditions, as in laundering and ironing. These conditions can exact a number of changes in wool. Even at 100 °C, there is slight evolution of hydrogen sulfide and ammonia, accompanied by the production of a few lanthionine and lysinoalanine residues. The evolution of ammonia may arise to some extent from the hydrolysis of amino acid residues containing amide groups. In addition, there is some destruction of cystine residues, and thiol and aldehyde groups are formed. One scheme proposed to account for this observation is as follows:

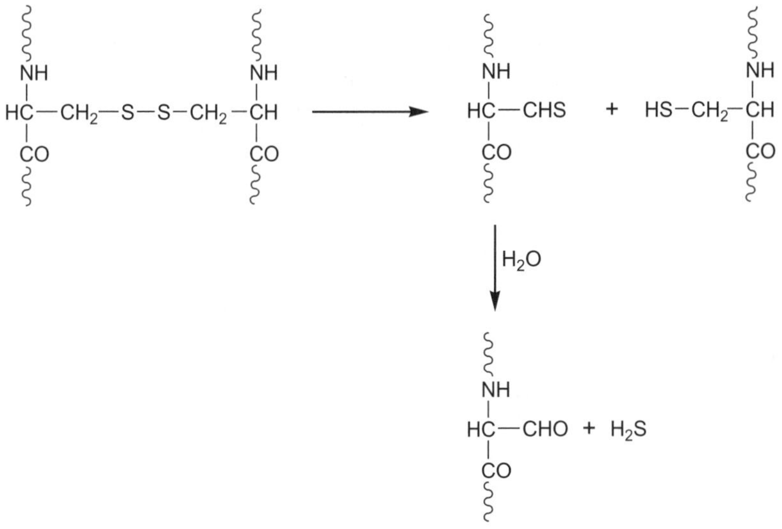

Hydrothermal treatment above 100 °C induces marked changes in the mechanical properties of wool, such as an increase in its elastic modulus, an effect attributed to reorganisation of the protein chains in the fibre matrix. Between 128 °C and 140 °C, an appreciable reduction in length of fibre is observed, a phenomenon known as supercontraction. The transition occurs at quite a sharp temperature, whose value depends on the rate of heating.

There is less damage to wool in the absence of water or steam, though it should be noted that 'dry' wool contains an appreciable proportion of water, as shown in Table 1.5. This absorbed water, in conjunction with atmospheric oxygen and water vapour, are the main influences on the thermal changes caused to wool. Below 140 °C, wool undergoes little damage, unless it is heated over a prolonged period. Above 140 °C, wool gradually yellows and appears scorched, and the mechanical performance of the fibres is reduced. Above 200 °C, cystine residues decompose, and their decomposition is accompanied by the formation of residues of cysteic acid, lanthionine, lysinoalanine and β-aminoalanine. There is also partial destruction of those amino acid residues containing hydroxyl groups. Above 250 °C, wool undergoes pyrolysis.

3.2.8.10   Action of Light.   Irradiation of wool with sunlight can cause either yellowing or bleaching, and indeed if sufficiently prolonged, can also cause loss of fibre strength. Maximum yellowing is observed at low wavelengths of sunlight, 290–320 nm (radiation from the sun of wavelengths <290 nm does not reach the earth's surface). Maximum bleaching occurs at 400–460 nm (blue light). The relative energies of the incident light over these two wavebands will determine whether yellowing or bleaching predominates. The intensity of the shorter waveband is more variable than that of the larger waveband, and depends on the season, the time of day and latitude. Yellowing and bleaching by light are promoted by the presence of water. Yellowing is also accelerated if the wool has already been chemically bleached, as described in Section 3.2.7.

The causes of yellowing have been the subject of considerable uncertainty. At one time, it was thought that damage to tryptophan residues in keratin chains was the prime cause of yellowing. Other amino acid residues that are partially decomposed are cystine, tyrosine and histidine. More recently, this conclusion has been questioned. Indeed, it has been suggested that small proline-rich proteins present in the epicuticle are responsible for much of the yellowing observed.

Bleaching using blue light has been tried as an alternative to bleaching by hydrogen peroxide, though both methods promote yellowing of wool on exposure to sunlight. The presence of water enhances the bleaching process. Thioglycolic acid also accelerates bleaching, but results in a loss of fibre strength. In addition, thioglycolic acid retained by the wool is difficult to remove, and the odour of thiol in the treated wool is evident. Less fibre damage is caused if thioglycolic acid is first complexed with zinc ions. Particularly effective bleaching occurs if wool is irradiated with blue light in the presence of dilute hydrogen peroxide solution, especially alkaline solution.

3.2.9 Setting

Setting refers to the stabilisation of a yarn or fabric in a desired conformation. Steam treatments are commonly used during industrial processing as a means of flat pressing wool fabrics and removing distortions in them. Pleats and creases can also be deliberately imparted to fabrics by steam setting. At a molecular level, setting involves the relaxation of stress through rearrangements in the conformations of keratin chains:

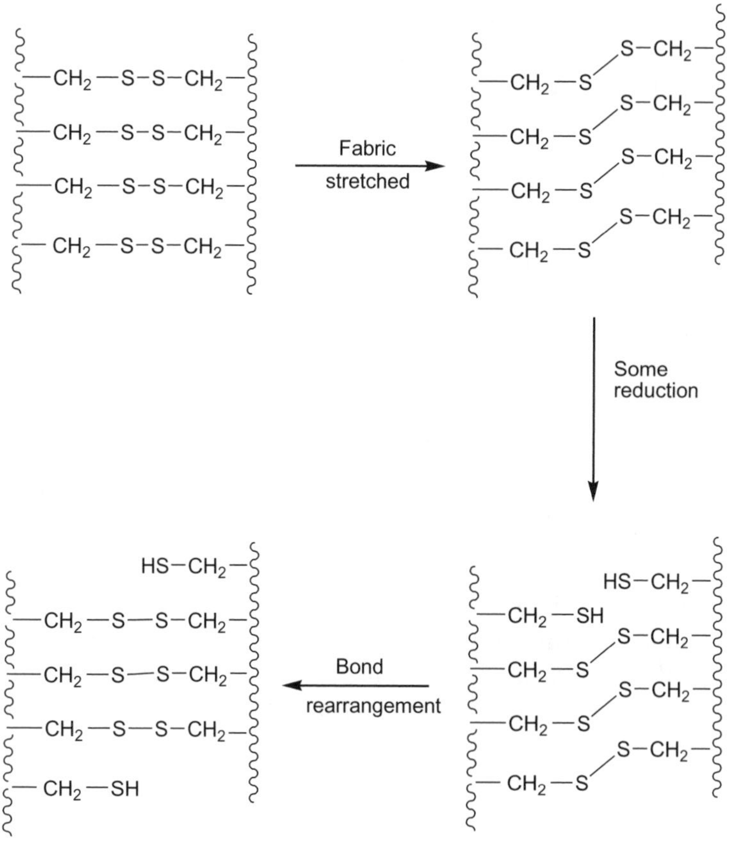

$$
\begin{array}{ccc}
-CH_2-S-S-CH_2- & & S-CH_2- \\
-CH_2-S-S-CH_2- & \xrightarrow{\text{Fabric stretched}} & -CH_2-S \\
-CH_2-S-S-CH_2- & & -CH_2-S \\
-CH_2-S-S-CH_2- & & -CH_2-S \\
\end{array}
$$

Some reduction

$$
\begin{array}{ccc}
HS-CH_2- & & HS-CH_2- \\
-CH_2-S-S-CH_2- & \xleftarrow{\text{Bond rearrangement}} & -CH_2-SH \\
-CH_2-S-S-CH_2- & & -CH_2-S \\
-CH_2-S-S-CH_2- & & -CH_2-S \\
-CH_2-SH & & -CH_2-S \\
\end{array}
$$

To achieve this relaxation of stress requires the rupture and renewed formation of interchain hydrogen bonds and disulfide bridges. Sufficient rearrangement of hydrogen bonds can be achieved by water. Rearrangement of disulfide bridges, however, is promoted through treatment with reducing agents, so that further thiol groups are introduced into the keratin chains. These thiol groups then induce rearrangement of the disulfide bridges by means of a thiol–disulfide interchange mechanism:

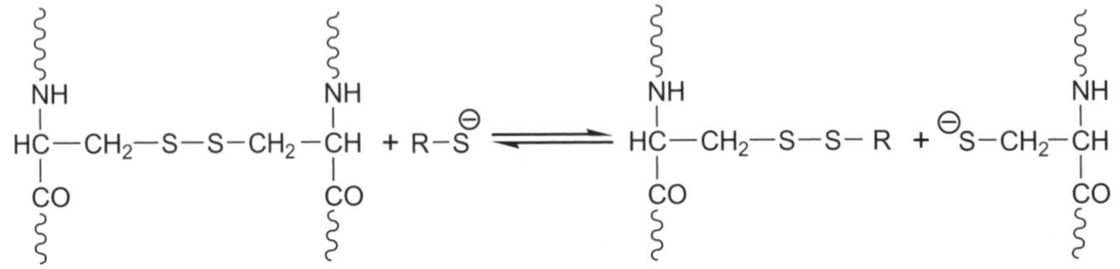

Sodium bisulfite and ammonium thioglycolate have been widely used as setting agents, but monoethanolamine sesquisulfite (MEAS; $(HO–CH_2CH_2–NH_3^+)_2\ SO_3^{2-}$) is nowadays more commonly applied. It reduces disulfide bridges by sulfitolysis.

3.2.10 Shrinkproofing

Almost all types of wool fabrics will shrink during use, unless some preventative treatment has been applied to them. Shrinkage is nearly always a consequence of felting (see Section 3.2.2). Two main types of treatment are available to prevent shrinkage. One is an oxidative method, which softens or even removes the cuticular scales. The other approach involves the deposition of a polymer, which masks the scale structure, or welds junctions between fibres, or even operates using both mechanisms. In many cases, an oxidative treatment is first applied, and then a suitable polymer is deposited on the fibre surfaces. Shrinkproofing treatments may be applied to wool fibres before they are spun into yarn, and also to woollen fabrics.

Shrinkproofing through oxidation has been traditionally achieved through chlorination, as outlined in Section 3.2.8.6. The most commonly used chlorinating agent is dichloroisocyanuric acid as illustrated below (or more usually its sodium or potassium salt):

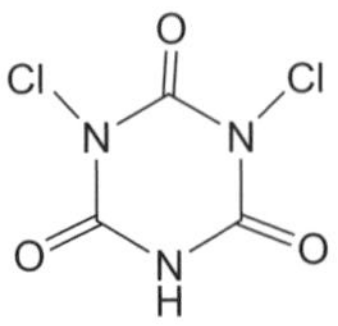

Potassium permanganate and permonosulfuric acid (Caro's acid; H_2SO_5) have also been used as oxidising agents. Oxidation converts disulfide bridges in the cuticular proteins to cysteic acid residues. Consequently, the fibre surfaces become strongly hydrophilic by virtue of the anionic sulfonic acid groups produced. The surface proteins swell, and the scales are softened. To ensure that oxidation is confined largely to the cuticle, oxidising agents can be dissolved first in strong salt solutions. Peptide linkages are also broken, principally where there are tyrosine residues. The degraded anioinic proteins that result may be progressively dissolved out of the fibres on repeated laundering; the shrink-resistant effect gradually wanes.

To prevent the leaching of degraded proteins from the fibre surfaces, they can be treated with a cationic polymer. The polymer forms complexes with these proteins; they are precipitated out of solution and are anchored more firmly to the fibre. Several cationic polymers are available commercially. One particularly effective polymer is Hercosett 57, a polyamide–epichlorohydrin polymer, produced by the Hercules Corporation. Other effective cationic polymers include Dylan GRC (Precision Processes Textiles Ltd.) and Basolan F (BASF). These polymers mask the degraded scale structures of the fibre surfaces.

Although shrinkproofing by chlorination and subsequent polymer treatment has been acknowledged to be effective on a commercial scale, there are now increasing concerns about the discharge of organic chlorine compounds in the waste liquor. The use of peroxy acid oxidising agents, or even suitable enzymes, may provide suitable alternatives. In addition, some textile technologists consider that suitable gas plasma treatments (discussed in Section 7.1) could be acceptable alternatives to chlorination, though different polymers will then need to be developed for the subsequent treatment stage. It would be desirable, however, to avoid the degradation stage altogether. Polymers have been developed which can be added to untreated wool fibres. This approach is increasingly being used for shrinkproofing wool fabrics. These polymers confer shrink resistance through inter-fibre bonding, though a disadvantage of this approach is a reduction in fabric flexibility.

Polymers that successfully impart shrink resistance to untreated wool include Synthappret LKF and Synthappret BAP (Bayer) and Basolan F (BASF). Synthappret LKF is a reactive polyether with terminal isocyanate groups. It is insoluble in water and is supplied as a solution in tetrachloroethene (perchloroethylene). There are, however, environmental concerns about the use of chlorinated solvents on a commercial basis. Synthappret BAP, which contains carbamoylsulfonate groups and is a bisulfite adduct of Synthappret LKF, is soluble in water. After application, the polymers are cured at elevated temperatures and the essential reactions occurring are as follows:

Synthappret LKF

$$\text{—NCO} \xrightarrow{\text{H}_2\text{O}} \text{—NH—CO—NH—} + CO_2$$

Synthappret BAP

$$2\ \text{—NH—CO—SO}_3^{\ominus} \xrightarrow{\text{H}_2\text{O}} \text{—NH—CO—NH—} + 2\overset{\ominus}{\text{H}}\text{SO}_3 + CO_2$$

3.2.11 Fibre Properties

The properties of wool fibres are shown in Table 3.4. The tenacity of wool fibres is low (especially when wet), compared with those of other textile fibres. However, the extent of stretching before wool fibres break is considerable, particularly for wet fibres. This high extensibility accounts in large measure for the resistance of woollen garments to hard wear.

The popularity of woollen products has declined over recent decades. A number of factors are responsible, mostly arising from changing consumer tastes. One factor is competition from synthetic fibre products. Another factor is the trend towards lighter-weight, less formal clothing. A third factor is improved heating at home and work. Blankets have been largely replaced by quilts, and fitted carpets in the home are nowadays far less common. Nevertheless, the development of softer, lighter-weight woollen fabrics is now bringing wool into casual clothing

Table 3.4 Properties of wool fibres.

Specific gravity	1.32.
Tenacity	9–15 cN tex^{-1}, but reduced to 7–14 cN tex^{-1} when wet.
Elongation at break	25–35% under ambient conditions, but 25–50% when wet.
Elastic recovery	High: $>50\%$ recovery from a 10% stretch.
Resilience	Very high.
Abrasion resistance	Apparel wool: quite low. Carpet wool: high.
Moisture regain	14–18%.
Launderability	Wool garments need to be washed with care – normally at 40 °C. Severe mechanical agitation in washing machines and tumble driers can induce felting.

and sportswear. The ability of wool to felt is utilised in the production of overcoats and baize playing surfaces for snooker and pool tables. Wool is also coming into use as a technical textile, *e.g.* for industrial felts, for use in agriculture, and as recycled wool for loft insulation.

3.3 SPECIALITY MAMMALIAN FIBRES

Figure 3.5 shows photographs of some mammals whose fibres are used in speciality textile fabrics, alongside corresponding micrographs of the fibres. The origins of a variety of speciality mammalian fibres are given in Table 3.5. The fibres differ from one another – and from wool fibres – in diameter and scale structure. Apart from mohair, these fibres are less durable than wool fibres. In cashmere fibres, the structure of the cortex is considerably different from that of fine wool. Although there is bilateral symmetry as in wool, there are also random cell arrangements, both structures occurring even in the same fleece. The cortex of cashmere contains fewer orthocortical cells and far more mesocortical cells (and in some cases fewer paracortical cells too) than the wool cortex, though it should be emphasised that there is nevertheless wide variation amongst cashmere fibres. The cortex of a mohair fibre consists of predominantly ortho- and mesocortex, with often $<10\%$ paracortex. Amongst llama fibres, vicuña and guanaco fibres exhibit bilateral structure in the cortex, whereas alpaca fibres do not. In camel fibres from the same fleece, the cortex consists of both bilateral and random cell arrangements.

The only significant differences in the amino acid compositions of these mammalian fibres are in the content of cystine and cysteic acid. For example, llama fibres generally possess considerably higher levels of cystine residues than do the other mammalian fibres. Moreover, the amounts of cysteic acid residue in cashmere and wool are notably lower than in the other fibres. It is, in fact, noteworthy that cysteic acid residues are so prominent in these other fibres, their enhanced presence probably being attributable to continuing photo-oxidation of the fibres as they grow. In

Table 3.5 Origins of some speciality fibres.

Fibre	Origin
Cashmere	Australia, China, Mongolia, Iran, Afghanistan
Mohair	South Africa, Turkey, USA, Argentina, New Zealand
Camel	China, Mongolia
Alpaca	Andes mountains
Llama	Peru
Vicuna	Peru

Figure 3.5 Photographs of mammals from which speciality protein fibres are obtained, together with SEM photographs of the fibres of (a) cashmere goat, (b) angora goat (mohair), (c) camel, (d) alpaca, (e) llama.

(e)

Figure 3.5 Continued.

addition, the extent of variation observed in the amino acid compositions of wool fibres is mirrored in other mammalian fibres too – for the same reasons. Thus, differences in amino acid composition cannot be exploited to identify the origin of mammalian fibres or to distinguish between them.

3.4 SILK

3.4.1 Introduction

Silk is produced as filaments, up to 0.5 km long, from secretions of the larvae of particular moths. Most silk is derived from the larvae of the moth, *Bombyx mori*, but some other silks come from the larvae of the Chinese Tussah moth (*Antheraea pernyi*) and the Indian Tussah moth (*Antheraea mylitta*).

The life cycle of *Bombyx mori* is *ca.* 50 days (see Figure 3.6). The newly hatched larvae feed exclusively on mulberry leaves for *ca.* 30 days. In this time, the mass of each silkworm will have increased by a factor of about 100, and its silk glands will have completely filled. The silkworm is now ready to begin spinning a cocoon. For this purpose it seeks a support and starts to extrude silk filaments from its glands by moving its head in a figure-of-eight pattern. Supports are generally made either of straw frames or plastic branch-like structures.

The cocoon is formed over a period of 3–6 days. Fibroin, which is contained in two of the glands inside the silkworm, is forced through two openings in its head. The two emerging filaments of fibroin are bound

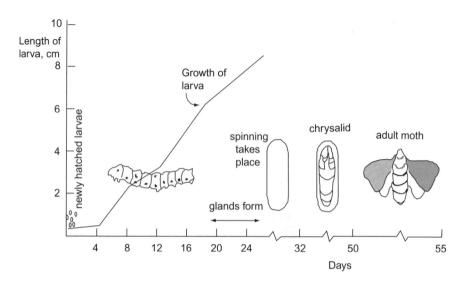

Figure 3.6 Life cycle of the *Bombyx mori* moth.

together by a protein gum, sericin, which is extruded from two adjacent glands. A single thread, of cross-section 15–25 µm, is thus formed. Tussah silks, however, are coarser, *ca.* 65 µm across. The cross-sections of the fibroin filaments from *Bombyx mori* secretions are approximately triangular, whereas tussah silks possess almost rectangular cross-sections. The length of thread with which the silkworm surrounds itself is generally 1–2 km. At this stage, the silkworm starts to transform into a chrysalid over 3–4 days, and then 10 days later into a moth. The moth can escape from the cocoon by secreting an enzyme which softens the cocoon and damages the silk filaments. The cocoons are therefore 'stifled', by being subjected to hot air (often at 110 °C) for several hours. The chrysalid is killed and the cocoon remains intact. Stifling is followed by drying to remove much of the moisture from the filaments, before they are reeled.

Sericin can be removed from the threads by a degumming procedure, in order to isolate the fibroin filaments. Traditionally, degumming was effected by treatment with soap solution for a number of hours. Other procedures utilise proteolytic enzymes, hot water at high pressures and dilute solutions of alkalis or acid. After degumming, the silk is bleached in hydrogen peroxide solution, and then rinsed.

3.4.2 Chemical Composition of Silk

Silk fibroin is a protein that consists of a large number of amino acid residues, as shown in Table 3.6. By far the most abundant residues are

Table 3.6 Amino acids present in silk fibroin, mol %.

No.	Amino acid	Structure of side chain (R)	Bombyx mori	Tussah
1	Glycine	H–	44.6	26.5
2	Alanine	CH_3-	29.4	44.1
3	Phenylalanine	$C_6H_5-CH_2-$	0.6	0.6
4	Valine	$\begin{array}{c} H_3C \\ CH- \\ H_3C \end{array}$	2.2	0.7
5	Leucine	$\begin{array}{c} H_3C \\ CH-CH_2- \\ H_3C \end{array}$	0.5	0.8
6	Isoleucine	$\begin{array}{c} H_3C-H_2C \\ CH- \\ H_3C \end{array}$	0.7	0
7	Proline (complete formula)	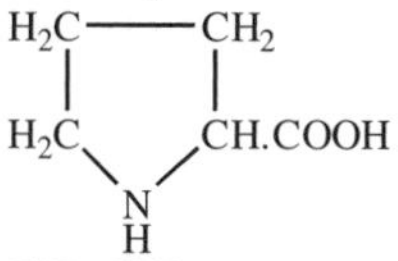 $\begin{array}{c} H_2C\text{-----}CH_2 \\ \mid \qquad \mid \\ H_2C \quad CH.COOH \\ \diagdown N \diagup \\ H \end{array}$	0.4	0.3
8	Serine	$HO-CH_2-$	12.1	11.8
9	Threonine	$\begin{array}{c} H_3C \\ CH- \\ HO \end{array}$	0.9	0.1
10	Tyrosine	$HO-C_6H_4-CH_2-$	5.2	4.9
11	Methionine	$CH_3-S-CH_2-CH_2-$	0.1	0
12	1/2-Cystine	$-CH_2-S-S-CH_2-$	0.2	0
13	Arginine	$\begin{array}{c} NH \\ \parallel \\ H_2N-C-N-CH_2CH_2CH_2- \\ H \end{array}$	0.5	2.6
14	Lysine	$H_2N-CH_2CH_2CH_2CH_2-$	0.3	0.1
15	Tryptophan	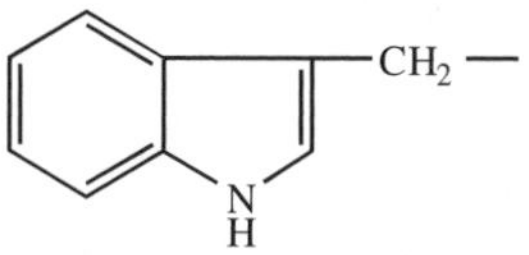 (indole)$-CH_2-$	0.1	1.1
16	Histidine	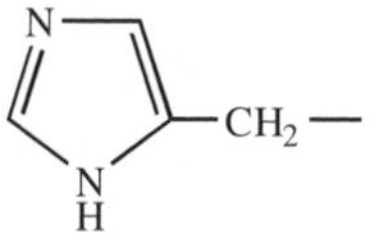 (imidazole)$-CH_2-$	0.1	0.8
17	Aspartic acid	$HOOC-CH_2-$	1.3	4.7
18	Glutamic acid	$HOOC-CH_2CH_2-$	1.0	0.8

glycine, alanine and serine; together they account for over 80% of the total residue content. In silk from *Bombyx mori*, glycine residues are over 50% more abundant than alanine residues, but in tussah silk from *Antheraea pernyi*, the relative content of the two amino acids is reversed. It is noteworthy too that, in contrast to wool and other mammalian hair fibres, the content of cystine is very small indeed.

Fibroin from *Bombyx mori* consists of an H-chain, of relative molar mass *ca.* 350 000, linked through a disulfide bridge to a much smaller L-chain, of relative molar mass *ca.* 25 000. Throughout most of the H-chain, glycine residues occupy alternate positions, and indeed Table 3.6 shows that the content of glycine is quite close to 50 mol%. It is also significant that the three most abundant residues in fibroin are the smallest side-chain groups amongst amino acids. As a consequence, there is little hindrance to crystallisation, and the degree of crystallinity is considered to be *ca.* 70%. Each crystal consists of layers of pleated sheets, in which each constituent fibroin chain is aligned in the opposite direction to its neighbour.

3.4.3 Bleaching of Silk

After degumming, silk often retains a small amount of pigment, which gives a yellow tint to the fibres. Fibre bleaching is therefore required. As in the case of wool, methods based on both oxidation and reduction are used. The most widely utilised bleaching agents that act by oxidation are hydrogen peroxide (at pH 8–9, in order to inhibit any alkaline hydrolysis of fibroin chains), sodium perborate ($NaBO_3$) and sodium persulfate ($Na_2S_2O_8$). A common reduction bleaching agent is sodium dithionite ($Na_2S_2O_4$), applied under neutral or mildly acidic conditions.

3.4.4 Chemical Reactions of Silk

3.4.4.1 Introduction. The chemistry of silk differs markedly from that of hair fibres. Whereas many of the reactions of hair fibres are based on cystine residues, the cystine content of silk is extremely small (see Table 3.6). Peptide linkages play a more prominent part in the chemistry of silk. Tyrosine and lysine residues also have significant roles.

3.4.4.2 Acid and Alkali. Treatment of silk with acid or alkali gives rise to hydrolysis of the fibroin chains. The degree of hydrolysis is highly dependent on pH. Hydrolysis is greater under acidic conditions than in alkaline conditions, but hot concentrated acids and alkalis both readily decompose silk. The lowest degree of hydrolysis occurs in the pH range of 4–8. The peptide bonds that are primarily attacked are those adjacent to serine and threonine residues. As with wool, it appears that ester groups are formed in the chains (see Section 3.2.8.3). These ester groups are more prone to hydrolysis than the peptide bonds from which they originate. Sulfuric acid also sulfonates the tyrosine residues in the fibroin chains.

Both acids and alkalis, under carefully controlled conditions, can induce a crepe effect, in which silk fabrics have a slightly wrinkled appearance. Such an appearance can be aesthetically desirable. Dilute organic acids, such as ethanoic, tartaric and citric acids, can enhance a rustling effect produced in silk fabrics, known as 'scroop'.

3.4.4.3 Oxidation. Oxidising agents, such as hydrogen peroxide, performic acid and peracetic acid, induce several different types of reaction. Peptide bonds, located at tyrosine residues in the fibroin chains, are broken, and amino groups at the ends of the chains can also be oxidised. Moreover, tyrosine residues themselves may be oxidised by potassium permanganate and chlorinating agents. 1,4-benzoquinone is released as a result of oxidation, and cross-linking is thought then to occur between free amino groups, such as those in lysine residues:

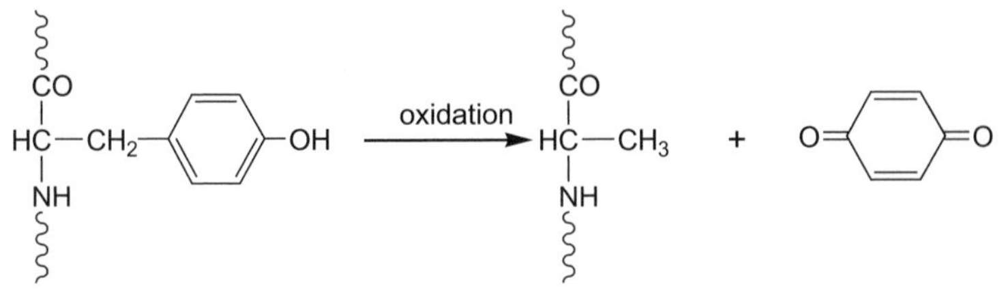

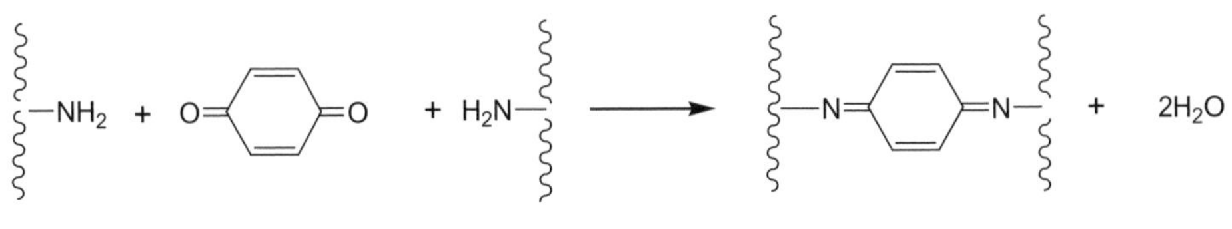

3.4.4.4 Cross-Linking. Cross-linking agents have been introduced into silk, in order to improve crease-resistance and washability. As is the case with wool, cross-links can be formed on treatment of silk with methanal and diepoxides. Interestingly, some lysinoalanine cross-links can also be formed on treatment of silk with alkali. Dehydroalanine residues (see Section 3.2.8.2) are formed from attack by alkali on the β-carbon atoms of, principally, serine. Lysinoalanine cross-links are then formed through the addition of lysine residues in other fibroin chains.

3.4.4.5   Reactive Dyes.   Reactive dyes can be applied to silk by virtue of the hydroxyl and primary amine groups present in the silk fibroin chains. As with wool and cellulosic fibres, reaction may occur by means of nucleophilic substitution and addition.

3.4.4.6   Action of Light.   Amongst natural fibres, silk is the most prone to photochemical degradation, and the fibres undergo yellowing. There is also a serious loss of fibre mechanical properties, such as tenacity and elongation to break. The underlying degradation mechanisms are complex and not well understood. It appears, however, that tyrosine residues are oxidised, and that there is fission of peptide bonds adjacent to tyrosine residues. Yellowing has been attributed to the decomposition of tryptophan, but some other residues are also likely to be degraded.

3.4.5   Silk Weighting

As a result of the loss of sericin during degumming, silk loses up to 25% of its weight. It is often commercially desirable to make up the weight loss. Traditionally, weighting was achieved with the addition of metal salts, notably tin salts, which possessed the added advantage of providing flame-resistance. However, the use of these salts is nowadays considered ecologically undesirable, and indeed silk fibres are rendered weaker, because of their increased susceptibility to hydrolysis and oxidation. Weighting with salts has now been superseded by grafting methacrylic acid or methyl methacrylate to the fibres and initiating polymerisation. This approach does not compromise the strength of the fibres.

3.4.6   Fibre Properties

The properties of *Bombyx mori* silk fibres are shown in Table 3.7. Their tenacity is moderate, though it is the highest of the protein fibres. The extent of stretching that *Bombyx mori* silk fibres can undergo before breaking is high, and is *ca.* 60% higher for tussah silks. Wet silk fibres can be stretched considerably further than dry ones. Tussah silks exhibit superior recovery to *Bombyx mori* silk above 12% stretch, but lower recovery below 12% stretch.

Table 3.7 Properties of *Bombyx mori* silk fibres.

Specific Gravity	1.33.
Tenacity	38 cN tex^{-1}, reduced by up to 20% when wet.
Elongation at break	~23% under ambient conditions, but up to 38% when wet.
Elastic recovery	Medium: *ca.* 50% recovery from a 10% stretch.
Resilience	High.
Abrasion resistance	Moderate.
Moisture regain	10–11%.
Launderability	Silk garments have to be gently washed with only mild agitation.

Silk production accounts for only *ca.* 0.2% of world textile fibre production overall. Silk is still perceived as a luxury fibre, closely associated with haute couture, despite haute couture's diminishing importance in the fashion industry. Silk is used too in scarves, ties, lingerie and furnishing fabrics. In more recent years, cheaper so-called 'sand-washed' silk was introduced into the fashion market, and made silk products more accessible in the high street. Sand-washed silk fibres possess abraided surfaces, and they are often weaker than conventional silk fibres. Outside the apparel sector, silk was also used in parachute fabric, although polyamide (nylon) has now replaced it. Silk also has some use as a medical textile, notably for ophthalmic sutures and the contact layers of wound dressings.

SUGGESTED FURTHER READING

1. W. S. Simpson and G. Crawshaw, *Wool: Science and Technology*, Woodhead Publishing Ltd., Cambridge, UK, 2002.
2. J. A. Maclaren and B. Milligan, *Wool Science, The Chemical Reactivity of the Wool Fibre*, Science Press, Marrickville, NSW, Australia, 1981.
3. R. R. Franck, *Silk, Mohair, Cashmere and Other Luxury Fibres*, Woodhead Publishing Ltd., Cambridge, UK, 2001.
4. L. N. Jones, D. E. Rivett and D. J. Tucker, in *Handbook of Fiber Chemistry*, ed. M. Lewin, CRC Press, Taylor & Francis Group, Boca Raton, USA, 2007 ch. 5.
5. A. Matsumoto, H. J. Kim, I. Y. Tsai, X. Wang, P. Cebe and D. L. Kaplan, in *Handbook of Fiber Chemistry*, ed. M. Lewin, CRC Press, Taylor & Francis Group, Boca Raton, USA, 2007, ch. 6.

CHAPTER 4

Regenerated Fibres

4.1 REGENERATED CELLULOSIC FIBRES

The linearity of the cellulose molecule which permits a high degree of crystallinity and intermolecular hydrogen bonding, results in cellulose neither melting nor dissolving in common solvents. During the second half of the nineteenth century, a number of attempts were made to produce cellulose-based fibres by chemical modification of the cellulose so that it could be dissolved, then spun as fine fibres from solution. The main goal was to produce fibres that would compete with silk in terms of fineness and handle. It has to be remembered of course that at that time, knowledge of organic chemistry was very basic and the range of chemicals available was rather limited, but despite this the degree of ingenuity shown by chemists was remarkable.

The earliest developments were based on nitrating cotton, and in 1846 a Swiss chemist called Schönbein produced the explosive nitrocellulose (gun-cotton). The first patent to describe the formation of fibres was taken out in 1855 by Audemars of Lausanne who nitrated fibres from the inner bark of mulberry trees. He dissolved the cellulose nitrate in a mixture of alcohol and ether, together with caoutchouc (the latex of rubber trees), and using a steel needle was able to pull out threads and wind them onto a spool. In 1883 Joseph Swan, an English inventor who was interested in developing carbon lamp filaments, patented a method for de-nitrating cellulose nitrate, but he did not develop his method to produce textile fibres. It was count Hilaire de Chardonnet in France, who finally developed a process to produce a fibre in commercial

The Chemistry of Textile Fibres
By Robert R Mather and Roger H Wardman
Published by the Royal Society of Chemistry, www.rsc.org

quantities. Chardonnet's silk was cellulose nitrate, which was obtained by nitrating cotton fibres in a mixture of nitric acid and sulfuric acid (as catalyst). A solution of the cellulose nitrate in a mixture of alcohol and ether (called collodion) was then wet spun into a bath of cold water. Cellulose nitrate is extremely flammable and Chardonnet developed a process to de-nitrate the fibres, based on treating them with ammonium hydrosulfide for about 20 hours at 65 °C, to regenerate cellulose. The whole process was patented in 1885 and the fibres were marketed as 'artificial silk' in 1889. The process was messy and inefficient, however, and is now obsolete. Nevertheless, Chardonnet's achievement was of considerable importance to the textile industry since it set the course for the development of other commercially viable rayon fibres, and Chardonnet has become known as the 'Father of Rayon'.

A different process was developed in England by Cross and Bevan, for which they were awarded a patent in 1893. This was the viscose process and was based on the extraction by dissolution, of the cellulose present in cotton. The process was still a long, messy one and it has to be said that if it were to be invented today, it is most likely that it would be considered as too much of an environmental risk to be commercially viable. The process involved the use of carbon disulfide, a chemical which is explosive, toxic and possesses a most awful smell. The fibres produced by the process, however, undeniably possessed properties that were quite unique in terms of handle, lustre, softness and absorbency, and for this reason it is still an important fibre today, both for apparel and non-apparel uses. It is only through manufacturers adapting their processes to recover the chemicals used in order to meet environmental legislation that manufacture has continued. Indeed, such has been the difficulties in meeting environmental standards with the viscose process, most European companies have closed down and there are only three left. The biggest manufacturer in Europe, Lenzing AG in Austria, has invested heavily since the 1980s in its process and recovery systems to meet the very strict demands of the Austrian government, whose environmental standards are far stricter even than those of the EU. Lenzing AG produces about 25% of the world supply of viscose, otherwise most of the world production is in China.

In recognition of the various types of regenerated cellulosic fibres and processes for producing them, the generic name 'rayon' was adopted in 1954 in America by the Federal Trade Commission (FTC). The FTC defines rayon fibres as those which are 'composed of regenerated cellulose, as well as manufactured fibres in which substituents have replaced not more than 15% of the hydrogens of the hydroxyl groups'. As will be seen later, some of the fibres in this class have been produced by modification of some of the hydroxyl groups in the cellulose. The term 'rayon' has not been formally adopted by BISFA in Europe, the names Viscose, Lyocell and Modal are used instead. The name 'Rayon' regularly appears on garment labels in Europe, however, and the word is generally understood to mean a manufactured cellulosic type fabric.

4.1.1 Viscose

The raw material for the manufacture of viscose is the cellulose present in beech wood. The trees are grown in managed forests, in countries such as Austria, Bavaria and Slovakia. The first step in manufacturing viscose is therefore that of isolating the cellulose from the wood. For this purpose, the logs are de-barked then chipped into small pieces of about 2–4 cm. The pulping process that is carried out next is designed to remove as much lignin, hemicelluloses and other extractable materials as possible, whilst avoiding degradation of the cellulose, though some controlled degradation is allowed in order to produce cellulose of the desired *DP*. In the bisulfite process, the wood chips are steamed under pressure with a mixture of magnesium oxide and sulfur dioxide (which form magnesium bisulphite) for about 8 hours at 150 °C and 8 bar pressure where the wood is digested. The resulting liquor has two components: non-cellulose materials such as sulfonated lignins and hemicelluloses, and cellulose pulp. The cellulose pulp is separated from the lignins and hemicelluloses which remain dissolved in the spent liquor. The spent liquor is incinerated with the recovery of heat and chemical products such as xylose, furfural and acetic acid. The pulp is further refined by washing, screening, caustic extraction and bleaching.

One difficulty of the process is that hemicelluloses, which are structurally very similar to cellulose, (see Section 2.2) are present in larger quantities in wood than in cotton, and not all are removed. Hemicelluloses are polysaccharides, but are made up of a variety of different simple sugars, in contrast to cellulose in which the sugar building blocks are glucose only. Hemicelluloses, of which there are two forms, β and γ, also have a lower degree of polymerisation than cellulose and their molecules are often branched. Both forms are soluble in alkali and can carry through into the viscose manufacture process, so techniques of dialysis of the steeping liquors together with nanofiltration techniques have been developed to remove them. Of the very small quantities of hemicelluloses that do carry through to the xanthate stage, the β-form becomes partly incorporated in the fibre on extrusion, but the γ-form remains in the spin bath system.

The next step is bleaching to remove any residual lignin. At Lenzing AG, this is carried out by a process that avoids the use of chlorine and employs a three-stage process instead, firstly using liquid oxygen, then ozone and finally hydrogen peroxide. The final product is 90–92% α-cellulose (the rest are hemicelluloses), in the form of sheets (approximately 0.5 m^2 in area and 2 mm thick) which are perfectly white in appearance. Of the original wood stock, about 40% is extracted as cellulose for use in the viscose production, 10–11% as secondary products (furfural, acetic acid and xylose) and the remaining materials are incinerated, producing steam and electricity.

The process for making viscose requires the cellulose to be dissolved. In order to achieve this, the cellulose sheets are shredded and their moisture content adjusted to 50%. At Lenzing, with its own wood pulping facility, the wood pulp is mechanically squeezed to this moisture content. The wood pulp flakes are then steeped in 18% sodium hydroxide solution, where the sheets swell as the alkali diffuses into them and reacts chemically with the hydroxyl groups of the cellulose to form alkali cellulose:

$$Cell-OH + NaOH \rightarrow Cell-O^-Na^+ + H_2O$$

The alkali cellulose is separated from the steeping lye by presses then shredded to obtain a bulky, reactive product. The shredded crumbs are next aged in air at ambient temperature for up to 24 hours, where oxidative degradation of the cellulose chains occurs, resulting in a decrease of molar mass. The alkali celluloses are then fed into churns and rotated under vacuum, where carbon disulfide (CS_2) is gradually introduced. The carbon disulfide reacts with the soda cellulose to form sodium cellulose xanthate, which is bright orange in colour.

$$Cell-O^-Na^+ + CS_2 \rightarrow Cell-O-\overset{\overset{\displaystyle S}{\|}}{C}-S^-Na^+$$

Side reactions can also occur:

$$3CS_2 + 6NaOH \rightarrow 2Na_2CS_3 + Na_2CO_3 + 3H_2O$$
$$Na_2CS_3 + 6NaOH \rightarrow 3Na_2S + Na_2CO_3 + 3H_2O$$

These side reactions are more significant at higher temperatures, but if the temperature is lowered the time required for complete xanthation is longer. A compromise is made and a temperature of between 25–37 °C is used, with times of 30–90 minutes. The sodium cellulose xanthate is dissolved in 1–2% sodium hydroxide solution at 8–12 °C to give an orange-brown spinning solution, the viscose 'dope'. This solution is then aged for a further 1–3 days until it reaches the correct viscosity and 'ripening index' for extrusion. This period also allows for the CS_2 to become evenly distributed throughout the cellulose. Before extrusion, the viscose dope is filtered to remove any small undissolved solids which might clog up the spinneret holes, and any air bubbles which would disrupt the flow of polymer through the spinneret are removed by de-aeration under vacuum. Finally before extrusion, additives may be added the to the dope, such as surface active agents to improve spinning performance and white finely dispersed titanium dioxide pigment if dull or matt fibres rather than the standard 'bright' fibres are required.

The fibres are produced by wet spinning. The spinneret, which is made of a gold–platinum alloy in which there are literally thousands of small holes (each of about 50 μm in diameter), is submerged in a coagulating bath containing 10% sulfuric acid, 18% sodium sulfate and 1% zinc sulfate.

During the extrusion of the sodium cellulose xanthate, the cellulose molecules near to, and in contact with, the walls of the orifices of the spinneret will experience a drag and will line up to a greater extent than the molecules in the centre. The composition of the coagulating bath is important (sometimes magnesium sulfate is added) because it controls the rate of generation and quality of the viscose. Typically the fibres develop with a 'skin-core' structure, whereby the orientation of the outer layer of the fibres is greater than that of the inner layer. The chemical reactions taking place in the coagulation bath can be represented as:

$$2\text{Cell-O-C(=S)-SNa} + H_2SO_4 \rightarrow 2\text{Cell-OH} + 2CS_2 + Na_2SO_4$$

The presence of zinc sulfate produces fibres of greater strength and with a serrated cross-section. It is thought that the divalent zinc ion forms a weak cross-link between adjacent cellulose xanthate anions:

$$\text{Cell-O-C(=S)-S}^- \text{---} Zn^{2+} \text{---} ^-\text{S-C(=S)-O-Cell}$$

This results in a retardation of the regeneration process. During the extrusion process, the zinc ions diffuse only slowly into the fibres and only into the outer layer, in contrast to the much more mobile hydrogen ion that penetrates throughout the fibre. The regeneration of the

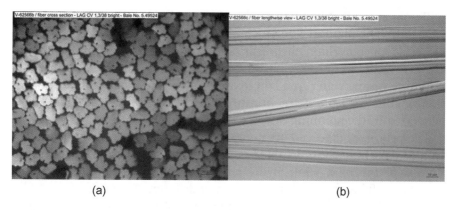

(a)　　　　　　　　　　　　　　　　　(b)

Figure 4.1 Microphotographs of viscose fibres: (a) cross-sectional view, (b) longitudinal view. (Photograph courtesy of Lenzing AG).

cellulose therefore occurs more slowly on the outer layer of the fibre, and because of the aligning effect of the spinneret described above, causes a more uniform, aligned arrangement of the cellulose molecules. As the core shrinks on regeneration of cellulose, the skin also contracts and becomes wrinkled, giving the fibres their characteristic jagged cross-sectional shape and striations along their length (see Figure 4.1).

During the extrusion stage the products of the side reactions formed in the xanthation stage, Na_2CS_3 and Na_2S, can react with the sulfuric acid present in the coagulation bath to give CS_2 and H_2S:

$$Na_2CS_3 + H_2SO_4 \rightarrow CS_2 + H_2S + Na_2SO_4$$
$$Na_2S + H_2SO_4 \rightarrow H_2S + Na_2SO_4$$

After extrusion the fibres are fed as a tow (the thousands of individual fibre filaments running parallel to each other) out of the bath, into a second hot water bath, *via* two rollers, called 'Godet' rollers. The second of the rollers rotates faster than the first, so that a stretch is induced into the fibres, which enhances their strength. The tow is usually cut into staple lengths of around 40 mm, formed into a fleece and finally thoroughly washed and dried. The fibres can be used in their initial 'bright' form, or they can be delustred to a 'dull' form, by the incorporation of the white pigment titanium dioxide into the fibres at the extrusion stage. These pigment particles show up as black dots in the fibres when viewed under a microscope.

A schematic representation of the manufacturing process is shown in Figure 4.6. From the foregoing description of the process, it is clear that a considerable amount of chemical waste and effluent will result.

Different manufacturers deal with the problem to varying degrees of efficiency. The plant at Lenzing in Austria is the world's most advanced in terms of recycling chemicals and producing minimum Chemical Oxygen Demand and Biological Oxygen Demand levels, with minimal pollution of air, water and soil. Key to this achievement is the application of four separate extraction processes to remove the CS_2 and H_2S generated in the spinning and washing baths.

The processes involved are complex. As the cellulose is formed in the spinning bath, carbon disulfide is slowly released and in the acid conditions of the bath some hydrogen sulfide (H_2S) is also formed at a faster rate. The spin bath is degassed and the mixture of H_2S and CS_2 incinerated to produce sulfur dioxide and sulfur trioxide, which when condensed with water yields sulfuric acid. This is called the 'wet sulfuric acid' process. From the second (stretching) bath the waste gas is richer in CS_2 than in H_2S. The two chemicals are removed by the Sulfosorbon process, which is an activated carbon adsorption process in which carbon impregnated with potassium iodide causes the H_2S to be oxidised to sulfur. The CS_2 is removed from the activated carbon by heating with steam, as liquid CS_2, and is used to extract the sulfur which was formed in the carbon. The CS_2/S mixture is distilled to separate the two components and finally the CS_2 is condensed.

The CS_2 is used again in the viscose process in the formation of sodium cellulose xanthate; the sulfur is incinerated to produce sulfuric acid, which is re-used in the spinning bath. A third extraction process, the Supersorbon process, is used for the removal of the CS_2 and H_2S from the later stages of the washing liquors, which are richer still in CS_2 than H_2S. The remaining H_2S is removed by treatment with NaOH in a soda washer, then the CS_2 is removed by adsorption onto activated carbon. The zinc salts remaining in the water are removed at the waste water treatment plant and are removed by precipitation as zinc sulfide (ZnS). This can be used to produce zinc sulfate again. Organic substances in the effluents from the viscose plant are degraded very effectively in a double-stage biological waste water treatment plant.

There is a considerable recovery of the chemicals which are re-used in the viscose production, so the main by-product is only sodium sulfate, which is sold to detergent manufacturers. In addition, acetic acid, furfural and xylose are the main by-products of the pulp mill, all of which are used elsewhere in the chemical industry.

Viscose fibres, like cotton fibres, comprise almost pure cellulose and so show very similar chemical properties in terms of their reactions with acids, alkalis and oxidising agents. Viscose differs quite substantially in its molecular structure, however, and has a less complex morphology than cotton. To begin with, the number of glucose units in wood cellulose is only about 1000 compared with up to 10 000 in cotton cellulose. During the manufacture of viscose, when the cellulose is converted into sodium cellulose xanthate and aged, some scission of the cellulose molecules occurs due to oxidative depolymerisation, so that in the cellulose of the final viscose fibres there are on average only around 270 glucose units. This considerably lower molecular mass of viscose cellulose manifests itself in a much lower strength than cotton fibres, and with behaviour opposite to that of cotton, the strength of viscose fibres decreases when they are wet. Viscose fibres swell considerably in water, their diameter increasing by about 35–40%. They have a water retention value of about 85–90%, a value more than double that of cotton.

Of course, since the extrusion process can be controlled, so the diameter of the viscose fibres produced can be perfectly uniform and also the fibres can be produced with any degree of lustre (without the need for a mercerisation process).

The properties of regular viscose fibres are given in Table 4.1. The main attributes of viscose fibres are their silk-like handle and shiny, lustrous appearance. Viscose is used extensively for both apparel and non-apparel applications. In apparel, it is widely used for linings because it is smooth and shiny so that garments such as jackets or overcoats slip easily over garments worn underneath. Other applications are those where its natural absorbency are useful, such as lingerie, blouses, dresses, and skirts. An interesting variant of viscose is Viloft, which has a flat cross-sectional shape (see Figure 4.2) which traps pockets of air, thereby giving insulation against the cold. Garments made from Viloft are used

Table 4.1 Properties of viscose fibres.

Fibre length	Can be varied according to need, but usually is about 40 mm.
Fineness	$\sim 20\,\mu m$.
Specific gravity	1.52.
Tenacity	25–30 cN tex^{-1}, but only about 18–20 cN tex^{-1}, when wet.
Elongation at break	Viscose is fairly extensible and stretches by about 15% (dry) and 25% (wet).
Elastic recovery	It does not recover well from stretching.
Resilience	Very good.
Moisture regain	12–13% – more absorbent than cotton.
Reaction to heat	Like cotton viscose has no melting point and it is very heat resistant but will yellow with a hot iron. It burns very readily when it gives a smell like burnt paper.
Sunlight	Viscose gradually loses strength on exposure to sunlight.
Launderability	Viscose is best washed at 40 °C. It can be dry cleaned but not tumble dried. It dries slowly and creases very easily.

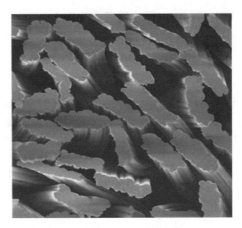

Figure 4.2 Cross-section of Viloft fibres. (Photograph courtesy of Lenzing AG).

for comfortable, absorbent, insulating underwear. In this respect it is very useful for base-layer garments for workwear.

4.1.2 Variants of Rayons

4.1.2.1 High Tenacity Rayons. Whilst viscose fibres do not possess the complex morphological structure of cotton, the cellulose molecules in viscose are not entirely of uniform arrangement either, due to the 'skin-core' effect described above. The degree to which the 'skin-core' structure exists can be controlled by adjusting the composition of the spinning bath. For example the use of higher concentrations of zinc sulfate (up to about 4%), or the addition of other salts such as magnesium sulfate, retard the regeneration of cellulose more consistently throughout the fibre, so that there is more of the 'skin' character than the 'core' character. Furthermore, the size and orientation of the cellulose crystallites can be influenced by stretching the fibres to the maximum extent possible without breaking them in a water bath at 90 °C. Less shrinkage occurs in the core and the fibres are smoother and also considerably stronger, especially in wet strength. The variants of viscose produced in this way are called 'high tenacity' rayons and have tenacities in the range 26–44 cN tex^{-1}. The high tenacity variants have been developed more for high-performance applications or industrial purposes than for apparel. For many years they were used for the manufacture of tyre cords, but their use declined with the development during the 1940s and 50s of nylon and polyester. Whilst this market is a highly competitive one, and is dominated by nylon, polyester and aramids, high tenacity rayons are still preferred for some types of tyres because they adhere better to

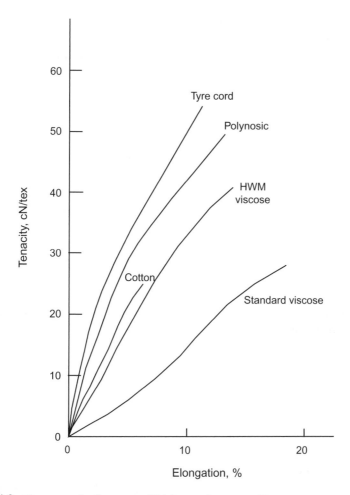

Figure 4.3   Stress–strain diagrams of high-tenacity rayon fibres.

rubber and have better heat resistance. The stress–strain curves of these fibre types are illustrated in Figure 4.3

4.1.2.2 Polynosic Rayons.   Another variant developed was the poly-nosic rayon type, the pioneering development of S. Tachikawa in Japan during the 1950s. Polynosic rayons differ from standard viscose in having a higher *DP*, leading to higher strength, and a lower degree of swelling in water which in turn gives a higher ratio of wet strength to dry strength. These improvements were brought about by modifications to the production method, chiefly:

• The ageing of the alkali cellulose crumbs stage was removed;

- The cellulose xanthate was dissolved in water instead of 3% sodium hydroxide;
- The sodium cellulose xanthate solution was not aged;
- Only a very low concentration of sulfuric acid, and no zinc sulfate was used in the extrusion bath; and
- Two- or three-stage stretching processes were applied to give a maximum stretch of about 300%.

Removing the ageing and ripening stages means that there is little reduction in the *DP* of the cellulose, with the polynosic and high tenacity variants having a *DP* values of as high as 800, in comparison to the *DP* of regular viscose of 270. The low acid concentration in the spinning bath leads to a much slower regeneration of the cellulose, enabling a higher degree of stretch to be imparted with the consequent increase in orientation and crystallinity. Polynosic fibres also possess a fibrillar structure, unlike standard viscose, and have rounder cross-sections and smoother surfaces. They produce 'lean' yarns which are difficult to dye to full shades. The fibres have the low elongation properties of cotton, but are quite brittle. Consequently, polynosic rayons have not been successful.

4.1.2.3 High Wet Modulus Rayons. A further development in the production of stronger variants of viscose, particularly variants with much higher wet strengths, was that of high wet modulus (HWM) rayon. Following the general principle that variants with higher wet strengths can be achieved by slowing down the rate of regeneration of cellulose in the extrusion bath, it was found during the 1950s that various compounds could act in this way, such as polyethylene glycols, alkylamines and phenols. The addition of these additives (known as 'modifiers') together with other adjustments made to the extrusion conditions (such as spinning speed and the temperature of the extrusion bath) also enabled greater degrees of stretch to be imparted. The modifiers act to assist in the formation of zinc cellulose xanthate, reducing the rate of regeneration of cellulose and causing a stronger de-swelling. In more recent years, considerable developmental work has led to yet further increases in tenacity. Significant amongst these developments has been the improvement in the quality of the cellulose from the wood pulp. The requirement is for α-cellulose, which is the highest quality cellulose for making textile fibres. The other forms, the β- and γ-celluloses, have lower molar masses and are poor quality celluloses for making textile fibres. They have a negative impact on pulp reactivity, the spinning stability of the dope and the mechanical properties of the resultant fibres. The introduction of a cold caustic extraction process, to remove short chain length celluloses, has enabled the high stretches to be applied to the fibres during spinning that are necessary for the production of the high tenacity grades of viscose.

4.1.2.4 Modal. Modal is a type of high wet modulus rayon that has a very precisely defined strength. The BISFA definition of modal fibres is a cellulose fibre having a high breaking force (*BF*) and a high wet modulus (*F_W*). The breaking force in the conditioned state and the wet modulus required to produce an elongation of 5% are given by:

$$BF \geq 1.3\sqrt{LD} + 2LD \tag{4.1}$$

$$Fw \geq 0.5\sqrt{LD} \tag{4.2}$$

where *LD* is the mean linear density (mass per unit length) in decitex. *BF* and *Fw* are expressed in centiNewtons. In addition to their characteristic strength, modal fibres are very soft to the touch and highly absorbent. They are used for garments worn next to the skin, such as lingerie, and for home textiles, such as towels, bath robes and bath mats. In these applications modal rayon is often blended with cotton. It is also used for fashion knitwear.

4.1.2.5 Flame Retardant Rayons. Whilst standard viscose burns very readily, its inherent absorbency, breathability and skin 'friendliness' can prevent heat stress and heat stroke. Lenzing AG produces a flame retardant version of viscose (FR viscose) for use in the manufacture of protective clothing for firefighters. This type of viscose is made by adding a flame-retarding agent, an organic compound containing phosphorus and sulfur, to the spinning dope, so that when the fibres are extruded the agent is permanently held within the fibres. The garments produced from FR viscose give protection from fire, radiant heat, electric arcs and molten metals. A blend of viscose FR/ Nomex/Kevlar (65:30:5) is also produced, which has enhanced strength at high temperatures. Another strategy for using viscose for flame retardant fibres is that used by Sateri Fibres which extrudes a mixture of sodium cellulose xanthate and water glass, $Si(OH)_4$, which polymerises to polysilicic acid on extrusion. The presence of the silicon within the fibres gives the flame retardant property.

4.1.2.6 Other Variants of Rayons. Viscose also has non-apparel applications, examples being interiors for seat covers, cushions, bedspreads and trimmings. More significantly it has become widely used in the manufacture of products for medical applications, such as wipes, disposables, feminine hygiene products, where it is used in non-woven form, for which the market is huge. The class of products called 'non-wovens' are assemblies of randomly orientated fibres that are not knitted or woven, but are held together by friction, adhesion or stitch bonding. Since the rayons (and Lyocell fibres – see Section 4.1.4) are cellulosic in nature they have a natural affinity for water and are much more suitable for this application than the more hydrophobic synthetic fibres. Nevertheless it is technically a very complex application, due in part to the different requirements of the products. For example a dry wipe must absorb a fluid rapidly and retain it, whilst a wet wipe must hold a liquid temporarily then release it on demand, say by squeezing. On the other hand, a tampon has to be capable of expanding immediately on wetting and retaining a large volume of liquid, even under pressure.

The key physical properties required of the fibres are water retention and water holding capacity. Water retention capacity is influenced by the fibre supermolecular structure, particularly the size and distribution of the voids over the fibre cross-section. Thus viscose fibres, with their skin and core structure, have small pores in the skin, but a mixture of small and large pores in their core. Lyocell fibres in contrast have a much more uniform distribution of pores across their cross-section. Water holding capacity is determined more by the surface characteristics of the fibres, such as crimp and cross-sectional shape, which influence the size of the fibre cavities within the non-woven fibre assembly, or any chemical finishes that have been applied.

Various strategies have been adopted to improve the performance of viscose for tampon and hygiene applications. The incorporation of water-soluble polymers (such as carboxymethyl cellulose, guar gum, alginate or sodium polyacrylate) into the viscose spinning dope prior to extrusion of the fibres increases the hydrophilicity of the fibres and can increase water retention value from 86% to 144%. Rayons containing these additives are called 'alloy rayons'. Alternatively, by extruding the dope through trilobal spinnerets, trilobal or star-shaped fibre cross-sections (see Figure 4.4) are formed, which influence the size of the void volumes between the fibres, increasing their absorptive capacity. The size and shape of the voids is also influenced by the crimp of the fibres and this property is affected by increasing the alkali content in the spinning dope. Table 4.2 shows typical values of water retention and water holding capacity for viscose fibre variants produced by these strategies.

For these types of products it is not just the biodegradability of the viscose that is important, but the flushability as well. The products have

Figure 4.4   Cross-section of Viscostar®. (Photograph courtesy of Lenzing AG).

Table 4.2   Water retention and water holding capacity for viscose fibre variants.

Fibre	Water retention, %	Water holding capacity, g g^{-1}
Standard viscose	86	22
Viscose + alginate	144	24
Viscose + carbamate	81	32
Viscose + carboxymethyl cellulose	126	22
Viscostar® trilobal viscose	80	22
Highly crimped viscose	86	26.5
Cotton	42	27

to be strong enough to be effective during use, but must disintegrate quickly when they are discharged into the sewerage system. This remains a challenge for the industry.

4.1.3   Cuprammonium Rayon

In 1857 a Swiss chemist, Schweizer, found that a good solvent for cellulose is a mixture of copper hydroxide and ammonium hydroxide solutions. The first cellulosic fibres from this solvent (known as Schweizer's reagent) were made in Germany by Fremery and Urban in 1891. At around the same time, however, a commercial process for the manufacture of viscose was developed by Bemberg, following the development of Thiele's stretch spinning technique, and to this day cuprammonium rayon is also known as 'Bemberg silk'. During the twentieth century, significant quantities of cuprammonium rayon were produced in Germany, England, America, Japan and Italy, but since the

1960s, production in all these countries, with the exception of Asahi Kasei Chemical Industries Co. in Japan, has ceased.

Schweizer's reagent is made by first mixing solutions of copper (II) sulfate and sodium hydroxide, from which copper hydroxide is precipitated:

$$CuSO_4 + 2NaOH \rightarrow Cu(OH)_2 + Na_2SO_4$$

In practice, however, this direct method is not used because the copper hydroxide is readily oxidised by air. Instead an indirect method is used, in which aqueous sodium carbonate is added to dilute copper sulfate solution, which results in the formation of basic copper sulfate:

$$4CuSO_4 + 3Na_2CO_3 + H_2O \rightarrow CuSO_4 \cdot 3Cu(OH)_2 + Na_2SO_4 + 3CO_2$$

Addition of sodium hydroxide then yields copper hydroxide:

$$CuSO_4 \cdot 3Cu(OH)_2 + 2NaOH \rightarrow 4Cu(OH)_2 + Na_2SO_4$$

Ammonium hydroxide is added to form the complex tetraamminediaquacopper dihydroxide, of formula $[Cu(NH_3)_4(H_2O)_2](OH)_2$. When the cellulose, usually from cotton linters, is dissolved in the solvent, the cuprammonium ions form a complex with the hydroxyl groups of the cellulose. The resulting solution, in which the cellulose concentration is about 9%, has a clear blue colour. After de-aeration and filtering it is extruded through a nickel spinneret into a slightly alkaline solution. On emerging into the spinning bath the fibres are forced down a funnel-shaped tube which causes them to stretch considerably (roughly 400%), before passing into a bath containing dilute sulfuric acid to complete the coagulation of the cellulose. Further washing of the cuprammonium yarns in water is required, to remove the copper and ammonium sulfates formed. Due to its toxicity the copper sulfate has to be fully recovered from the washing liquors.

The properties of cuprammonium rayon fibres are shown in Table 4.3. Since the fibres are regenerated cellulose, they have similar characteristics to those of viscose. However cuprammonium rayon filaments are

Table 4.3 Properties of cuprammonium rayon fibres.

Specific gravity	1.54.
Tenacity	21–28 cN tex^{-1}, and about 18–20 cN tex^{-1} when wet.
Elongation at break	Very similar to viscose, about 15% (dry) and 25% (wet).
Elastic recovery	It does not recover well from stretching.
Resilience	Very good.
Moisture regain	11%.

finer than those of viscose and their strength, particularly wet strength, is also higher.

4.1.4 Lyocell Fibres

The process for the manufacture of Lyocell fibres was developed during the 1970s when patents were filed by Akzo Nobel. The company granted both Courtaulds and Lenzing free licence to develop the process for the manufacture of staple fibres. Courtaulds started the development at Coventry in the late 1970s and it commenced full-scale production in 1992 at its plant in Mobile, Alabama. Lenzing started its development work independently in the mid-eighties and commenced full-scale production in Heiligenkreuz in Austria in 1997. It has been estimated that some £100 m was spent on the research, which aimed to produce cellulosic fibres from renewable resources such as wood pulp, by extruding it from solution in an organic solvent.

In 1989, BISFA adopted the generic name 'Lyocell', defined as 'cellulosic fibres obtained by an organic solvent spinning process'. This is a carefully worded definition, because it distinguishes Lyocell fibres from viscose or cuprammonium fibres in that the Lyocell fibres are formed from a solution without the formation of a derivative. In 1995, the FTC in America also adopted the generic name 'Lyocell', but as a sub-category under 'Rayon'.

When Courtaulds commenced production of their Lyocell fibre in 1992 they marketed it under the trade name Tencel®. During the 1990s, the company was bought by Akzo Nobel and the fibres division was subsequently sold to CVC, which formed Acordis. The Tencel® Group was sold to Lenzing AG in Austria, in 2004. Lenzing AG currently produces Tencel® in Austria, but also at sites in Grimsby in the UK and Mobile in the USA. Lenzing is by far the largest producer of Lyocell fibres, though there are two other producers with pilot plants, Smart Fiber AG in Germany (which operates under licence from Lenzing and produces speciality Lyocells such as Seacell® and Smartcell® on a small but commercial basis) and Hyo Sung in Korea (which produces pilot plant quantities of Lyocell for tyre cords). In addition there are companies in some other countries with experimental scale operations, such as Shanghai Lyocell Development Group.

Like viscose, Tencel® is produced from wood pulp but the process is chemically much less complex because the cellulose in the wood pulp is dissolved directly in an organic solvent. The trees are obtained from responsibly managed forests in Brazil and South Africa. As in viscose manufacture, the wood is debarked and chipped, and the chippings are then treated in the same way to produce sheets of virtually pure cellulose. Unlike the viscose or cuprammonium rayon fibre manufacture, the cellulose is not converted into a chemical derivative prior to extrusion.

The solvent used for the manufacture of Lyocell is *N*-methylmorpholine-*N*-oxide (NMMO), which is one of a group of compounds known as amine oxides. NMMO is highly polar and very soluble in water, and aqueous solutions can readily form hydrogen bonds with the hydroxyl groups of cellulose and so dissolve it. At temperatures of around 100 °C, an aqueous solution of NMMO is a very effective solvent for cellulose, and especially the premium quality α-celluloses that have high molecular weight and are required for the best quality fibres.

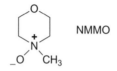

The water content and temperature of the NMMO play an important role in the dissolution of cellulose (see Figure 4.5). If the water content is too high (>15%), the cellulose will not dissolve due to the domination of water–NMMO hydrogen bonding. In practice it is found best to add the wood chips to 50–60% aqueous NMMO solution, and then reduce the water content of the solution by evaporation under reduced pressure to give a spinning dope of composition NMMO:water:cellulose (76:10:14). Temperatures of between 90–120 °C are used. Cellulose degrades rapidly in NMMO, so it is necessary to include an antioxidant (usually *n*-propyl gallate) at a concentration of between 0.01–0.1% to prevent degradation. At this stage the spinning dope contains some undissolved particles of cellulose which have to be removed by filtering prior to extrusion. The fibres are formed by the dry-jet wet spinning process, in which the spinning dope is extruded through a spinneret which is located between 50–100 mm above the spinning bath, at a speed of about 100 m min^{-1}. The spinning bath contains either water or a very dilute solution of NMMO. After extrusion the fibres are thoroughly washed in water to remove the NMMO.

Whilst NMMO is claimed to be non-toxic and biodegradable, it is very expensive, and a closed-loop process for Lyocell production has been developed that is so effective that some 99.5% of the NMMO is recovered which can be re-used in the process. Schematic representations of the key stages in Lyocell and viscose fibre production are shown in Figure 4.6. The whole process for the manufacture of Lyocell fibres is

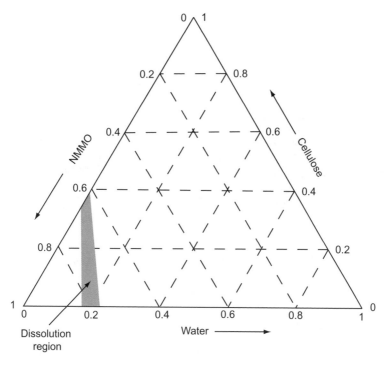

Figure 4.5 Schematic phase diagram for NMMO, water and cellulose at 100 °C.

much shorter (about 8 hours) than that for viscose, where the need for the various ageing stages extends the process time to over 40 hours.

The resulting Lyocell fibres also have very different structural properties from viscose. To begin with, the *DP* of the cellulose in Lyocell fibres is about 600, over twice that of viscose fibres, and Lyocell fibres are considerably more crystalline, with a higher degree of orientation. Modal fibres (a type of high wet modulus rayon) have structural features in between those of viscose and Lyocell (see Figure 4.7). Another feature that distinguishes Lyocell fibres from those of viscose are their much smoother and more circular cross-sectional shapes (see Figure 4.8).

As a consequence of these structural characteristics, Lyocell fibres have greater tenacity and greater wet strength than viscose fibres. Due to their enhanced strength, Lyocell fibres of very small diameter (fineness < 1 dtex) can be produced, giving the advantage of fineness to the touch. A summary of the key tensile properties of the man-made cellulosic fibres, together with those of cotton, is shown in Table 4.4.

An important property of Lyocell fibres is their propensity to fibrillate due to their high degree of polymer chain orientation and lack of lateral cohesion. The fibres develop micro-fine surface hairs called 'fibrils' under

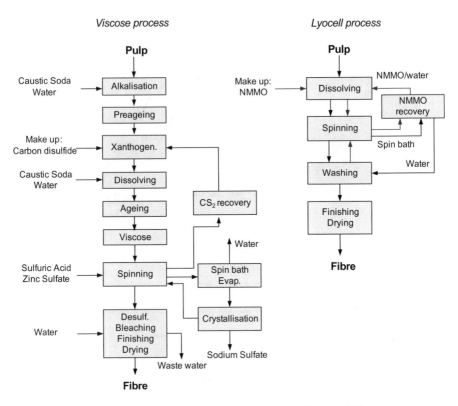

Figure 4.6   Key stages in the manufacture of viscose and Lyocell fibres.

conditions that cause abrasion, particularly under warm aqueous conditions such as washing processes. The SEM photograph in Figure 4.9 shows the fibrillation that has occurred in some fibres on dyeing a Tencel® fabric. It shows that it is mainly the surface fibres of the yarns that are susceptible to fibrillation; the fibres within the yarns remain smooth.

This tendency of the fibres to fibrillate can be both an advantage and a disadvantage. When the fibres are first processed, for example during dyeing operations, the fibrils cause the loose fibres on the fabric surface to form pills which are aesthetically undesirable. The fibrils can cause fabrics that are dyed to heavy depths of shade to have a 'frosted' appearance; since the fibrils are so fine (about 1–4 µm), they are almost transparent and have a white appearance. This is called primary fibrillation and the offending fibrils can be removed by treatment with enzymes. When fibrillation occurs in a subsequent wet processing operation, secondary fibrils which are shorter and finer are formed. On woven fabrics, this imparts a different surface characteristic, called a 'peach-skin' handle, which is an aesthetically pleasing handle for casual wear. Fibrillation is

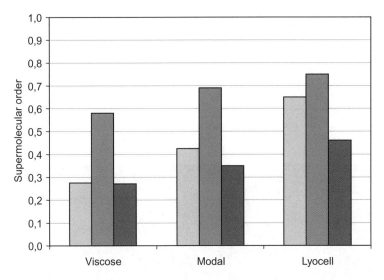

Figure 4.7 Degree of polymerisation [$10^3\,\mathrm{g\,mol^{-1}}$] (▢), orientation factor (▨) and crystallinity index (▩) in viscose, modal and Lyocell fibres. (Courtesy of Schmidtbauer, Lenzing AG).

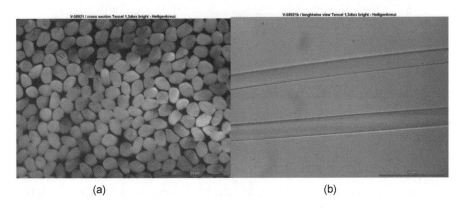

(a)   (b)

Figure 4.8 Microphotographs of Tencel® fibres: (a) cross-sectional view, (b) longitudinal view. (Courtesy of Lenzing AG).

not always required and it can be avoided by careful choice of the dyeing method. Indeed today most Lyocell is processed in 'open width' dyeing machines which prevent the wet abrasion, giving a clean, flat fibre surface rather than a peach surface. Fabrics dyed in this way have found applications in markets such as formal shirtings and bed linen.

The property of fibrillation is also useful in the manufacture of non-woven technical products, especially those made by the hydroentanglement process, where the fibrils serve to enhance the strength of the final product.

Table 4.4 Key tensile properties of cotton and man-made cellulosic fibres.

	Cotton	Viscose	HWM	Polynosic	Modal	Cupro	Lyocell
Tenacity, cond, (cN tex^{-1})	20–24	25	45	38	35	20	36
Tenacity, wet (cN tex^{-1})	26–28	13	30	30	20	10	29
Elongation, cond, %	7–9	20	12	8	13	15	14
Elongation, wet %	12–14	23	15	9	15	25	16

Note: 'Cond' = measured under standard conditions of 65% R.H and 20 °C.

Figure 4.9 Fibrillated fibres in Tencel® yarns. (Photograph Courtesy of U Syed, Heriot-Watt University).

Since fibrillation is not always a desirable attribute, a cross-linked version of Lyocell is also produced (such as Lenzing's Tencel A100®) in which the cross-linking agent trisacryloylhexahydrotriazine (TAHT) is used. The reactions which take place are as follows:

The A100 version has a higher dyeability than standard Tencel®, but it is not really suitable for blending with cotton because it is not stable to high levels of alkali so cannot be mercerised, which is an important requirement for cotton blends. In addition, A100 dyes more heavily than cotton and so it is either used alone or in blends with synthetic fibres. Another variant, called Tencel LF, is cross-linked using sodium hydroxydichlorotriazine and is more suitable for blending with cotton because it is more stable in alkali and can be mercerised, and it has a dyeability closer to that of cotton.

Lyocell fibres are used for shirts, blouses, sportswear and leisure garments, and bedding materials such as mattresses and quilts. It is claimed to be a naturally hygienic fibre which restricts bacterial growth because of its highly absorbent nature, which absorbs moisture and perspiration from the body. Being cellulosic Tencel® is biodegradable, and because it is produced from wood which is a renewable resource, it possesses excellent environmentally-friendly credentials.

It is possible to produce variants of Lyocell fibres that possess specific properties by adding chemical modifiers to the NMMO spinning dope prior to extrusion. Examples of such modifiers are biologically active agents and flame retardants. For example Lyocell fibres intended for medical purposes have been produced by incorporating Triclosan as a bactericidal agent, or by incorporating silver compounds. Triclosan is a chlorinated phenoxy compound which is EPA-registered and used as an additive in toothpaste, as well as being approved for use in fibres. The incorporation of mineral nano-powders, such as ZnO, TiO$_2$ or BaSO$_4$, into the spinning dope also yields fibres that provide UV protection from the sun's rays. Tencel C® is a fibre with anti-bacterial properties made by incorporating chitosan into standard Tencel. SeaCell® produced by Smartfibre AG contains seaweed which is an anti-inflammatory and skin protective additive. A version of SeaCell® with enhanced antimicrobial properties contains silver as well. The company's Smartcell™ range of fibres, also produced by the Lyocell process, have various functionalities, such as odour- and bacteria-reducing qualities and electrical conductivity.

4.1.5 Other Solvent-Based Processes

Numerous solvent systems for cellulose have been investigated over the years, though they have not become commercially viable processes, either because they cause degradation of the cellulose upon dissolution, or they are corrosive or difficult to use in ways that are environmentally acceptable. Examples of such solvent systems are calcium thiocyanate–water and ammonia–ammonium thiocyanate. However two processes that have been developed which show more promise are the Akzo process, in which cellulose powder is extruded from solution in phosphoric acid into a bath of acetone, and the Carbamate process, in which alkali cellulose from wood pulp is treated with urea in xylene to form cellulose carbamate, which is then dissolved in dilute NaOH and wet spun. The carbamate process is also claimed to be environmentally friendly. Neither of the processes is commercially significant at present but the carbamate process in particular could become so in the future.

4.1.6   Cellulose Acetate Fibres

The commercial development of the acetate derivatives of cellulose is an interesting story. The acetate of cellulose was first prepared in 1869 by the French chemist Paul Schutzenberger by reaction of cellulose with acetic anhydride. The reaction was not easy to carry out, and it was not until 1894 that it was found that the reaction could be greatly facilitated by the use of sulfuric acid as catalyst. The product was the triacetate, in which each of the three hydroxyl groups of the glucose rings of the cellulose are acetylated. The triacetate is soluble in dichloromethane.

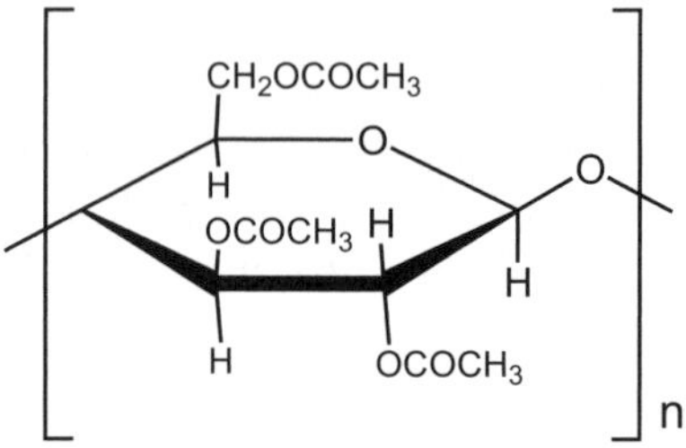

It was further discovered that if the cellulose triacetate is partially hydrolysed, so that on average 2.4 of the three hydroxyl groups on each glucose ring are acetylated, the product is soluble in acetone. This product is named cellulose diacetate but is more commonly known as cellulose acetate. During the early twentieth century two Swiss chemists, Henri and Camille Dreyfus, produced a 'dope' of this cellulose acetate dissolved in acetone. The primary use of this dope at the time was for the wings of fighter aircraft used in the 1914–18 war. The wings were made of a wooden skeleton, covered with fabric. On applying the cellulose acetate dope, the acetone evaporated leaving the cellulose acetate coated on the fabric, the result of which was to make the fabric impervious to air. Such was the demand for the dope during the war period that the Dreyfus brothers were persuaded to manufacture it in England, where they established a company called British Cellulose Co Ltd at Spondon, near Derby.

The end of the war saw a very sharp fall in demand for the dope for aircraft wings and the Dreyfus brothers were left with a huge manufacturing capacity, but with little market for it. After a considerable amount of research work, Henri Dreyfus established a process for extruding the cellulose acetate from the acetone solution as fibres. Much of the research work involved the development of the dry spinning process that was necessary to extrude these fibres. The fibres were marketed under the trade name 'Celanese' in the UK in 1921 and by his brother Camille from 1924 in the USA.

To manufacture cellulose acetate, it is necessary firstly to completely acetylate cellulose and then to partially hydrolyse it until the degree of substitution is 2.4. It is not possible simply to acetylate directly to a degree of substitution of 2.4 and confer solubility in acetone. The cellulose is sourced either from cotton linters (the very short staple length cotton – see Section 2.2.1) or from wood pulp. The process involves the following stages:

1. Pre-treatment of the pure cellulosic material in a small amount of glacial acetic acid at about 35 °C for one hour, which enables the acid to diffuse through the cellulose material and cause a degree of swelling, which facilitates the acetylation reaction.
2. The pre-treated cellulose is added to the acetylating mixture of acetic acid, acetic anhydride (in theoretical excess) and sulfuric acid (as catalyst). Efficient stirring is necessary to ensure uniform acetylation, as well as cooling, since the reactions that occur are highly exothermic and high temperatures could cause degradation of the cellulose. The reaction is allowed to continue until a clear, viscous dope has formed, (usually after about 8 hours) at which point acetic acid is added to 'kill' the excess acetic anhydride and stop the reaction. Although the sulfuric acid is described as a catalyst, this is not strictly correct, since it does actually take part in the reaction, forming sulfate linkages with the cellulose and some of it is actually consumed.
3. The dope from the acetylation reaction is run into water to give an approximately 85–90% solution and left for about 20 hours at a temperature of about 60–80 °C. It is important to balance the temperature and acid concentration so that degradation of the

cellulose chains is minimised, whilst allowing the hydrolysis re-action to take place. Once the required reduction of acetyl groups (to about 2.4 per glucose ring) has been achieved, the solution is dropped into an excess of water where the cellulose acetate is precipitated as flakes. It is important that all of the sulfate radicals are removed during the hydrolysis stage since any remaining can adversely affect the quality of the final fibres, in terms of colour and dyeability. At this stage the *DP* is about 350–400.

4. The flakes of cellulose acetate are thoroughly washed in a counter-current flow of water and then dried. The acetic acid that is present in the water washings is directed to a recovery plant.

5. Unlike the manufacture of other regenerated cellulosic fibres, such as viscose and Lyocell which are continuous processes, the above stages in the manufacture of cellulose acetates are batch processes. Before the fibre extrusion stage, several batches are blended to ensure a consistency of final product.

6. The spinning dope is prepared by dissolving the cellulose acetate in about three times its weight of acetone–water (95:5), to give a solution of about 30% cellulose acetate. Powerful stirring is required and the dissolution takes about 24 hours to complete. Before extrusion, white titanium dioxide pigment can be added if a delustred version is required rather than the standard 'bright' fibre, or very finely dispersed carbon black pigment (both at about 2% loading) if black fibre is required. After careful filtering and de-aeration, the dope is heated to reduce its viscosity then dry spun.

7. The air leaving the extrusion unit is laden with acetone, which is adsorbed onto activated charcoal from which it can be recovered for re-use.

Similar processes are used for the production of cellulose triacetate fibres, the essential difference being that Stage 3 of the above process (the hydrolysis stage) is of course not necessary and for extrusion, a dope of the polymer in dichloromethane containing about 10% methanol is used.

Both the acetate and triacetate fibres are not very crystalline, in contrast to the starting cellulosic material, so consequently they are not particularly strong fibres. By contrast, they are noted, especially the acetate, for their superb handle and silk-like quality. They form fabrics that drape very well and because of their smoothness they are widely used for linings in jackets. As might be expected, since it contains no hydroxyl groups, the triacetate is less hydrophilic than the acetate, so is

Table 4.5 Properties of acetate and triacetate fibres.

Property	Cellulose acetate	Cellulose triacetate
Tenacity	15 cN tex^{-1} and 10 cN tex^{-1} when wet.	13 cN tex^{-1}, but only about 9 cN tex^{-1} when wet.
Elongation at break	25%.	20–28%.
Elastic recovery	Highly elastic up to an extension of 5%.	It does not recover well from stretching.
Specific gravity	1.30.	1.32.
Moisture regain	6.5%.	4.5%, more hydrophobic than the acetate.
Reaction to heat	Softens easily, so care is necessary when ironing. Melts at 230 °C.	Melts at 290 °C and does not soften quite so easily as acetate. Can be heat-set.
Sunlight	Good resistance.	Good resistance.

not as absorbent. The acetate is thermoplastic and softens readily on heating. Ironing fabrics made of acetate is tricky since it melts at 230 °C but long before that point is reached it turns sticky on an iron. The triacetate however is more resistant to heat, melting at about 290 °C. It can be heat-set at about 170 °C so it will hold pleats and is resistant to creasing during washing. The heat-set triacetate fabrics are more crystalline, have a crisper handle, but are more hydrophobic. The properties of both acetate and triacetate are given in Table 4.5.

An interesting consequence of the introduction of these two fibres on to the market during the 1920s was the need to develop a new type of dye with affinity for them. Very few of the dyes existing at the time would build up to satisfactory depths of shade without washing out. This led the dye manufacturers to develop the class of dyes known as 'disperse dyes', which possess only marginal solubility in water, but could be applied by a normal dyeing operation to give fast colours. These dyes were also to become important for the coloration of nylon and polyester fibres when these were introduced during the 1940s and 50s.

These two fibre types were highly successful when they originally came on to the market during the 1920s, their brightness, lustre and handle being their main advantages, and serving as direct low-cost alternatives to silk. They became especially popular in Japan, the main uses being for women's wear, such as blouses, skirts and dresses where the easy-care 'wrinkle-free' characteristics were important. The importance of the acetate and triacetate fibres reached its peak in the early 1970s, but in the years since the synthetic fibres have taken over many of their applications and their popularity has declined considerably. Also the toxicological problems associated with using dichloromethane have caused production of the triacetate in the USA and much of Europe to cease, though it is still produced in some other countries. What has become a significant market for acetate fibres during this period, however, is in the manufacture of cigarette filters. For this application fibre tows are used which comprise bands of thousands of fibre filaments held together loosely by a crimp that is given to them in production. A 'Y'-shaped fibre cross-section is used, which has a greater surface area than a circular cross-sectional shape and is more efficient for absorbing smoke. None of the synthetic fibres can match cellulose acetate fibres for this application.

4.2 REGENERATED PROTEIN FIBRES

Fibres made from regenerated proteins have not been anything like as successful commercially as regenerated cellulosic fibres, mainly because they are relatively weak. The molecular chains of proteins do not align easily to form the ordered, crystalline structures necessary for strength, and whilst the fibres have a soft handle they have to be used in blends with other fibre types.

In the USA the generic name 'Azlon' was adopted by the FTC for any regenerated naturally occurring protein fibre, but no generic name has ever been assigned to this class of fibres in Europe. Different sources of proteins can be used, such as peanuts, corn, soybeans and milk. Of these, fibres produced from casein were the most produced and reached their maximum popularity during the 1930s and 1940s. Thereafter, as the production of synthetic fibres such as nylon, polyester and then acrylic increased and came to dominate the market, the demand for casein fibres rapidly declined and many products were discontinued during the 1950s. More recently interest in protein fibres has been renewed and in 2001 soybean protein fibre manufacture commenced under the trade name Soysilk®. It is made by extracting proteins from the residue of soybeans left in the manufacture of tofu. The interest in the fibre centres round the fact that it is produced from a renewable resource, by an eco-friendly process, and is biodegradable. The fibre is claimed to have a 'cashmere' feel and to be smoother than cashmere, and to have the lustrous appearance of silk. It is also claimed to have antibacterial properties, making it suitable for underwear and summer wear. The fibres are also used for the manufacture of soft toys and for hand knitting yarns.

SUGGESTED FURTHER READING

1. C. Woodings, *Regenerated Cellulosic Fibres*, Woodhead Publishing Ltd., Cambridge, UK, 2000.
2. R. Kotec, in *Handbook of Fiber Chemistry*, ed. M. Lewin, Taylor & Francis Group, Boca Raton, USA, 2007 ch. 10.
3. H. L. LaNieve, in *Handbook of Fiber Chemistry*, ed. M. Lewin, Taylor & Francis Group, Boca Raton, USA, 2007 ch. 11.
4. K. Bredereck and F. Hermanutz, *Rev. Prog. Coloration Textiles*, 2005, **35**, 59–75.

CHAPTER 5
Synthetic Fibres

5.1 PROCESSING OF SYNTHETIC FIBRES

5.1.1 Introduction

The production of synthetic fibres from polymer granules consists of a number of consecutive processes. The granules must first be converted to a liquid form, either as a melt or as a solution. The liquid is then extruded (spun) through the tiny holes of a spinneret (illustrated in Figure 5.1), to form continuous filaments. The holes are usually circular, with the aim of forming filaments of circular cross-section. However, filaments with non-circular cross-sections, *e.g.* triangular, cross-shaped and even hollow (as shown in Figure 1.3), can be produced from spinneret holes with corresponding profiles. For successful extrusion, the viscosity of the fluid has to be carefully adjusted. If the viscosity is too low, the extruded liquid breaks into drops. If the viscosity is too high, too great a pressure is required to force the liquid through the spinneret holes. The viscosity of a melt can be adjusted by raising or lowering its temperature, although too high a temperature may trigger some decomposition of the polymer. The viscosity of a polymeric solution can be controlled by polymer concentration. As soon as the continuous filaments are formed, they need to be converted back to the solid phase, either by cooling (from the melt) or by removal of solvent (from solution). The spun filaments are then gathered together to form a multifilament yarn.

After extrusion, the filaments need to be stretched under carefully controlled conditions (a process known as drawing), in order to be

The Chemistry of Textile Fibres
By Robert R Mather and Roger H Wardman
© Robert R Mather and Roger H Wardman 2011
Published by the Royal Society of Chemistry, www.rsc.org

Figure 5.1 Diagram of a spinneret.

strengthened sufficiently for textile applications. For some of these applications, notably apparel, the drawn multifilament yarn is texturised, in order to confer desirable aesthetic properties, improve comfort and enhance thermal insulation. Texturising confers a significantly larger volume to the yarn, by inserting bends and loops into the constituent filaments, so that there is considerable air space within the yarn. Texturised yarns may be cut into staple fibres: these are fibres of a desired length, typically 4–20 cm. The staple fibres are then packed into bales, in some cases for subsequent mixing with a natural fibre, and then conversion into blended yarns (see Chapter 9). Alternatively, slivers of parallel staple fibres are made, for conversion into yarn.

The yarns, however, may possess a significant degree of heat shrinkage, whose magnitude depends on the degree of crystallinity and the overall orientation of the polymer chains. Heat shrinkage can be overcome in many cases by heating the filaments to a sufficiently high temperature. The intermolecular bonds linking adjacent polymer chains are weakened, and their mobility increases accordingly. This relaxation process leads to re-formation of intermolecular bonds and stabilisation of the fibre structure. In addition, heat setting induces further crystallisation, which also contributes to a more stable fibre structure.

5.1.2 Melt Spinning

Melt spinning is the simplest extrusion process, in that no addition and subsequent removal of solvent is required. It is, therefore, the preferred method for those polymers that can melt without thermal degradation and that are thermally stable over the range of temperatures required for the correct extrusion viscosity. Melt spinning is used for producing polyamide ('nylon'), polyethylene terephthalate ('polyester') and polypropylene fibres, all of which are widely used commodity textiles. The principal drawback of melt spinning is the danger of thermal decomposition of the molten polymer during extrusion, due to oxidation. Thermal oxidation normally involves free radical chain reactions. Resistance to thermal oxidation can be conferred through the addition of antioxidants to the polymer. The antioxidants are oxidised preferentially, so remain effective unless they are completely consumed. Many antioxidants terminate the oxidation process by forming free radicals that are insufficiently reactive towards oxygen to continue radical chain reactions. Other antioxidants remove hydroperoxide radicals ($\cdot OOH$) that are often formed during thermal oxidation, and so prevent the start of fresh oxidation cycles. Thermal oxidation of the polymer melt may also be prevented by extrusion in an atmosphere of nitrogen.

Figure 5.2 shows schematically the melt extrusion process. Polymer granules are fed into a screw extruder, which contains a series of heated chambers. After the granules have melted, the molten polymer is compressed, and the screw action propels it towards the metering pump. The metering pump is a gear pump, which provides a constant flow of molten polymer through the holes of the spinneret (see Figure 5.1). However, before the polymer reaches the spinneret, it passes through a filter pack, to remove unwanted solid impurities that may be present. As the filaments emerge from the spinneret, the polymer melt is no longer constrained by the walls of the holes. As a result, 'die-swell' is observed: this is a dilation of the extruded filament at its exit from the spinneret. Die-swell limits the minimum cross-section that can be obtained under a particular set of extrusion conditions.

After emergence from the spinneret, the individual filaments are pulled through a long cooling chamber, where they solidify, towards a take-up roller. Cooling is achieved by a transverse flow of air, whose velocity can be regulated, in order to control the cooling rate. By the time the filaments reach the take-up roller, they have been gathered into multifilament yarn. The yarn is treated with a liquid spin finish just before it reaches the roller. Spin finish prevents filament abrasion and the build-up of static electricity. Take-up velocity is generally in the

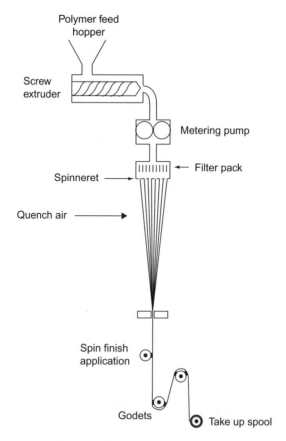

Figure 5.2 Schematic diagram of the melt spinning process.

range of 1000–6000 m min^{-1} and is much higher than the velocity at which the filaments emerge from the spinneret. The consequent stretching of the filaments induces some ordering of the constituent polymer chains.

Single thicker filaments, known as monofilaments, may also be extruded from a spinneret with just one hole. These filaments are often cooled by passage through a water bath. Take-up velocity is much lower, often 30–100 m min^{-1}.

5.1.3 Solution Spinning

Solution spinning is adopted for those polymers that cannot be melt extruded. There are two main types of solution spinning, the choice depending on the method by which the solvent is removed from the extruded filaments. In dry spinning, the solvent is removed by evaporation; in wet spinning, it is removed by coagulation in another liquid, which is compatible with the spinning solvent yet is not a solvent for the filaments.

Synthetic fibres produced by dry spinning include acrylic, polyvinyl chloride and elastomeric fibres. Figure 5.3 shows schematically the dry spinning process. Polymer chip is dissolved in a suitable solvent, in most cases to a concentration of 15–25%, depending on the desired viscosity for extrusion. After filtration, the spinning solution is fed into a metering pump at a predetermined temperature and pressure (often 10–20 bar). It then passes to the spinneret, and at this stage the temperature is often raised. For example, the temperature of a solution of polyacrylonitrile in dimethyl formamide is typically raised from 90 °C to 140 °C. The

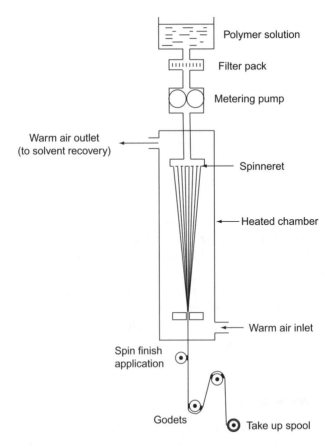

Figure 5.3   Schematic diagram of the dry spinning process.

spinning solution emerges from the tiny spinneret holes to form filaments and, as with melt-spun filaments, die-swell is often observed. The extruded filaments pass through a heated chamber, and much of the solvent evaporates. Some residual solvent remains in the filaments, however, in order to assist the subsequent drawing process. An oil-based spin finish is then applied to the filaments before they are wound on a roller. Spinning velocities range from 100 to 1000 m min^{-1}.

The profiles of the cross-sections of the extruded filaments are more difficult to control, than they are for melt spinning. The problem arises from the diffusion of solvent to the surface of each filament and the subsequent evaporation of solvent from the surface. If the rates of diffusion and evaporation are about equal, then the cross-sectional profile will be largely unaltered. If, however, evaporation is significantly faster than diffusion, a hard skin of dry polymer forms and the cross-section becomes distorted. Figurre 5.4 shows the kind of distortion that may occur to filaments extruded from circular spinneret holes.

The other main method of solution spinning is wet spinning, illustrated in Figure 5.5. This method is applied to polymers that cannot be melt extruded, and which are insoluble in solvents that can be readily evaporated. Synthetic fibres formed by wet spinning include some acrylics, polyvinyl chloride and polyvinyl alcohol. Compared with dry spinning, wet spinning is often conducted at lower temperatures, and the concentration of dissolved polymer is 10–30%. The solution is highly viscous. After the solution has been outgassed to remove any air, it is filtered and then passed through a metering pump to the spinneret, which is located in the coagulation bath. Spinnerets are normally constructed from precious metals. The number of holes in a spinneret can be

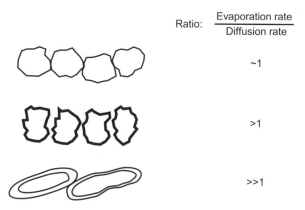

Figure 5.4 Schematic representation of fibre distortion on extrusion from a spinneret.

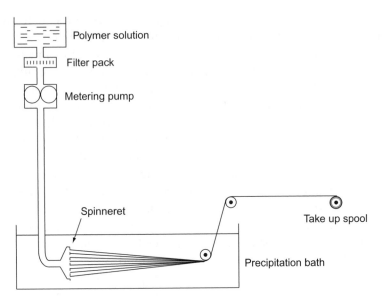

Figure 5.5   Schematic diagram of the wet spinning process.

as high as tens of thousands! The cross-section of each hole is generally within the range of 0.05–0.38 mm.

As the viscous polymer solution emerges from the spinneret holes, the polymer comes out of solution in the coagulation bath in the form of continuous filaments. The solid filaments formed, however, must be in a gel state, in which some solvent is still present and also some liquid from the coagulation bath. In this state, the filaments are highly elastic, and subsequent drawing is therefore facilitated. The filaments are taken up by a roller at a higher velocity than that at which they emerged from the spinneret, so that they are stretched $1\frac{1}{2}$–3 times. Spinning velocities are low, generally 100–1000 m min^{-1}, so the use of spinnerets with many thousands of holes is necessary to achieve productivity. The filaments are then drawn, while they go through the subsequent processing steps of washing and drying. The degree of stretch achieved may be up to *ca.* 30 times.

The cross-sectional profiles of the wet-spun filaments can vary (see Section 1.4), even from the same profile of spinneret hole. For example, a circular spinneret hole may yield a variety of filament cross-sections, ranging from circular to much flatter, depending on the rate of coagulation and the nature of the additives in the coagulation bath.

A procedure has also been devised, which combines dry spinning and wet spinning. It is known as dry-jet wet spinning and is used, in particular, for processing some types of aramid fibres. As noted in

Chapter 4, it is also used for spinning Lyocell fibres. The process is discussed later in Chapter 6.

5.1.4 Drawing

Freshly extruded filaments that have not been subsequently drawn (stretched) have poor mechanical performance. However, drawing of the filaments by factors of five to ten times or more can greatly improve mechanical properties. Drawing confers to the polymer chains a much greater degree of alignment in the direction of the filament axis. Drawing may be performed either directly after the extrusion process or as a separate operation, starting with undrawn filaments wound on bobbins. Moreover, it may consist of several stages. Important control parameters are draw ratio, draw temperatures, draw speeds and the number of drawing stages. (Draw ratio is the ratio of the linear densities of the filaments before and after drawing.) The drawing process is irreversible.

Undrawn filaments deform during drawing either inhomogeneously, with a marked necking effect, or homogeneously, in which the deformation is uniform. In the initial part of inhomogeneous drawing, the undrawn filament becomes thinner at particular points. As drawing proceeds, the length of the thinner region progressively increases, until drawing is complete. Figure 5.6 shows the type of stress–strain curve observed for inhomogeneous drawing. From A to B, the relation between stress and strain is almost linear, and deformation is elastic and practically reversible. Beyond B, necking occurs, and the filament deformation is plastic and irreversible. At C, neck propagation is complete. The stress then rises sharply with small increases in strain, until the filaments rupture at D.

Whether drawing occurs homogeneously or inhomogeneously depends on the structure of the undrawn filaments and the drawing conditions. Neck formation generally occurs at lower draw temperatures, higher draw speeds and higher orientation of the polymer chains in the undrawn filaments. It is also more likely to occur, the larger the cross-section of the filaments. Neck formation is considered to arise from variations in stress along a filament, arising from deformation during drawing. In turn, these variations may be attributable to small inhomogeneities along the length of the filament, such as slight variations in cross-sectional area.

The lower the drawing temperature, the more likely are the introduction of fresh stresses and structural defects into a filament. These

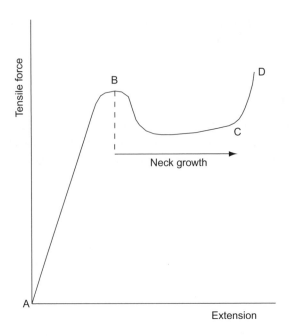

Figure 5.6   Stress–strain curve for inhomogenous drawing.

stresses and defects can be substantially removed if the filaments are heated to a sufficiently elevated temperature.

5.2   POLYAMIDE FIBRES

5.2.1   Introduction

Polyamides are polymers whose constituent monomeric units are joined by the amide linkage, –NH–CO–. Those polyamides that are predominantly aliphatic are members of a generic class often called 'nylons'. Polyamides in which ≥ 85% of the amide linkages are adjacent to aromatic structures are known as 'aramids'. This section is concerned with 'nylons', which in most countries are in fact called 'polyamides'. Aramids are the subject of Section 6.2.

Polyamide fibres were the earliest synthetic fibres to become major commercial products. Wallace Hume Carothers is almost invariably cited as the pioneer of polyamide fibres. He joined E. I. du Pont de Nemours in 1928 at their base in Wilmington, Delaware, USA. However, it was not until 1934 that a member of Carothers' group, D. D. Coffman, produced nylon 9 fibres, formed from the polymerisation of ethyl 9-aminononanoate ($H_2N(CH_2)_8COOC_2H_5$). (In nylon terminology, the numerals that distinguish each type of nylon represent the number of carbon atoms lying between successive amide groups in the polymer chain.) A patent was filed by DuPont the following year, which claimed a number of similar polyamides, including nylon 6. These polyamides had been produced from the polymerisation of both 6-aminohexanoic acid ($H_2N(CH_2)_5COOH$) and ethyl 6-aminohexanoate ($H_2N(CH_2)_5COOC_2H_5$).

Nylon 6,6, the polyamide from DuPont that finally made real commercial impact, was first produced by Gerard Berchet in early 1935, and only a few months later, a decision was made by Elmer Bolton, the head of the DuPont laboratories, to exploit the new fibre commercially. Nylon 6,6 is produced by the copolymerisation of 1,6-hexandioic acid (adipic acid) and 1,6-diaminohexane (hexamethylene diamine). In 1939, nylon 6,6 fibres were commercially produced, with the first nylon stockings also being sold that year. The first production in the UK of polyamide fibres was at British Nylon Spinners in 1941 (under licence from DuPont), and much of the production at that time was for parachute fabrics.

It is ironic to note that in 1930, Carothers and Berchet had reported that caprolactam does not undergo polymerisation. Nevertheless, Paul Schlack, who was working at I. G. Farbenindustrie in Germany, demonstrated in 1938 that caprolactam could be polymerised, and that the product, nylon 6, could form strong fibres. Large-scale production of nylon 6 fibre, known as 'Perlon', began at Landsberg in 1943. Production of the fibre in Germany was stopped at the end of the Second World War in 1945, and was only resumed in 1950.

In the early 1950s, polyamide almost completely dominated the synthetic fibres market. By 1970, polyester, acrylic and polypropylene fibres had been introduced into the market, and world-wide production of synthetic fibres had risen by more than 50 times. Nevertheless, polyamide fibres still accounted for *ca.* 40% of synthetic fibre production. Since 1975, polyester has overtaken nylon, and is now by far the most used synthetic fibre. Polyamide, however, still has an important commercial presence.

This section will concentrate on nylon 6,6 and nylon 6, because they are the two most important commercial polyamides. Nylon 6,6 can be represented as:

$$-[NH-(CH_2)_6-NH-CO-(CH_2)_4-CO]_n$$

Nylon 6 can be represented as:

$$-[NH-(CH_2)_5-CO]-_n$$

For both polyamides, the constituent monomeric units each contain six carbon atoms. As shown in Section 5.2.2, the monomers can be conveniently synthesised from aromatic compounds. Other commercial polyamides are produced on a much smaller scale. These include nylon 4,6, nylon 6,10 and nylon 11.

5.2.2 Manufacture of Monomers

5.2.2.1 1,6-Hexandioic Acid (Adipic Acid).   1,6-Hexandioic acid can be synthesised from benzene. The first stage is hydrogenation to cyclohexane, which can then be oxidised in air to yield a mixture of cyclohexanol and cyclohexanone. Conversion of the mixture to 1,6-hexandioic acid can be achieved with concentrated nitric acid in the presence of a copper–vanadium catalyst:

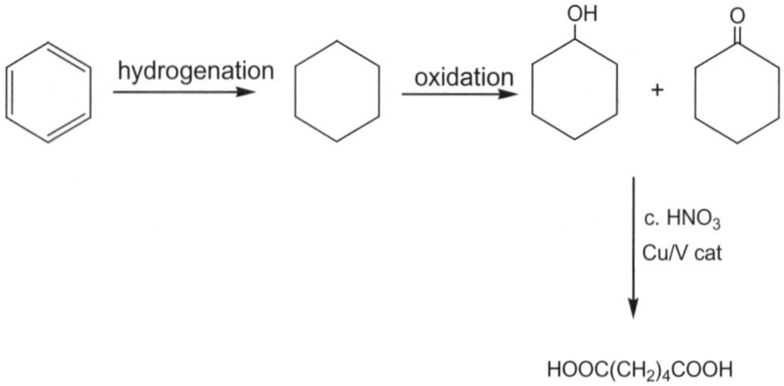

Another starting material is phenol, which can be directly hydrogenated to yield cyclohexanol and some cyclohexanone.

An alternative route is the treatment of butadiene with carbon monoxide and methanol, to give the dimethyl ester of 1,6-hexandioic acid. The ester is then hydrolysed:

$$CH_2 = CH - CH = CH_2 + 2CO + CH_3OH$$
$$\rightarrow CH_3O-OC-(CH_2)_4-CO-OCH_3 \rightarrow HOOC-(CH_2)_4-COOH$$

The ester may also be formed by hydrogenation of dimethyl hexenoate, itself produced through the dimerisation of methyl propenoate (methyl acrylate):

$$2CH_2 = CH-CO-OCH_3 \rightarrow CH_3O-OC-CH_2-CH = CH-CH_2-CO-OCH_3$$
$$\rightarrow CH_3O-OC-(CH_2)_4-CO-OCH_3$$

5.2.2.2 1,6-Diaminohexane (Hexamethylene Diamine). A number of methods are available for synthesizing 1,6-diaminohexane on a commercial scale, but in nearly all cases the last stage involves the reduction of 1,4-dicyanobutane (adiponitrile) in the presence of a cobalt or nickel catalyst:

$$NC–(CH_2)_4–CN \rightarrow H_2N–(CH_2)_6–NH_2$$

The processes differ, therefore, in their routes to 1,4-dicyanobutane.

Originally, 1,6-hexandioic acid was the starting material. Treatment with ammonia effects conversion to 1,6-hexandiamide (adipamide), which on dehydration over a phosphorus or boron oxide catalyst yields 1,4-dicyanobutane:

$$HOOC–(CH_2)_4–COOH \rightarrow H_2N–OC–(CH_2)_4–CO–NH_2$$
$$\rightarrow NC–(CH_2)_4–CN$$

Other processes utilise butadiene as the starting material. In one of these, treatment with hydrogen cyanide yields 1,4-dicyanobutane:

$$CH_2{=}CH–CHCH_2 \rightarrow NC–(CH_2)_4–CN$$

\Alternatively, butadiene is treated with copper (ɪ) cyanide in the presence of iodine. The product, 1,4-dicyanobut-2-ene, is then hydrogenated:

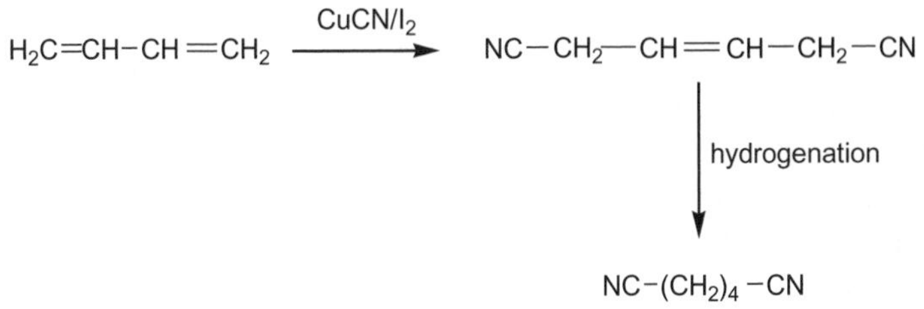

$$H_2C{=}CH–CH{=}CH_2 \xrightarrow{\text{CuCN/I}_2} NC–CH_2–CH{=}CH–CH_2–CN$$

$$\downarrow \text{hydrogenation}$$

$$NC–(CH_2)_4–CN$$

1,4-Dicyanobutane can also be synthesised through the reductive electrolytic dimerisation of cyanoethene (acrylonitrile):

$$2CH_2{=}CH–CN + 2H_2O + 2e^- \rightarrow NC–(CH_2)_4–CN + 2OH^-$$

1,6-Diaminohexane can also be synthesised from tetrahydrofuran (THF). Treatment of THF with hydrochloric acid yields 1,4-dichlorobutane, which can be converted by sodium cyanide to 1,4-dicyanobutane (adiponitrile). Reduction of 1,4-dicyanobutane yields 1,6-diaminohexane. 1,4-dicyanobutane can also be hydrolysed to 1,6-hexandioic acid, but it is cheaper to produce 1,6-hexandioic acid by the routes outlined in Section 5.2.2.1.

5.2.2.3 Caprolactam. Caprolactam can also be synthesised from a mixture of cyclohexanol and cyclohexanone (see Section 5.2.2.1). As cyclohexanone is the preferred starting compound, the mixture can be passed over a copper catalyst; the cyclohexanol is converted to cyclohexanone by dehydrogenation. Alternatively, a process has recently been reported by which cyclohexanone can be synthesised directly from phenol. The reduction is achieved with palladium and a Lewis acid catalyst, such as aluminium trichloride. The cyclohexanone is then converted to an oxime by treatment with hydroxylamine (NH_2OH), and the oxime is cyclised in oleum by a Beckmann rearrangement to yield caprolactam:

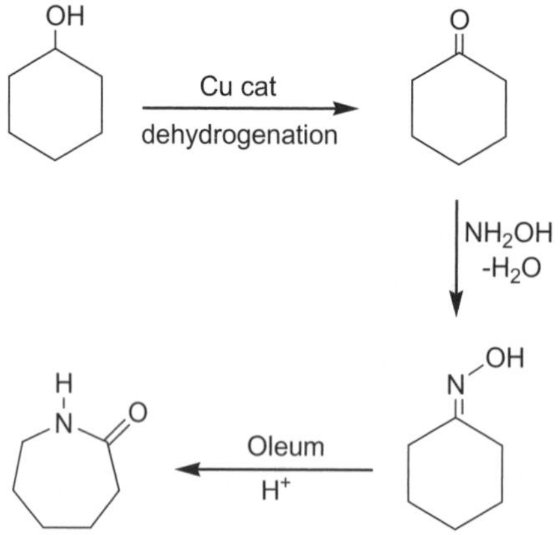

A number of techniques have been used to generate hydroxylamine in a form that can be readily applied. In one approach, hydroxylamine sulfate is used. Alternatively, hydroxylamine phosphate is used.

Another commercially viable route to caprolactam involves butadiene, carbon monoxide, ammonia and hydrogen as feedstocks. The process consists essentially of four consecutive stages:

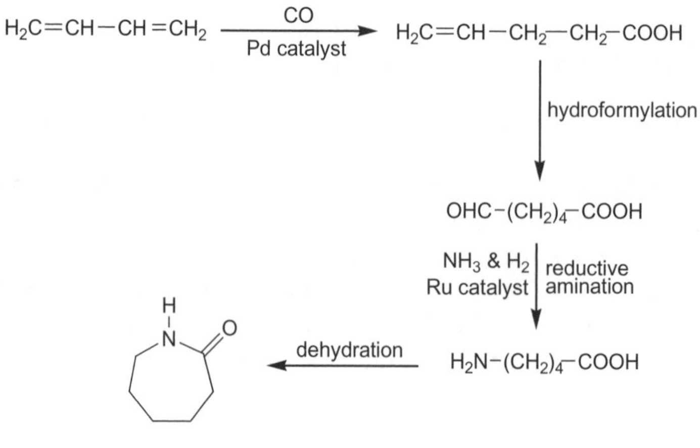

Caprolactam can also be synthesised from toluene. Toluene is oxidised to benzoic acid, which in turn is hydrogenated in the presence of palladium or platinum to cyclohexanoic acid. The reduced acid is then treated with nitrosyl sulfuric acid ($HO \cdot SO_2 \cdot O \cdot NO$) in the presence of oleum:

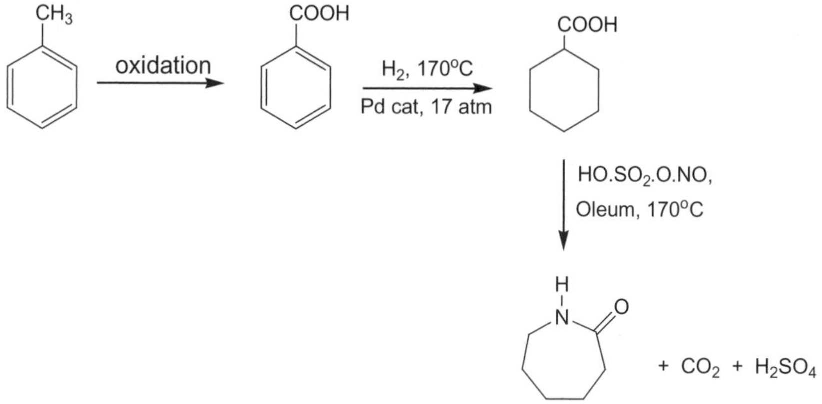

5.2.3 Production of Nylon 6,6

Nylon 6,6 is an alternating copolymer of 1,6-hexandioic acid and 1,6 diaminohexane. The initial stage in the production of nylon 6,6 involves the formation of so-called 'nylon salt' from equimolar quantities of the two compounds in methanol:

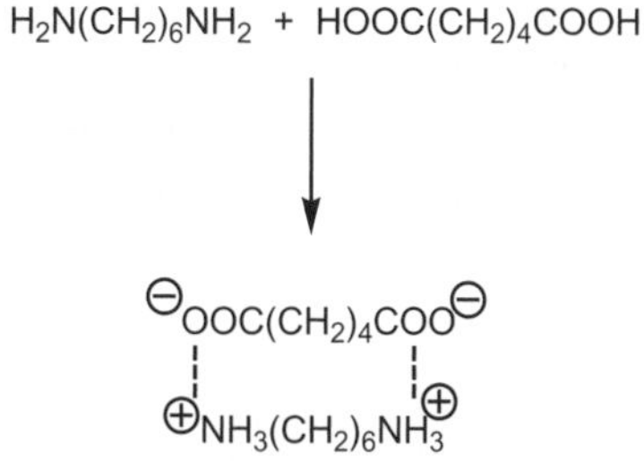

Nylon "salt"

The salt is precipitated from the methanol and subsequently dissolved in water to a level of 60%. The solution is transferred to an evaporator, where the salt concentration is raised to >80%.

After a small quantity of monofunctional agent, such as ethanoic acid, has been added, step growth polymerisation is carried out in an atmosphere of nitrogen. Oxygen is excluded, because the polymer product is prone to degradation by oxygen at the reaction temperatures used (see Section 5.2.7.4). As steam is also a product of the reaction, the molar mass of nylon 6,6 achievable is governed by the pressure of the steam during the reaction. Initially, nylon 6,6 of low molar mass (*ca.* 4000) is formed at a temperature of 260–280 °C and a pressure of 1.8 MPa. The pressure is then released by allowing the steam to escape, and the molar mass of the nylon 6,6 increases to *ca.* 12 000. A still higher molar mass can be achieved by conducting the final stages of the polymerisation at reduced pressure. Thus, careful control of pressure can lead to the desired molar mass. The presence of the monofunctional agent also serves to control molar mass by terminating growth of the polymer chains. The number average molar mass required for nylon 6,6 fibres is in the range of 12 000–22 000, depending on the fibre properties required. The level of cyclic oligomers formed in the process is usually <2%, a level usually considered acceptable in nylon 6,6 fibres.

Nowadays, nylon 6,6 is almost invariably produced by a continuous process, and on formation the product is often led straight to a melt extruder. Production processes have also been devised in which nylon salt is polymerised in the dry form. Although less energy is expended on removing water, particle size has to be carefully controlled.

5.2.4 Production of Nylon 6

For the processing of nylon 6 into fibres, caprolactam is normally polymerised by a hydrolytic process at 250–270 °C at atmospheric pressure. In a completely dry form, caprolactam cannot be polymerised. Acid, base or even merely water is required to initiate polymerisation. In industrial processes, water is generally used, as it is a cheap initiator! As with the production of nylon 6,6, a small amount of monofunctional agent is often present, to control the molar mass of the final polymer.

The initial step in the production of nylon 6 is hydrolysis of caprolactam to 6-aminohexanoic acid:

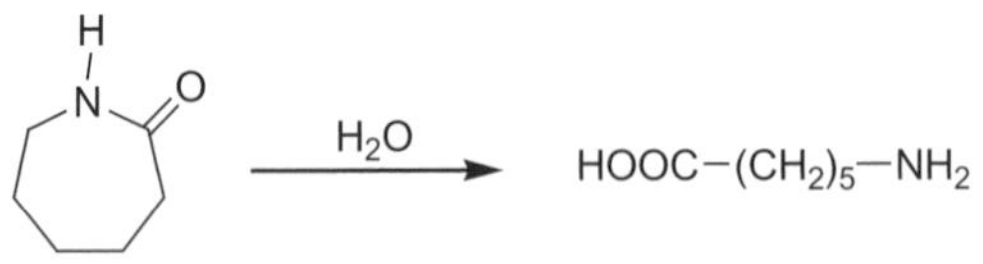

$$\text{caprolactam} \xrightarrow{\ H_2O\ } HOOC-(CH_2)_5-NH_2$$

Addition polymerisation then occurs, until an equilibrium mixture of polymer (*ca.* 90%) and caprolactam (*ca.* 10%) is attained:

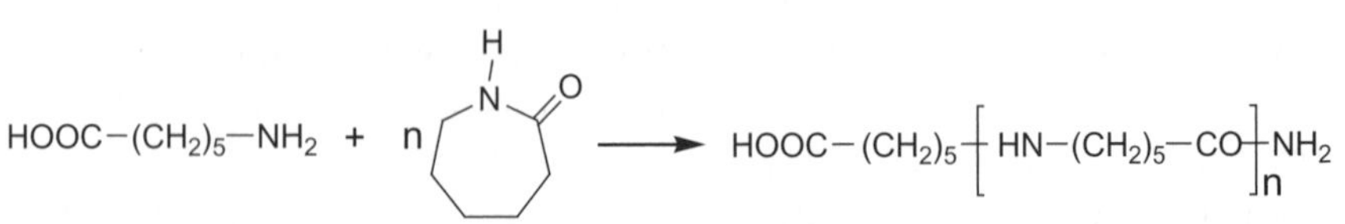

$$HOOC-(CH_2)_5-NH_2 + n\,\text{(caprolactam)} \longrightarrow HOOC-(CH_2)_5-[HN-(CH_2)_5-CO]_n-NH_2$$

Step growth polymerisation then follows, involving the polymer molecules already formed, to yield nylon 6. The number average molar mass of the nylon 6 produced is generally *ca.* 20 000 or so.

The nylon 6 produced contains up to 10% of cyclic oligomers, consisting of unreacted caprolactam, cyclic dimer:

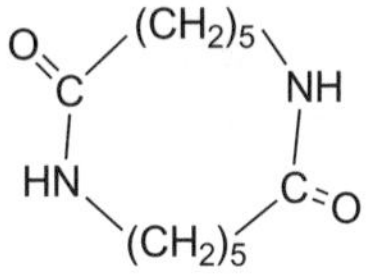

and even cyclic trimer and tetramer. These oligomers are removed either by extraction with hot water or by vacuum extraction. Boiling methanol also dissolves away the oligomers from the polymer.

5.2.5 Fibre Extrusion

Polyamide fibres are melt extruded. During extrusion, the polymer is maintained in an atmosphere of dry nitrogen, to prevent oxidative degradation (see Section 5.2.7.4). Typical extrusion temperatures are 280–290 °C for nylon 6,6 and 250–260 °C for nylon 6. The spinneret holes are typically 0.1–0.4 mm in diameter, and extrusion velocity is normally 1000–1200 m min^{-1}. For nylon 6,6 particularly careful control of extrusion conditions is required, because of its susceptibility to degradation. Nylon 6 is less sensitive to extrusion conditions. However, exposure of nylon 6 to elevated temperatures during extrusion does bring about some regeneration of free caprolactam.

After extrusion, the polyamide filaments pass through a cooling chamber of up to 5 m in length. The filaments then need to be moistened before they are finally wound or perhaps fed directly to a drawing unit. Polyamide filaments are able to absorb a small amount of water (it can be noted that moisture regain is 4.0–4.5% under standard conditions of 20 °C and 65% relative humidity), and in consequence become slightly longer. Therefore, if the polyamide yarn is wound without moisturizing, it would become slightly slacker on absorption of moisture. It would then be difficult to unwind the yarn for subsequent processing, or even properly to maintain the yarn on the winder.

Nylon 6,6 can be moistened by passage through a steam chamber prior to winding. Alternatively, the yarn is moistened by a spin finish. However, the presence of free caprolactam excludes the use of a steam chamber for moistening nylon 6 yarn, as the constituent filaments would become too sticky. Conditioning with spin finish is, therefore, preferred.

The extruded filaments are then drawn, usually to a draw ratio of 4–5. In general, cold drawing is adopted for nylon yarn to be used in apparel and carpets. Stronger yarns that are required for more demanding applications are drawn at higher temperatures. The drawn yarn may then be texturised, and finally heat set. Heating of nylon 6,6 filaments to 180–200 °C and of nylon 6 filaments to 160–180 °C induces an annealing process in them, and a permanent set is achieved. In the presence of steam, lower temperatures suffice: 120–130 °C for nylon 6,6 and 100–105 °C for nylon 6.

5.2.6 Physical Structure

As with most synthetic fibres, polyamide fibres possess crystalline and amorphous regions and a number of intermediate phases. The number and arrangement of these phases within the fibre is determined by the chemical structure of the polyamide in question and the processing conditions by which the fibre was made. In general, nylon 6,6 fibres crystallise more quickly after extrusion, and so nylon 6 fibres tend to possess the more open structures.

Polyamides can exist in several crystal forms. For nylon 6,6 and nylon 6, the most energetically stable crystalline structures are their α-forms, which for both polymers comprise stacks of sheets of planar polyamide chain segments. Within each sheet are hydrogen bonds linking adjacent chain segments. In nylon 6,6 the chain segments run parallel to one another in a triclinic cell (see Figure 5.7a). The distance between neighbouring chains is 0.42 nm, and between adjacent sheets is 0.36 nm. In nylon 6, however, the sheets of the α-form are characterised by an antiparallel alignment of the chain segments in a monoclinic cell (see Figure 5.7b). The interchain distance is 0.44 nm, and the intersheet distance is 0.37 nm. The interchain and intersheet distances are, therefore, very similar for both types of polyamide fibre.

Although the α-crystalline form is the most stable, the γ-form can occur in freshly extruded polyamide filaments. At low spinning speeds, distinct crystalline forms are absent in nylon 6,6 filaments, but the γ-form is predominant at higher speeds. Extruded nylon 6 filaments contain both α- and γ-forms. The higher the spinning speed, the greater the proportion of the γ-form present. Annealing the filaments at elevated temperatures converts the γ-form to the α-form. Thus, although the α-form is energetically more favoured, the constraints on the polyamide chains, as the filaments are cooled, tend to promote formation of the γ-form.

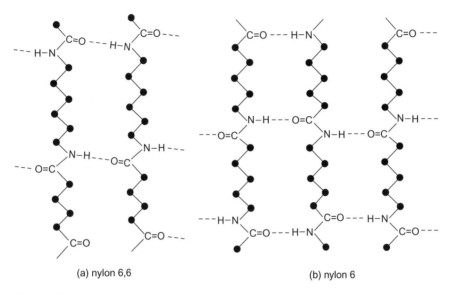

(a) nylon 6,6   (b) nylon 6

Figure 5.7   Structures of (a) nylon 6,6, (b) nylon 6.

5.2.7   Chemistry of Polyamide Fibres

5.2.7.1   Introduction.   The chemical properties of polyamide fibres are largely determined by the amide groups along the polymer chains. The remainder of the chains consists of methylene ($-CH_2-$) groups, which are far less reactive. Polyamide fibres are generally resistant to alkalis and to most common organic solvents. They do, however, dissolve in concentrated methanoic acid, phenol and *m*-cresol. They are also attacked by strong acids and are prone to thermal oxidation and photodegradation.

5.2.7.2   Hydrolysis.   Polyamide fibres are hydrolysed by superheated steam ($\geq 150\,^\circ C$) and by strong acids, such as concentrated hydrochloric acid and sulfuric acid. Depolymerisation back to the constituent comonomers occurs. Nylon 6 is more readily hydrolysed than nylon 6,6, most likely because the more open structure of nylon 6 allows easier penetration of acids and steam.

5.2.7.3   Pyrolysis.   At elevated temperatures in the absence of oxygen, polyamide fibres are degraded, with the accompanying formation of ammonia, carbon dioxide and water. The liberation of ammonia is the result of a deamination reaction between primary amine end-groups:

$$—CH_2—NH_2 + H_2N—CH_2— \longrightarrow —CH_2—NH—CH_2— + NH_3$$

The liberation of carbon dioxide and water arise from more complex reactions, involving carboxylic acid end-groups. The reaction sequence is:

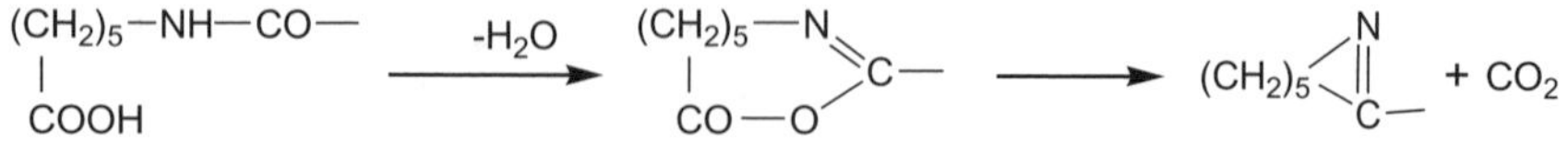

In the case of nylon 6,6, cyclopentanone is also released; 1,6-hexandioic acid end-groups tend to cyclise, with the loss of carbon dioxide and water:

cyclopentanone

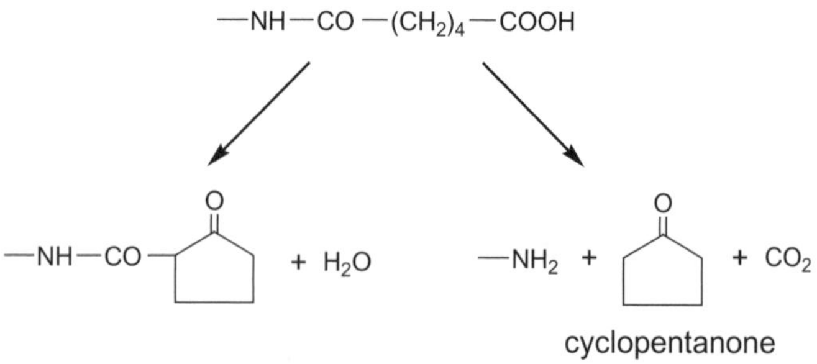

Knowledge of pyrolysis mechanisms is important for the control of the extrusion of polyamides. In addition, polyamide fibres are used extensively in tyres, where the fibres are embedded in rubber and are inaccessible to air.

5.2.7.4 Thermal Oxidation. Oxidation of polyamide fibres is undesirable in that it causes their discoloration and considerably reduces their strength. It occurs by free radical mechanisms. The process is initiated by small quantities of hydroperoxides and carbonyl compounds, produced during processing of the polyamide fibres. The free radicals formed from these species can abstract hydrogen atoms from the methylene groups adjacent to the amide nitrogen atoms along the polyamide chains. Reaction with oxygen then follows to yield a peroxide radical, which can in turn abstract a further hydrogen atom from the polymer chain. A hydroperoxide is formed, together with another polymer radical, which then combines with further oxygen in a chain reaction. The hydroperoxide decomposes to form two new radicals, and so chain branching is initiated:

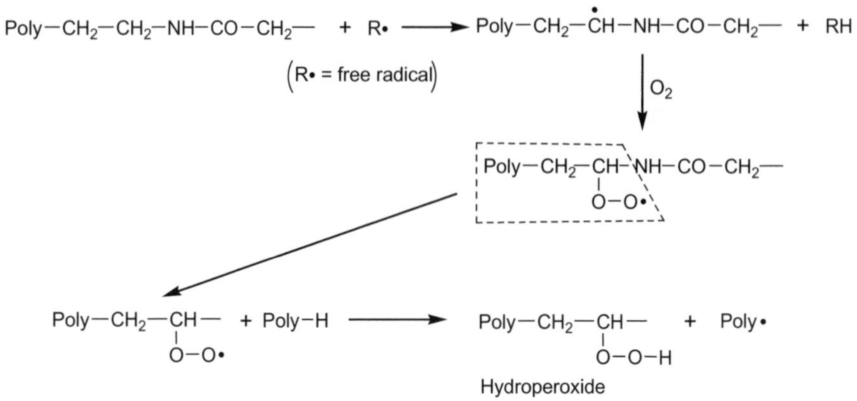

Poly—CH$_2$—CH$_2$—NH—CO—CH$_2$— + R• $\longrightarrow$ Poly—CH$_2$—$\overset{\bullet}{C}$H—NH—CO—CH$_2$— + RH

(R• = free radical) $\downarrow$ O$_2$

Poly—CH$_2$—CH—NH—CO—CH$_2$—
     |
     O—O•

Poly—CH$_2$—CH— + Poly—H $\longrightarrow$ Poly—CH$_2$—CH— + Poly•
     | |
     O—O•   O—O—H
                                            Hydroperoxide

Then:

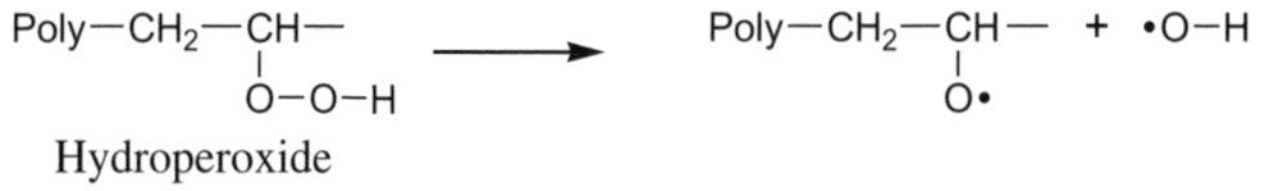

Poly—CH$_2$—CH— $\longrightarrow$ Poly—CH$_2$—CH— + •O—H
     | |
     O—O—H   O•
Hydroperoxide

or:

2 Poly—CH$_2$—CH— $\longrightarrow$ Poly—CH$_2$—CH— + Poly—CH— + H$_2$O
        | | |
        O—O—H   O•   O—O•
Hydroperoxide

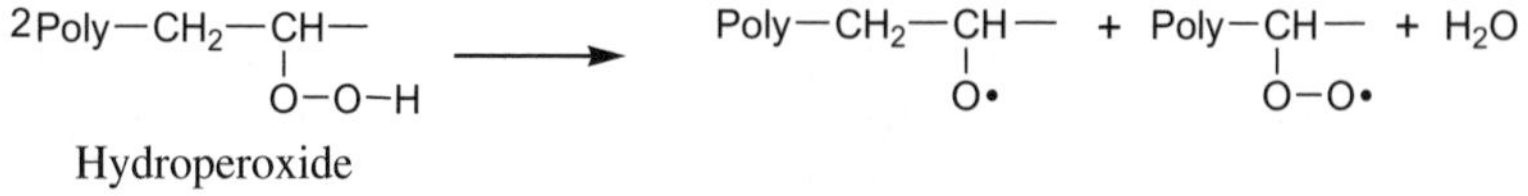

In addition, the polymeric oxygen centred radicals formed can be cleaved, and the polyamide fibres accordingly become weaker.

Discoloration of polyamide fibres on oxidation is attributable to the formation of azomethine (–CH=N–) groups in structural units with conjugated bonds. The azomethine groups are formed from reaction of aldehydes, produced by oxidation, with primary amine groups in the polyamide chains:

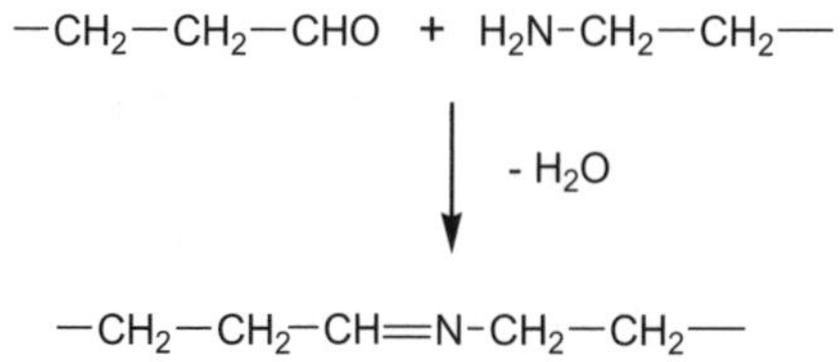

—CH$_2$—CH$_2$—CHO + H$_2$N-CH$_2$—CH$_2$—

$\downarrow$ - H$_2$O

—CH$_2$—CH$_2$—CH=N-CH$_2$—CH$_2$—

5.2.7.5 Photodegradation. Exposure to light, and in particular to ultraviolet radiation, can also degrade polyamide fibres. The extent of photodegradation depends on the nature of the radiation and on the presence of compounds, such as colorants, in the fibres that may sensitise them to the radiation. The absorption of light by these additives leads to the formation of free radicals that initiate oxidation. As in the case of thermal oxidation, hydrogen atoms are extracted from methylene groups next to the nitrogen atoms in the polyamide chains. Aldehydes and carboxylic acids are formed during photodegradation. The aldehydes then tend to condense, to form larger structures with conjugated double bonds.

5.2.7.6 Stabilisation of Polyamide Fibres. To prevent the thermal degradation and photodegradation of polyamide fibres, stabilisers are incorporated into them. These compounds must also meet a number of other processing criteria. For example, they must themselves possess sufficient stability to withstand the conditions under which the fibres are processed. Moreover, they must not migrate within the fibres, nor must they colour the fibres. They must not adversely react with any other additives present in the fibres.

Stabilisers can act in a number of ways. Antioxidant stabilisers, for instance, can stabilise polyamide fibres against both thermal degradation and photodegradation. They function either by scavenging radicals or by decomposing peroxides. Some radical scavengers react with the free radicals formed in the polyamide chains, to yield stable products that cannot lead to further chain degradation. These scavengers are sterically hindered phenols, such as 2,6-di-*tert*-butyl-*p*-cresol:

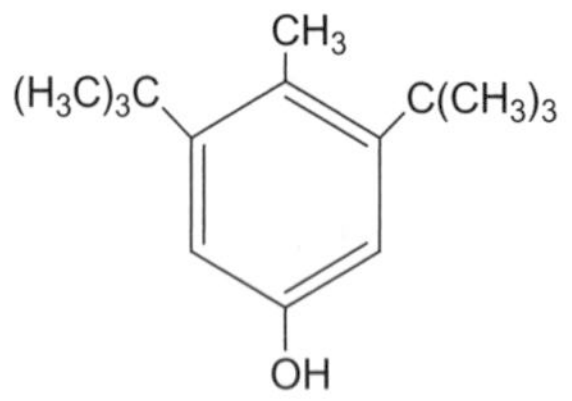

or secondary aromatic amines. A chain reaction is involved, which gives rise to a neutral product and a stable radical in which there is extensive electron delocalisation. In the case of 2,6-di-*tert*-butyl-*p*-cresol, delocalisation leads to formation of a quinone structure, which can then deactivate another free radical by a termination reaction:

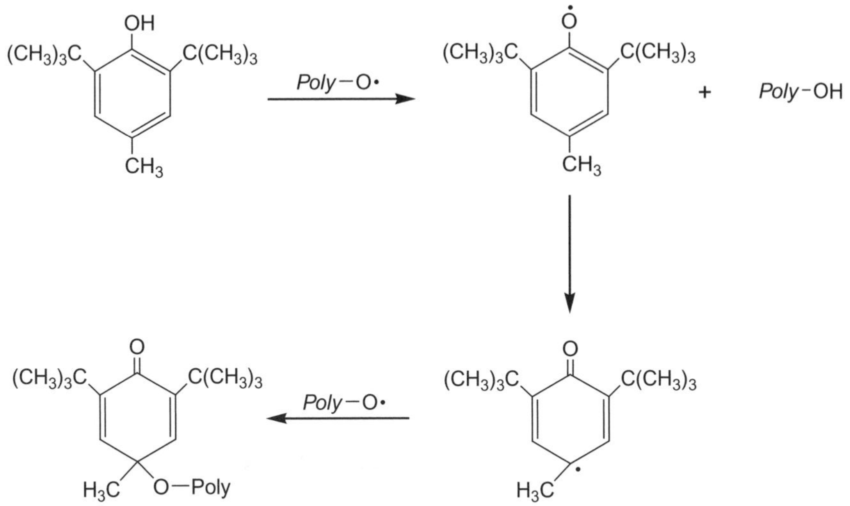

Peroxide decomposers for polyamide fibres are predominantly metal cations, such as Mn^{2+}, Cu^{2+} and Cu^{+}. Mn^{2+} ions hinder photodegradation, and Cu^{+} ions improve the thermal resistance of the fibres. Examples of their reactions with radicals formed in the polyamide chains are:

$$Poly^{\bullet} + Cu^{2+} \rightarrow Poly^{+} + Cu^{+}$$

$$PolyOO^{\bullet} + Cu^{+} \rightarrow PolyOO^{-} + Cu^{2+}$$

Care has to be taken that the concentration of the metal ion is above a minimum of *ca.* 0.01%. Very low concentrations of peroxide decomposers may catalyse oxidation of the polyamide fibres.

Photodegradation may be impeded by the presence of ultraviolet absorbers. These compounds preferentially absorb those wavelengths that can cause fibre degradation. Ultraviolet absorbers include 2-hydroxybenzoquinone and derivatives of 2-hydroxyphenyl benzotriazole. Photodegradation may also be countered by the presence of hindered amine light stabilisers (HALS). Most of these stabilisers contain 2,2,6,6-tetramethyl piperidyl groups:

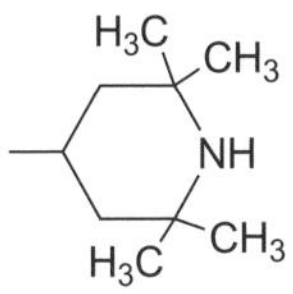

HALS function as free radical scavengers. The active species are nitroxyl radicals, formed by photo-oxidation of the amine group:

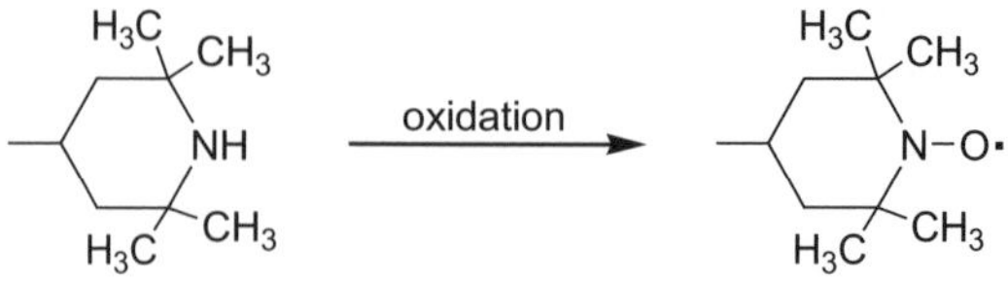

During scavenging, the nitroxyl groups are mostly regenerated.

5.2.8 Fibre Properties

The properties of polyamide fibres are shown in Table 5.1. The mechanical properties of nylon 6,6 and of nylon 6 are quite similar. Fibres of higher tenacity are produced for more demanding industrial (non-apparel) applications.

In the apparel market, polyamide fibres are used mostly in tights, swimwear and lingerie. In combination with elastane fibres (see Section 5.6.1), they are used in skiwear and sportswear. They are still used extensively as carpet fibres, although carpets in the home are less common than hitherto. Wool/nylon blends (often *ca.* 80 : 20) are often used; nylon is resilient and hard-wearing, whilst wool provides warmth. High tenacity fibres find extensive application in tyre cord; not only are they strong, but they adhere well to rubber. Hence, they are also used to reinforce rubber in drive belts, conveyor belts and hoses. Other major applications are in climbing ropes (light-weight, water resistant and abrasion resistant) and vehicle air-bags (good elasticity).

Table 5.1 Properties of polyamide fibres.

Specific gravity	1.14.
Tenacity	40–60 cN tex^{-1} normally, but up to 90 cN tex^{-1} for high tenacity yarns.
Elongation at break	20–30% but up to 40% for nylon 6; 15–20% for high tenacity yarns.
Elastic recovery	Very good: 100% from an 8% stretch (high tenacity yarns, 100% recovery from 4% stretch).
Resilience	High.
Abrasion resistance	High.
Moisture regain	4.0–4.5%.
Launderability	Polyamide garments should not be washed in hot water, and only subjected to gentle agitation. Otherwise, the fabric distorts and wrinkes, due to excessive breakage of intermolecular hydrogen bonds on absorption of water.

5.3 POLYESTER FIBRES

5.3.1 Introduction

Polyesters are polymers whose monomeric units are joined by the ester linkage, –O–CO–. By far the most important polyester in commercial terms is polyethylene terephthalate (PET). Indeed, production of PET fibres far outstrips that of any other synthetic fibre, and most of the following discussion of polyester fibres is focused on PET fibres. Nevertheless, other polyester fibres have had some commercial impact. Amongst these are polybutylene terephthalate (PBT), polytrimethylene terephthalate (PTT) and, to a lesser extent, polycyclohexanedimethylene terephthalate (PCT) polyester. Polyethylene naphthalate (PEN) has also been marketed as a fibre.

The first polyester fibres were produced in the DuPont laboratories by W. H. Carothers and some of his colleagues. In 1930, J. W. Hill condensed octadecandioic acid ($HOOC-(CH_2)_{16}-COOH$), with propane-1,3-diol ($HO-CH_2-CH_2-CH_2-OH$). A product of molar mass *ca.* 12 000 was obtained, from which fibres could be pulled. After this success, Carothers and his group made a number of other fibre-forming polyesters. However, because they were synthesised from aliphatic co-monomers, their melting points were all too low for practical use as textiles, and many dissolved readily in dry-cleaning solvents. Meanwhile the group was finding success with polyamide fibres, and it concentrated more and more on their production and commercialisation.

Nevertheless, the publications of the Carothers group on polyesters caught the attention of J. R. Whinfield, who worked at the Calico Printers Association in Accrington, UK. During the Second World War, he began a research programme on the formation of polyester fibres, and with J. T. Dickson succeeded in 1941 in producing a fibre-forming polyester from ethane-1,2-diol (ethylene glycol; $HO-CH_2-CH_2-OH$) and terephthalic acid:

$$HOOC-\langle\!\!\!\bigcirc\!\!\!\rangle-COOH$$

This polyester was PET.

After evaluation by UK government laboratories, further development of the fibre was undertaken by Imperial Chemical Industries (ICI) from 1943 and by DuPont in the USA from 1944. After the end of the Second World War in 1945, DuPont acquired the production rights in the USA and ICI in the rest of the world. ICI named their product 'Terylene', whilst DuPont initially called their product 'Fiber V', and from 1951

'Dacron'. Both the ICI and DuPont brands became very successful, and from the mid-1970s, production of PET fibres overtook that of polyamide fibres. PET fibres now dominate the synthetic fibre industry.

Other polyester fibres have only a modest commercial presence in comparison with PET fibres. PBT and PTT fibres were also produced by Whinfield and Dickson in 1941. PBT is synthesised from butane-1,4-diol ($HO–CH_2–CH_2–CH_2–CH_2–OH$) and the dimethyl ester of terephthalic acid. Whilst PBT enjoys some application as a carpet fibre, it is used much more as a polymer for injection moulding. PTT fibres have very good properties, but for many years one of the comonomers, propane-1,3-diol, was too expensive for economic commercial production of the fibre. (The other comonomer is terephthalic acid.) However, propane-1,3-diol can nowadays be sourced economically (see Section 5.3.2.5), and so PTT fibres are becoming more widespread. PCT originated from Eastman Kodak in 1958. As a fibre, it is used in carpets, but it is used far more extensively for injection moulding and as plastic sheeting. PEN was first produced at ICI in the late 1940s. It is used primarily in high-performance polyester films and blow-moulded containers. In 2002, Honeywell Performance Fibers commercialised PEN fibre.

The structures of the polyesters described in this section are:

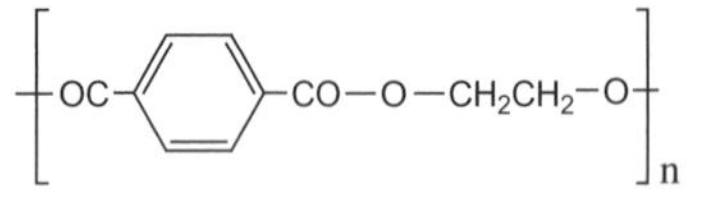

Polyethylene terephthalate (PET)

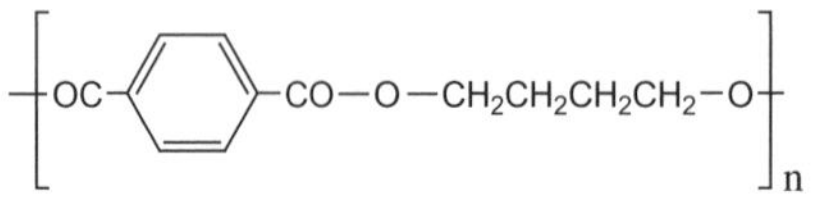

Polybutylene terephthalate (PBT)

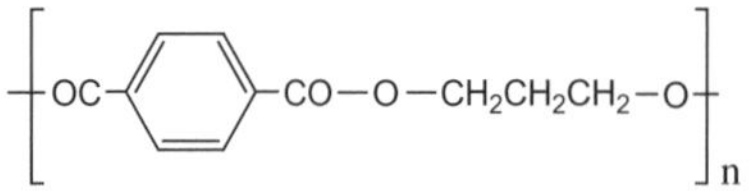

Polytrimethylene terephthalate (PTT)

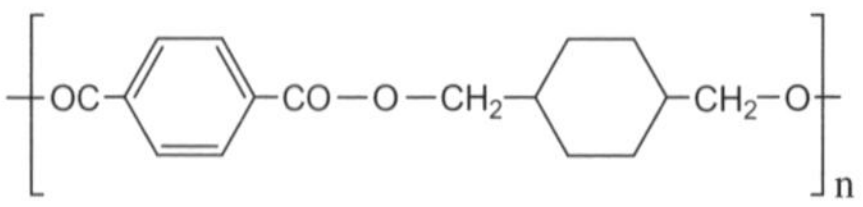

Polycyclohexanedimethylene terephthalate (PCT)

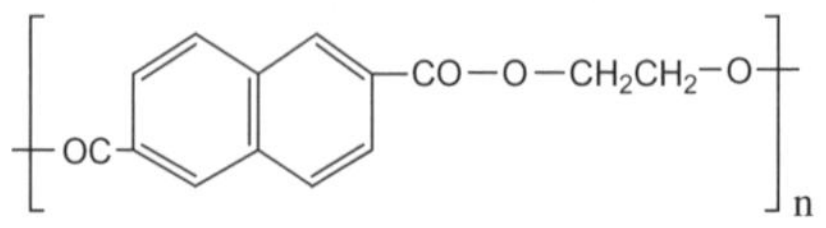

Polyethylene naphthalate (PEN)

Finally, an important emerging class of polyesters consists of those that are biodegradable. These polyesters have particular importance as temporary implants and scaffolds *in vivo*. They are discussed fully in Chapter 7.

5.3.2 Manufacture of Monomers

5.3.2.1 Terephthalic Acid. Terephthalic acid is produced commercially by oxidation of a solution of *p*-xylene in ethanoic acid, using oxygen, and catalytic quantities of cobalt and manganese salts, activated by bromide anions. The product which is precipitated from solution is, however, far from pure, the major impurities being *p*-toluic acid and 4-carboxybenzaldehyde:

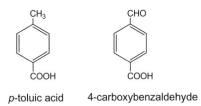

p-toluic acid   4-carboxybenzaldehyde

Originally these impurities prevented the use of terephthalic acid as a comonomer for synthesising polyesters, as its quality was too poor. However, a process for the removal of the impurities was developed by Amoco. The crude terephthalic acid is dissolved in superheated water at elevated pressures, and exposed to a small volume of hydrogen over a hydrogenation catalyst, such as a noble metal. All the 4-carboxybenzaldehyde is reduced to *p*-toluic acid. After carefully controlled cooling, the purified terephthalic acid crystallises from solution, leaving behind the dissolved *p*-toluic acid. The terephthalic acid so produced is sufficiently pure for polymerisation.

5.3.2.2 Dimethyl Terephthalate. Although dimethyl terephthalate is no longer normally used as a comonomer for the polymerisation of PET, it is still used for the synthesis of PBT. Dimethyl terephthalate is produced from the esterification of terephthalic acid with methanol at elevated pressure above 200 °C. The ester can be purified by distillation under vacuum and subsequent recrystallisation from methanol.

5.3.2.3 Ethane-1,2-diol. Ethane-1,2-diol (ethylene glycol) was at one time well known as antifreeze, before being very largely replaced by the less toxic propane-1,2-diol (propylene glycol). Ethane-1,2-diol is produced from ethene by a two-stage process. The first step comprises oxidation of ethene with air at about 350 °C. The product, epoxyethane (ethylene oxide), when treated with water, produces ethane-1,2-diol:

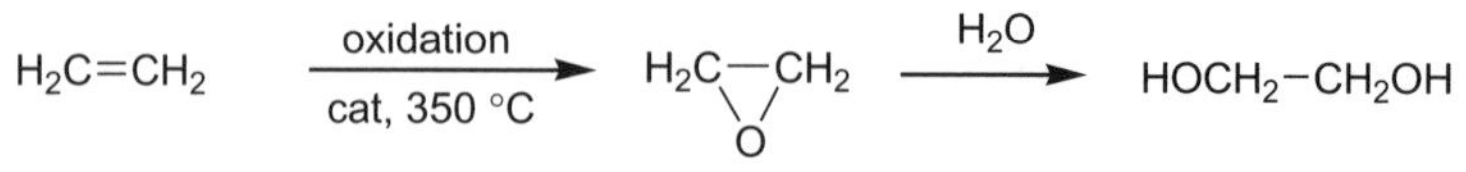

$$H_2C{=}CH_2 \xrightarrow[\text{cat, 350 °C}]{\text{oxidation}} H_2C\overset{\displaystyle\diagup\diagdown}{\underset{O}{}}CH_2 \xrightarrow{H_2O} HOCH_2{-}CH_2OH$$

An alternative route starts from 'synthesis gas', a mixture of carbon monoxide and hydrogen, which is produced from the gasification of coal. Hydrogenation of carbon monoxide over rhodium or ruthenium catalysts in tetrabutylphosphonium bromide, leads to high yields of ethane-1,2-diol.

5.3.2.4 Butane-1,4-diol. The principal procedure for synthesising butane-1,4-diol is from the reaction of ethyne with methanal:

$$CH{\equiv}CH + 2HCHO \rightarrow HO{-}CH_2{-}C{\equiv}C{-}CH_2{-}OH$$
$$\rightarrow HO{-}CH_2{-}CH_2{-}CH_2{-}CH_2{-}OH$$

The first step is catalysed by a mixture of bismuth and copper(I) ethynide (cuprous acetylide). Hydrogenation can be achieved using a Ranay nickel catalyst.

An alternative route starts with the conversion of epoxypropane (propylene oxide) to 2-propen-1-ol (allyl alcohol), which on hydroformylation using a rhodium-based catalyst yields 4-hydroxybutanal:

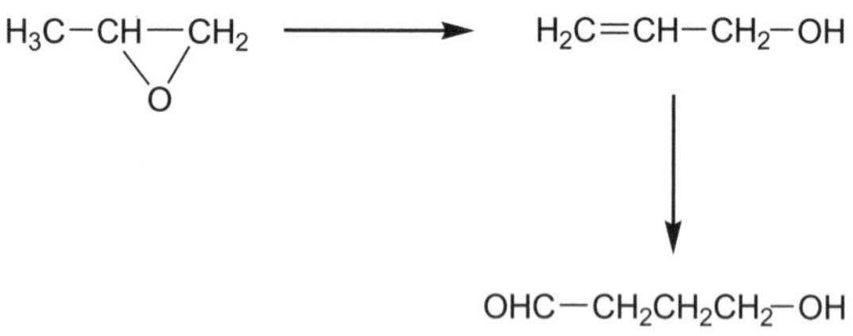

$$H_3C{-}CH\overset{\displaystyle\diagup\diagdown}{\underset{O}{}}CH_2 \longrightarrow H_2C{=}CH{-}CH_2{-}OH$$

$$\downarrow$$

$$OHC{-}CH_2CH_2CH_2{-}OH$$

4-Hydroxybutanal is then reduced to butane-1,4-diol under carefully controlled conditions that avoid unwanted side-reactions.

5.3.2.5 Propane-1,3-diol. Propane-1,3-diol can be synthesised by a variety of routes. For example, it may synthesised from the addition of water under pressure to propenal (acrolein), and the subsequent hydrogenation of the product, 3-hydroxypropanol, over a Ranay nickel catalyst:

$$CH_2{=}CH{-}CHO + H_2O \rightarrow HO{-}CH_2{-}CH_2{-}CHO$$
$$\rightarrow HO{-}CH_2{-}CH_2{-}CH_2{-}OH$$

The unwanted dimerisation of propenal, however, may accompany the hydration reaction to some extent. An alternative method of producing propane-1,3-diol is the hydroformylation of epoxyethane.

Fermentation processes may also yield propane-1,3-diol economically. Corn sugars, which consist largely of glucose, are converted by suitable yeasts to propane-1,2,3-triol (glycerol). Various bacteria are capable of reducing propane-1,2,3-triol to propane-1,2-diol.

5.3.2.6 Cyclohexanedimethanol. Cyclohexanedimethanol is synthesised by hydrogenation of dimethyl terephthalate over a palladium catalyst and subsequent conversion of the product, cyclohexane-1,4-dicarboxylate, with a copper(II) chromite catalyst to cyclohexanedimethanol:

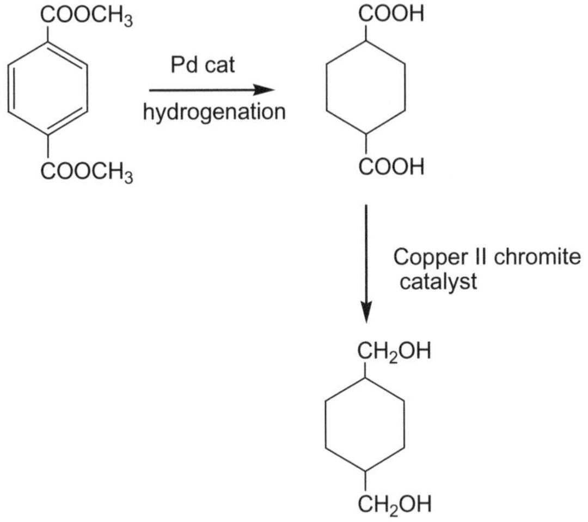

Cyclohexanedimethanol exists in *cis* and *trans* isomeric forms. The desired ratio of the two forms for commercial PCT fibre is *cis*–*trans* (1 : 2).

5.3.2.7 Naphthalene-2,6-dicarboxylic Acid. Naphthalene-2,6-dicarboxylic acid can be synthesised by a variety of methods. For example, it can be synthesised from *o*-xylene and 1,3-butadiene:

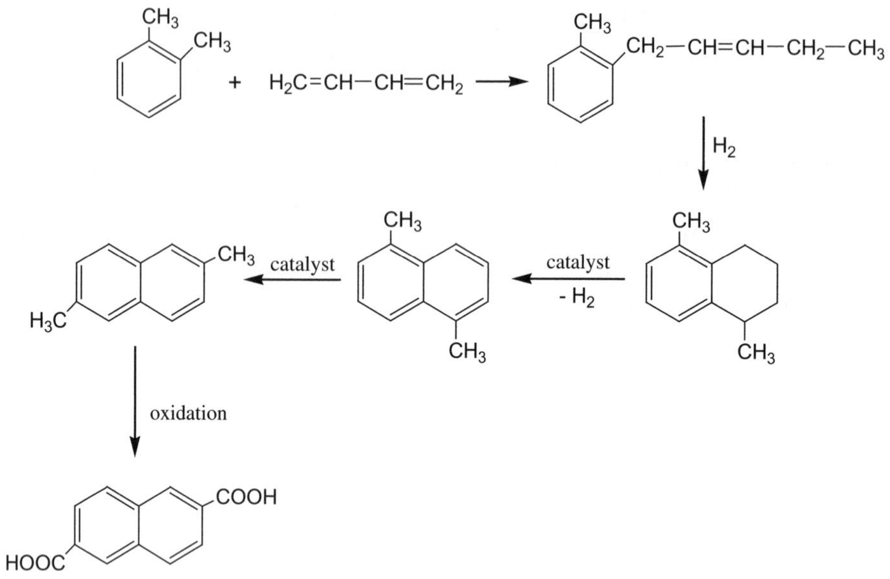

Alternatively, it can be synthesised from naphthalene and propene:

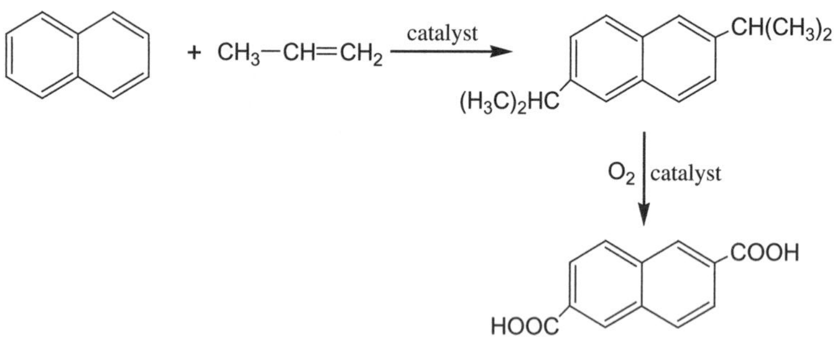

5.3.2.8   Dimethyl Naphthalene-2,6-dicarboxylate. Dimethyl naphthalene-2,6-dicarboxylate can be produced from crude naphthalene-2,6-dicarboxylic acid by a process that is closely analogous to that used for producing dimethyl terephthalate.

5.3.3 Production of Polyesters

5.3.3.1 Production of PET. The preferred route for producing PET is direct esterification from purified terephthalic acid and ethane-1,2-diol. It is important, however, that the two comonomers are in a 1:1 equimolar ratio. To ensure this stoichiometry, an initial reaction is conducted first to produce *bis*hydroxyethyl terephthalate:

HOOC—⟨benzene⟩—COOH   +   2HO—CH_2CH_2—OH

↓

HO—CH_2CH_2—OOC—⟨benzene⟩—COO—CH_2CH_2—OH   + $2H_2O$

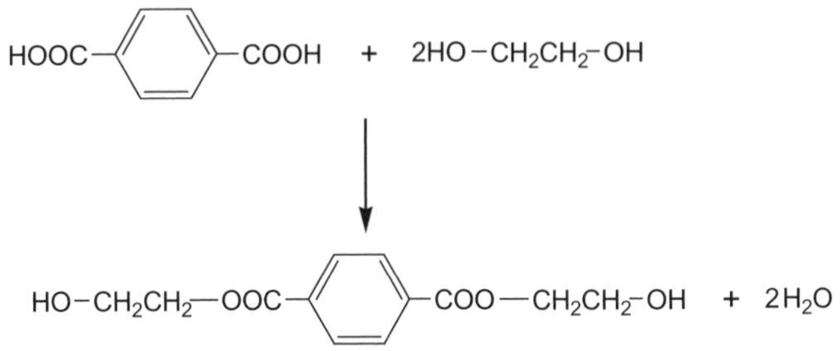

Before the production of pure terephthalic acid was commercially viable, *bis*hydroxyethyl terephthalate was produced from ester interchange between dimethyl terephthalate and ethane-1,2-diol:

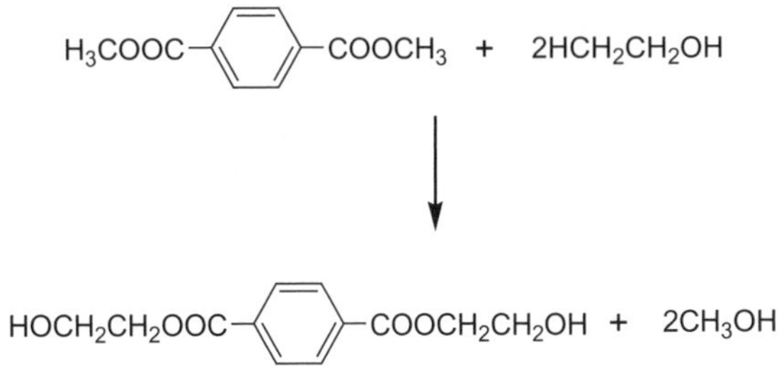

H_3COOC—⟨benzene⟩—$COOCH_3$ + $2HCH_2CH_2OH$

↓

$HOCH_2CH_2OOC$—⟨benzene⟩—$COOCH_2CH_2OH$ + $2CH_3OH$

Step growth polymerisation of *bis*hydroxyethyl terephthalate then follows:

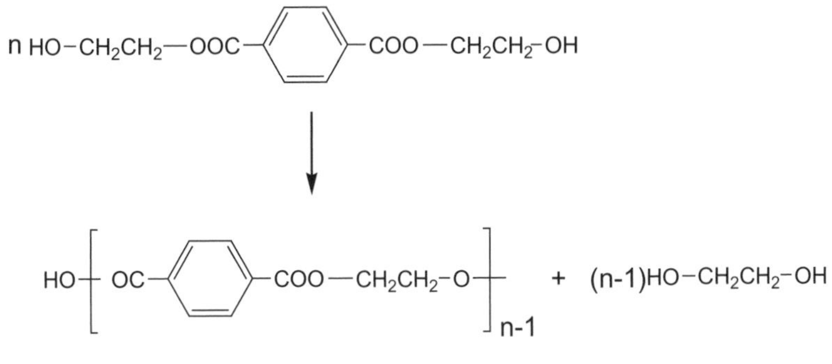

n HO—CH_2CH_2—OOC—⟨benzene⟩—COO—CH_2CH_2—OH

↓

HO—[OC—⟨benzene⟩—COO—CH_2CH_2—O—]$_{n-1}$ +   (n-1)HO—CH_2CH_2—OH

There are a number of advantages in producing PET from purified terephthalic acid rather than from its dimethyl ester. Not only can PET grades of higher molar mass eventually be formed, if required, but also the direct esterification reaction is catalysed by the carboxylic acid groups that are present. Stronger acids are sometimes added as additional catalysts.

There are, however, several problems associated with producing PET. During direct esterification, there is a side reaction by which an ether is formed from reaction between two molecules of ethane-1,2-diol, to yield dihydroxydiethyl ether (diethylene glycol; $HO–CH_2–CH_2–O–CH_2–CH_2–OH$). The reaction is acid-catalysed but can be suppressed to some extent through the addition of a small quantity of sodium hydroxide or an organic quaternary hydroxide. The formation of the ether can also be reduced if direct esterification is carried out in the temperature range, 280–290 °C.

The step growth polymerisation of *bis*hydroxyethyl terephthalate is a reversible reaction. It is, therefore, important to remove the ethane-1,2-diol produced in the reaction, in order to drive the polymerisation process forward. Ethane-1,2-diol is removed under vacuum and recycled as a starting material for the direct esterification process.

The catalyst, traditionally used for the step growth polymerisation step, has been antimony trioxide (Sb_2O_3). A major drawback with antimony trioxide is its susceptibility to reduction by products from PET thermal degradation. Metallic antimony is formed as a fine suspension within the PET, giving rise to a grey tinge in the polymer. Moreover, concerns over the toxicity of antimony are increasingly rendering it less attractive as a catalyst. Suitable alternatives that have the effectiveness of antimony trioxide are catalysts comprising titanium dioxide and either silica or zirconium dioxide.

During the polymerisation, the PET produced is subject to some thermal degradation. The chemistry underlying the degradation processes will be discussed in Section 5.3.6.4. In addition, cyclic oligomers (polymers of only a few monomeric units) are also formed, up to a level of 2–3%. The oligomers are mainly trimers. They can cause significant problems during melt spinning, by coating the spinneret plates and even breaking down the molten filaments after they have emerged from the spinneret. These oligomers can also interfere with the dyeing of PET fibres.

As with nylon 6,6, PET is nearly always produced by a continuous process. The product can then be led straight to a melt extruder.

5.3.3.2 Production of PBT. PBT is produced using ester interchange between dimethyl terephthalate and butane-1,4-diol, because heating terephthalic acid with butane-1,4-diol leads to the substantial formation of tetrahydrofuran, which is highly volatile. Catalysts based on titanium are used for both the ester interchange and the subsequent step growth polymerisation. Thermal degradation of the product is less marked than with production of PET, although a little tetrahydrofuran is released during the process, due to breakdown of 4-hydroxybutyl end groups:

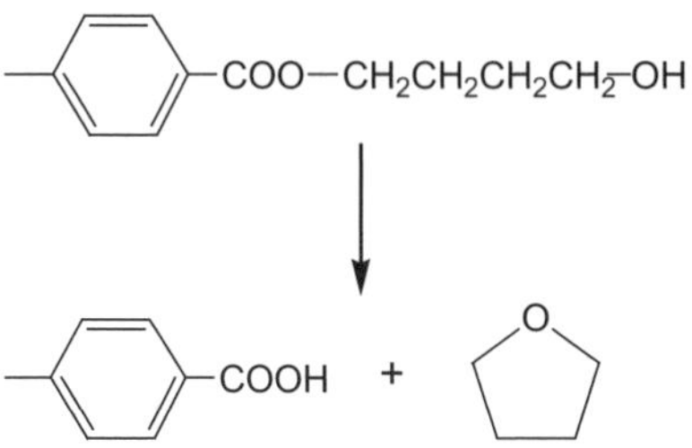

5.3.3.3 Production of PTT. PTT is mostly produced by the direct esterification route, using terephthalic acid and propane-1,2-diol. Catalysts based on tin or titanium are generally used. An unusual feature of PTT production is that it is a 'batch-to-batch' process. Oligomers and polymers of low molar mass left behind from the previous batch are adopted as the reaction medium for the succeeding batch. When direct esterification is complete, the mixture of oligomers produced is partially transferred to the polymerisation vessel, so some is left behind for the next direct esterification.

As with the production of PET, not only are oligomers produced (though they are mainly dimeric in this case), but so too is an ether, formed by reaction between two molecules of propane-1,3-diol. Care also has to be taken to minimise thermal degradation of the PTT produced.

5.3.3.4 Production of PCT. PCT is produced using the ester interchange route. Titanium based catalysts are used for both stages of the production process. The proportion of *cis* to *trans* cyclohexane ring isomers in the product is *ca.* 1:2.

5.3.3.5 Production of Polyethylene Naphthalate. Polyethylene naphthalate can be produced from the reaction of ethane-1,2-diol with either 2,6-naphthalene dicarboxylic acid or dimethyl naphthalene-2,6-dicarboxylate. The processes are analogous to the production of PET using terephthalic acid or dimethyl terephthalate, respectively.

5.3.4 Fibre Processing

Polyester fibres are melt extruded. Before extrusion, the polymer is rigorously dried at elevated temperatures, to prevent hydrolysis of the polymer melt in the extruder. The moisture content is reduced to <30 ppm. Extrusion is carried out in an atmosphere of dry nitrogen at a temperature normally *ca.* 20 °C above the melt temperature of the polymer. Thus, PET filaments are extruded at 280–290 °C, and PBT and PTT filaments are extruded at 250–260 °C. PCT is extruded at 305 °C. Since terephthalate polyester chains start to show signs of degradation above 260 °C, thermal degradation can be a significant problem in the melt extrusion of PCT.

The spinneret holes in the extruder are generally 0.2–0.4 mm in cross-section. Extrusion velocity varies according to the strength of yarn finally required. For so-called low-oriented PET yarn, extrusion velocity is 500–1500 m min^{-1}. At the other extreme, production of so-called fully oriented PET yarn requires velocities of above 6000 m min^{-1}. In structural terms, the categories of yarn differ from one another in the degree of orientation of the polymer chains, and hence in the strength of the yarns. After extrusion, the polyester filaments pass through a cooling chamber, before being finally wound.

Extruded PET filaments are generally amorphous, but drawing brings the PET chains closer to one another, with the result that some crystallinity is induced in the filaments. Drawing is normally performed at 60–90 °C, with a draw ratio between $1\frac{1}{2}$ and 6, depending on the strength of the yarn required. The drawn yarn may then be texturised, and finally heat set, usually at *ca.* 130–150 °C. Heat setting is necessary with PET filaments, to introduce further crystallinity into the yarn and stabilise it against thermal shrinkage during later processing and use.

In contrast to extruded PET filaments, PBT and PTT filaments crystallise readily after extrusion, and their crystallinity increases even more after the extruded yarn has been wound. The yarn dimensions alter accordingly. If, as in the extrusion of nylon 6,6, the yarn is steamed prior to winding, crystallisation is accelerated and indeed is almost complete before the yarn is wound. Drawing is carried out in a similar way to the drawing of PET yarns. Extruded PCT yarns possess virtually no crystallinity with low chain orientation. They are drawn at 120 °C.

5.3.5 Physical Structure

The proportion of crystalline and amorphous phases (and also intermediate phases) in polyester fibres depends very much on the type of polyester and the conditions under which the fibres were processed. Moreover, there is quite a wide divergence of crystalline cell structures.

The unit cell of a PET crystal is triclinic, and its dimensions vary slightly depending on processing conditions. Thermal treatment under tension, for example, extends the cell's c-axis (this is the axis most nearly parallel to the fibre axis). The a- and b-axes are slightly diminished. If no tension is imposed, the c-axis contracts, and the a- and b-axes expand.

PBT crystals are monoclinic. However, when the filament is subject to strain, there is a transformation from a contracted α-unit cell to an extended β-unit cell, which is also monoclinic. This transformation is accompanied by changes in conformation about the carbon–carbon bonds along the PET chains. PTT crystals, however, are triclinic.

PCT contains both the *cis-* and *trans-*variants of the cyclohexane ring. Each homopolymer possesses an individual unit cell structure, so it is perhaps not surprising that, after extrusion, the filaments produced have virtually no crystallinity.

PEN possesses two types of unit cell, both of which are triclinic. Normally, the extended α-form is produced in PEN fibres, but under certain conditions a slightly extended β-form is produced instead.

5.3.6 Chemistry of Polyester Fibres

5.3.6.1 Introduction. The chemical properties of polyesters are mostly governed by the ester groups in the polymer chains. Polyesters are resistant to acids, except strong acids at high temperature, but are more susceptible to alkalis. They are resistant to many common solvents, but dissolve in 3-methyl phenol (*m*-cresol), trifluoroethanoic acid and 2-chlorophenol. The polymer chains are prone to degradation above *ca.* 260 °C. Polyester fibres are more resistant to sunlight than polyamide fibres are.

5.3.6.2 Hydrolysis. When polyester fibres are hydrolysed by either acids or alkalis, depolymerisation back to the constituent comonomers occurs. However, the rate of hydrolysis and the parts of the fibre affected depend on the rate of diffusion of the acid or alkali into the fibre. Thus, in the case of many alkalis, such as sodium hydroxide solution, diffusion is slow and only the fibre surfaces are attacked. Fibre cross-section is, therefore, reduced, but the molar mass of the bulk of the fibre is largely unaltered. Acids generally cause hydrolysis throughout the fibre. The molar mass falls, but there is little change in fibre cross-section.

Polyester fibres are also hydrolysed at high temperature by moisture. It has already been noted, for example, that polyester chip must first be rigorously dried before admittance to an extruder. Significant hydrolysis of fibres can also occur above 150 °C. The reaction is catalysed by the carboxylic acid end groups, and since hydrolysis produces further carboxylic acid groups, the process is autocatalytic.

5.3.6.3 Thermal Degradation. As noted in Section 5.3.6.1, most polyesters start to show signs of degradation at 260 °C. Thus, PET degrades quite slowly in the absence of oxygen below its melting point (*ca.* 260 °C), but above 300 °C, degradation becomes substantial. Random scission of $-CH_2-O-$ bonds occurs by a β-elimination process, to yield free carboxylic acid and vinyl ester end-groups:

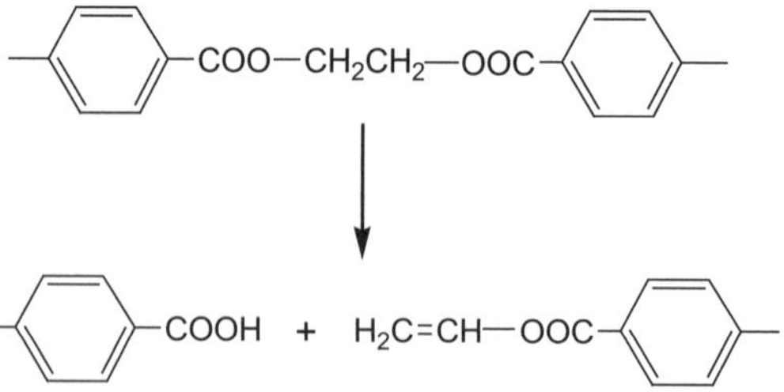

The vinyl ester end-groups can then degrade, to yield carbon monoxide, carbon dioxide or ethyne. Some of these processes are:

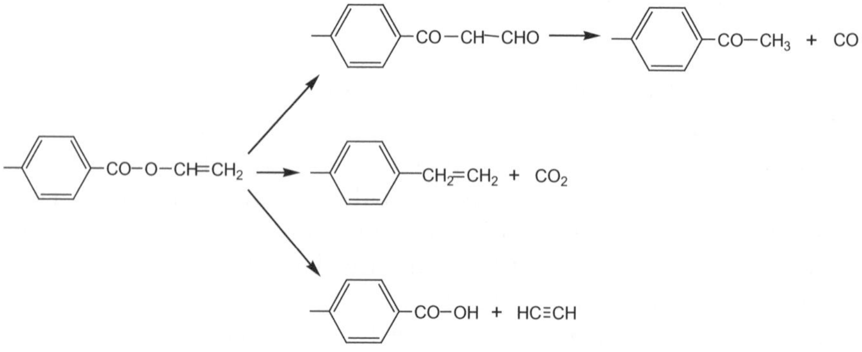

Alternatively, they react with other chain ends to form acid anhydride or ester links, with evolution of ethanol (acetaldehyde):

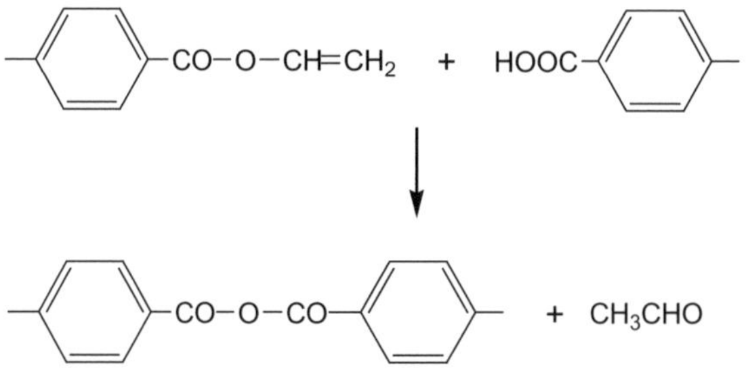

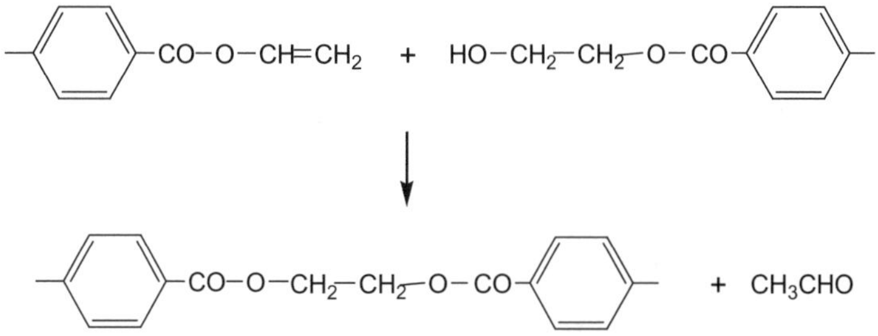

$$\text{—C}_6\text{H}_4\text{—CO—O—CH=CH}_2 \quad + \quad \text{HOOC—C}_6\text{H}_4\text{—}$$

$$\downarrow$$

$$\text{—C}_6\text{H}_4\text{—CO—O—CO—C}_6\text{H}_4\text{—} \quad + \quad \text{CH}_3\text{CHO}$$

$$\text{—C}_6\text{H}_4\text{—CO—O—CH=CH}_2 \quad + \quad \text{HO—CH}_2\text{—CH}_2\text{—O—CO—C}_6\text{H}_4\text{—}$$

$$\downarrow$$

$$\text{—C}_6\text{H}_4\text{—CO—O—CH}_2\text{—CH}_2\text{—O—CO—C}_6\text{H}_4\text{—} \quad + \quad \text{CH}_3\text{CHO}$$

PBT is not as thermally stable as PET. The main products of its decomposition are butadiene and terephthalic acid. Pyrolysis of PTT yields 2-propen-1-ol (allyl alcohol; $CH_2{=}CH{-}CH_2{-}OH$) and propenal (acrolein; $CH_2{=}CH{-}CHO$). The pyrolysis of PCT yields vinylidene–cyclohexane units. PEN fibres undergo analogous pyrolysis reactions to those of PET fibres.

5.3.6.4 Photodegradation. Compared with many other types of synthetic fibre, polyester fibres possess a high resistance to sunlight. In particular, PET fabrics show a clear superiority to most other commonly used fabrics, when exposed behind glass to sunlight. Nevertheless, PET fibres do undergo some photodegradation in ordinary daylight. In the absence of oxygen, scission of the PET chains yields carboxylic acid groups and carbon monoxide, and is followed by cross-linking of chains:

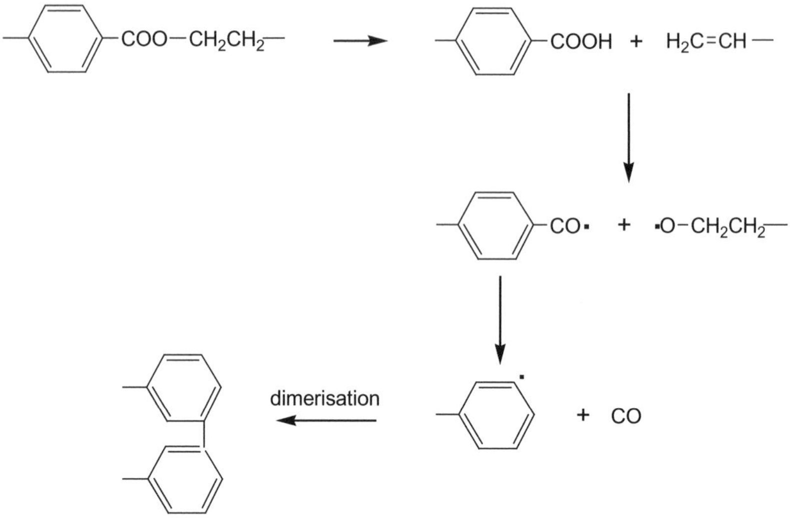

In the presence of oxygen, chain scission yields carbon dioxide. Hydroxylation of aromatic rings in the chains also occurs.

5.3.7 Fibre Properties

The properties of PET fibres are shown in Table 5.2.

PET fibres are used in a whole host of applications: apparel, household, medical and industrial. In apparel, PET fibres are often blended with other fibres. Blends with cotton fibres are used, for example, in shirts, skirts, dresses, overalls and sportswear. Blends with wool are used in suits, trousers, dresses *etc.* 100% PET fibre is used in ties, pockets and

Table 5.2 Properties of PET fibres.

Specific gravity	1.38.
Tenacity	35–56 cN tex^{-1} normally, but up to 100 cN tex^{-1} for high tenacity yarns.
Elongation at break	15–40%; 5–10% for high tenacity yarns.
Elastic recovery	Inferior to polyamide fibres; 80% from an 8% stretch.
Resilience	High.
Abrasion resistance	High.
Moisture regain	0.4%.
Launderability	PET garments can withstand vigorous washing treatments, because of their high strength and abrasion resistance. The conditions for laundering blends of PET and other fibres are largely governed by the other fibres.

linings. PET fibres are used extensively in curtain nets, where their resistance to sunlight behind glass is a major asset. PET/cotton blends are used extensively in bedsheets and pillowcases. Medical applications include sutures, heart valves, artificial tendons and ligaments, orthopaedic bandages and surgical hosiery. Industrial applications include vehicle tyre reinforcement, ropes, tents and sailcloths. Special PET fabrics are also used for reinforcement and drainage to stabilise soil in road construction.

PTT fibres exhibit very high resilience, owing to the elastic nature of the unit cells in PTT crystals. The fibres, therefore, find use as carpet face yarns. They are also used in stretch apparel, such as sportswear, and in upholstery. The applications for PBT fibres include carpets, sportswear and hosiery. PCT fibres are also used for carpet yarns, as well in cushions and pillows. These fibres have also been used in some high temperature filtration and insulation applications. PEN fibres are used in tyre cords and high-performance sailcloth.

5.4 ACRYLIC AND MODACRYLIC FIBRES

5.4.1 Introduction

Acrylic fibres are based on polyacrylonitrile (PAN), produced by the polymerisation of cyanoethene (acrylonitrile), a substituted vinyl compound in which the substituent X=CN.

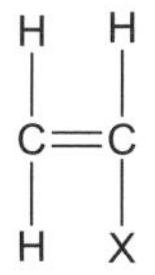

$$
\begin{array}{cc}
H & H \\
| & | \\
C & \!\!=\!\! C \\
| & | \\
H & X
\end{array}
$$

The polymerisation of acrylonitrile to polyacrylonitrile was first achieved by a French chemist Moreau in 1894, but the polymer formed was not soluble in any of the solvents available at the time and it decomposed before it melted. Since the polymer was so difficult to process, it was not investigated any further for many years. Interest in the polymer was re-kindled in the late 1930s when it became an important constituent for making synthetic rubbers, the first of which was acrylonitrile–butadiene rubber produced by I. G. Farbenindustrie in Germany in 1937. Around this time it was discovered that polyacrylonitrile would dissolve in solvents that are strongly polar, the

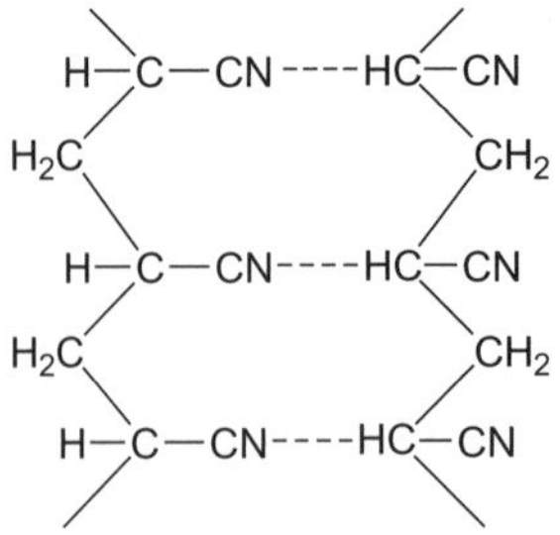

Figure 5.8   Hydrogen bonding in PAN fibres.

reason being that adjacent polymer chains are very strongly hydrogen bonded to each other (see Figure 5.8). Consequently solvents able to break hydrogen bonds, such as *N,N*-dimethylformamide, dimethyl sulfone, and *m*-nitrophenol would dissolve it. Aqueous solutions of inorganic compounds, such as sodium thiocyanate and lithium bromide were also found to dissolve it.

The first acrylic fibre was marketed under the trade name 'Orlon' in 1944 by Du Pont in America, who developed a process for dry spinning the polymer from solution in *N,N*-dimethylformamide. In 1952 another American company, Chemstrand, launched a different version of the fibre under the trade name 'Acrilan', though this was produced by a wet spinning extrusion process. All of the acrylonitrile produced in America was made by a company called American Cyanamid and in 1958 it launched its own version of an acrylic fibre called 'Creslan'. The 1950s saw rapid growth in the manufacture of acrylic fibres, with production also commencing in European countries, especially France, Germany, Italy and England. In England, Courtaulds established a manufacturing plant at Grimsby, the trade name for its fibre being 'Courtelle'. The fibre was wet spun from solution in sodium thiocyanate and a process for dyeing the fibres continuously directly after extrusion (before they had dried) when they were very porous and receptive to dye molecules, was developed (the 'Neochrome' process). There followed a rapid increase in acrylic fibre production in the Far East, particularly in Japan, China and Korea, during the 1970s but production peaked in the late 1970s. Much of the production went into yarns for knitwear, where acrylic fibres were seen as an alternative to wool, and into carpet yarns. In the years since, however, there has been a substantial decline in the popularity of acrylic fibres and much of the manufacturing capacity in both America and Europe has now disappeared. For example, the plant at Grimsby now produces the Tencel® Lyocell fibre type instead (see Section 4.1.4), a reflection of the trend towards eco-friendly fibres. The only manufacturer in Europe now is at Dolan GmbH, which is now a subsidiary of Lenzing Plastics. The decline in manufacture in America and Europe has largely arisen from the shift in production to the Asian countries, especially China, with Japan still an important manufacturer. Over the last five years, however, a decrease in world demand has resulted in a decline in production levels, even in China. The production of acrylic fibres worldwide fell by 20% between 2007–2008, partially due to the economic recession, but also to the economics of manufacture. The problem for acrylic fibres is that acrylonitrile is a relatively expensive raw material and on a cost-performance basis, acrylic fibres offer no significant advantages over the other main synthetic fibres, nylon and polyester, or indeed of wool.

Acrylic fibres based solely on acrylonitrile have a number of un-desirable properties:

1. Poor solubility in industrial solvents;
2. High melting point, with degradation ($\sim$330 °C), making melt spinning impossible;
3. High glass transition temperature ($\sim$105 °C), making it difficult to dye at the boil; and
4. Low saturation absorption of disperse dyes.

Although solvents for the polyacrylonitrile have been found to enable the manufacture of acrylic fibres, the fibres formed still pose difficulties for commercial application. Modifications to the polymers are made, the essential feature of which is to copolymerise the polyacrylonitrile with a small percentage of another vinyl monomer. Whilst precise details of the processes have never been disclosed, it is generally believed that the comonomers used are typically vinyl acetate (ethenyl acetate), with methyl acrylate (methyl propenoate) or methyl methacrylate (methyl-2-methylprop-2-enoate) also being used, added in a proportion of about 10%, with 90% of the acrylonitrile. The presence of, for example the large acetate group, opens up the molecular structure of the final polymer, aiding dissolution in solvents. Additionally the glass transition temperature is much lower at $\sim$75–80 °C, enabling dye molecules to penetrate the fibres more easily at temperatures below 100 °C.

Vinyl acetate   Methyl acrylate   Methyl methacrylate

In addition to these 'neutral' comonomers, some acrylic fibre variants also contain a third comonomer which is either acidic or basic in character, to confer different dyeing properties, such as affinity for basic dyes or acid dyes respectively. Examples of the acidic types of comonomer are acrylic acid ($H_2C\!\!=\!\!CH \cdot COOH$), sodium styrene sulfonate or sodium methallyl sulfonate ($H_2C\!\!=\!\!CH \cdot CH_2OSO_2ONa$). The basic types are typically vinyl pyridine (2-ethenyl pyridine) or ethylene imine (aziridine).

Vinyl pyridine   Ethylene imine

Clearly there are many possible variants of acrylic fibres, depending on both the nature of the comonomers used and their proportions. Because of the generally accepted need to include one or more comonomers in the fibre to overcome the difficulties of fibres made from pure poly-acrylonitrile, the Federal Trade Commission (FTC) in America defined acrylic fibres as those containing at least 85% by weight of acrylonitrile comonomer. In 1960, the FTC defined the term 'modacrylic' as those fibres containing at least 35% but less than 85% by weight of acrylonitrile comonomer. In Europe BISFA defines modacrylic fibres as those containing at least 50% but less than 85% by weight of acrylonitrile, though the International Standards Organisation specifies the wider 35–85% range.

Modacrylic fibres became important for their high flame resistance, achieved by the use of halogen-containing comonomers, such as vinylidene chloride, vinyl chloride or vinyl bromide in the polymer. They are used typically in home furnishings where flame resistance is required.

5.4.2   Manufacture of Acrylonitrile

There are a number of synthetic routes for the manufacture of acrylonitrile, but the most commercially viable process nowadays is based on the heterogeneous vapour phase catalytic oxidation of propene in the presence of ammonia, called *ammoxidation*. The overall reaction can be represented as:

$$CH_2\!\!=\!\!CH\!-\!CH_3 + NH_3 + \tfrac{3}{2}O_2 \rightarrow CH_2\!\!=\!\!CH\!-\!CN + 3H_2O$$

Different processes have been developed over the years, the two most important being the Distillers process and the Sohio process. In the Distillers process, propene is firstly oxidised in air to form acrolein and water, which are then reacted with ammonia in the presence of molybdenum oxide and air to yield acrylonitrile. The reaction sequence is:

$$CH_2\!=\!CH\!-\!CH_3 + \tfrac{3}{2}O_2 \rightarrow CH_2\!=\!CH\!-\!CHO + 3H_2O$$

$$CH_2\!=\!CH\!-\!CHO + NH_3 \rightarrow CH_2\!=\!CH\!-\!CH\!=\!NH + H_2O$$

$$CH_2\!=\!CH\!-\!CH\!=\!NH + \tfrac{1}{2}O_2 \rightarrow CH_2\!=\!CH\!-\!CN + H_2O$$

Various degradation reactions occur simultaneously, leading to the formation of by-products such as HCN, CH_3CN, N_2, CO and CO_2. These compounds have to be removed from the product stream. The main difficulty is the separation of the acetonitrile from the acrylonitrile, which have very close boiling points (81 °C and 77 °C respectively) and a series of azeotropic distillations are required to separate the acrylonitrile at the required purity. In the BP Sohio process, which is a one-stage process, the ammoxidation is carried out in air at 2–3 atmospheres pressure and a temperature of 425–510 °C. Various catalytic systems have been employed, a much used one being bismuth phosphomolybdate doped by additives of cobalt, nickel and potassium. The BP Sohio process has come to dominate acrylonitrile manufacture and it is licensed extensively.

Future developments are focused on the ammoxidation of propane and various companies are exploring the commercialisation of the process. It is claimed that only minor modifications are required to the existing propene feedstock plant. The reaction can be represented overall as:

$$CH_3CH_2CH_3 + 2O_2 + NH_3 \rightarrow CH_2\!=\!CH\!-\!CN + 4H_2O$$

though it is believed propene is formed in an intermediate stage.

5.4.3 Polymerisation of Acrylonitrile

The polymerisation of acrylonitrile is usually carried out by a free radical method, either in bulk, in solution, by emulsion polymerisation or by aqueous dispersion polymerisation. The difference between emulsion and aqueous dispersion polymerisation is that in the former, the acrylonitrile and comonomers are first emulsified, when micelles of about 100 molecules are formed. Radicals, which are generated in the aqueous phase, are adsorbed onto the surface of the micelles where they promote the polymerisation. In aqueous dispersion polymerisation the polymer chains become insoluble in water as they grow in size whereupon they aggregate to form nuclei. These nuclei also contain monomer and the radicals that continue to be formed in solution interact at their surface, thereby maintaining the polymerisation. For the production of acrylic fibres, the solution and aqueous dispersion methods are most used, especially the latter. A redox initiator is usually employed. Emulsion polymerisation is used more for the production of modacrylic fibre types. Bulk polymerisation is rarely used commercially.

In solution polymerisation, the solvents used are usually dimethylformamide, dimethylsulfoxide or aqueous solutions of an inorganic salt, such as sodium thiocyanate, in which the polymer is dissolved so that once the unreacted monomer has been removed the spinning dope is formed directly and is ready for extrusion. Thermally activated initiators are used, such as azobisisobutyronitrile (AIBN), ammonium persulfate or benzoyl peroxide, and the polymerisation proceeds according to the classical sequence of initiation, chain propagation and termination.

Initiation:	$I + M \rightarrow P_1^*$
Chain propagation:	$M + P_n^* \rightarrow P_{n+1}^*$
Chain termination by radical combination:	$P_n^* + P_m^* \rightarrow P_{n+m}^*$
Chain termination by radical disproportionation:	$P_n^* + P_m^* \rightarrow P_n + P_m$

Free radical transfer reactions to the solvent can also occur:

$$P_n^* + S \rightarrow P_n + S^*$$

which restrict the formation of high molar mass polymer.

In aqueous dispersion polymerisation, the method most widely used for acrylic fibre manufacture, the redox initiator system used is usually ammonium or potassium persulfate (oxidizing agent), sodium bisulfite (reducing agent) and ferric or ferrous iron catalyst. The reactions that create the free radicals are:

$$S_2O_8^{2-} + Fe^{2+} \rightarrow Fe^{3+} + SO_4^{2-} + SO_4^{\bullet-}$$
$$HSO_3^- + Fe^{3+} \rightarrow Fe^{2+} + HSO_3^{\bullet}$$

The sulfate and sulfonate radicals then react with the monomer molecules in the manner described above. A consequence of initiation using this initiator system is that sulfate ($-SO_4^-$) and sulfonate ($-SO_3^-$) groups remain at the ends of the polymer chains which provide sites for the adsorption of cationic (basic) dyes. Acrylic fibres therefore have an inherent substantivity for these types of dyes and this is a property of key importance.

In order to produce an acrylic fibre that can be dyed to full shades, it is necessary that the polymer has a molecular weight distribution that provides sufficient dye sites (at the ends of the polymer chains). Whilst there will be more dye sites available in a polymer of low number average molar mass M_n, if the value of M_n is too low the rheological properties will not be satisfactory for extrusion and the fibre-forming properties of the polymer will be inferior. Clearly it is necessary to strike a balance and if sufficient dye sites are not provided by the initiator residues there is the option of including a sulfonated comonomer in the polymer that enhances dye substantivity by providing sulfo groups along the length of the polymer chain. Typically the M_n values for acrylic fibre polymers lie in the range $40\,000–60\,000\,\text{g mol}^{-1}$, corresponding to about 1000 repeat units depending upon the comonomer composition.

5.4.4 Fibre Extrusion

Most acrylic fibres are wet spun (some 85%), usually from solution either in sodium thiocyanate with the PAN containing methyl acrylate as comonomer, or in solution in dimethylacetamide with the PAN containing vinyl acetate as comonomer. The remainder is produced by dry spinning, usually from solution in dimethylformamide when the comonomer is usually methyl acrylate. The productivity of dry spinning is much lower than of wet spinning, so manufacturing costs are therefore much higher. The wet spinning processes have all the typical characteristics described in Section 5.1, with the dope being extruded through the spinnerets submerged in the coagulation bath containing a dilute aqueous solution of the spinning dope solvent. The spinnerets have between 10 000–60 000 holes, each of between 0.05–0.38 mm in diameter, and the individual filaments emerging from each hole are drawn from the bath together as large 'tows'. From there the tows are passed through water baths with a counter-current flow of water for effective washing.

Some manufacturers have the facility for gel-dyeing in which the undried filaments, in the form of the tow, are fed into a bath of water-soluble cationic dye. In the never-dried state, the fibres are still highly swollen and receptive to dye, and the dyeing process at this stage of fibre production is highly efficient. The dye molecules become trapped within the fibre matrix so consequently high levels of wash fastness are obtained. Some manufacturers add pigment to the dope prior to extrusion, as an alternative but equally effective means of coloration of the fibres. Whilst coloration of the fibres at this stage of manufacture has the benefits of simplicity and of providing excellent fastness properties, the manufacturers have to be sure that they will be able to sell the colours they are producing. Coloration by this method, so early in the production chain, is not suitable for the fibres intended garment industry where quick response to fashionable colours is required.

A feature of the fibre filaments produced by the wet spinning process is their extremely porous nature. It is normal for the wet tow to be passed over two pairs of heated rollers to induce drawing and orientate the fibres, during which the voids collapse as the fibres dry out. Depending on the extrusion conditions, fibres of different cross-sectional shapes and microvoid structures can be formed, which in turn influences the physical properties of the final processed fibres. The rate at which solvent diffuses out of the filaments in relation to the rate non-solvent diffuses in, during extrusion, is the key factor. Both rates increase with the temperature and with the speed of extrusion. Two mechanisms of solidification can occur: 'gelation' which is a gradual transition of the dope into a gel, and 'phase separation' which is a more distinct separation of the solvent from the polymer. Gelation dominates at lower temperatures and to maximise the formation of a homogeneous structure, it is preferable that a gelation stage precedes the phase separation stage. It has been found that higher temperatures (up to 70 °C) of the spin bath lead to the formation of larger voids (macrovoids) and although they collapse during drying, their harder surfaces remain, giving a lower degree of internal homogeneity and adversely affecting strength. If the solvent diffuses out of the filaments quicker than non-solvent diffuses into them, contraction takes place and the cross-section becomes more kidney-bean shaped. Wet spinning of the polymer dissolved in an organic solvent also produces kidney-bean shaped fibres, due to a skin formed initially around the filaments. As the filaments shrink, the skin folds to a bean shape. This behaviour is very much the case in dry spinning (when there is no non-solvent present to diffuse in) and the fibres formed have very characteristic dog-bone cross-sectional shapes.

After extrusion, all acrylic fibres are thoroughly washed to remove all traces of solvent, drawn to increase orientation, dried and relaxed. The relaxation stage is an important part of the manufacturing sequence. After drawing and drying, the fibres have reasonable strength but they are not very extensible and they fibrillate easily. Dyeability is also poor.

Relaxation of the fibres is usually carried out in either water or in steam, the presence of moisture being required to enable lower temperatures to be used. No tension is applied and during the process shrinkage of up to 40% can occur. It is thought that the water plasticises the fibre molecules, enabling cracks and fissures from the collapsed voids to fuse together. Certainly the treated fibres have improved abrasion resistance and resistance to fibrillation.

The final stage in the manufacture is to impart a crimp into the fibres, the purpose of which is to promote bulkiness and cohesion between fibres, when the staple fibres are converted into yarns. The crimp imparted can be substantially lost, however, during dyeing and laundering processes.

5.4.5 Chemistry of Acrylic Fibres

Acrylic fibres are noted for their chemical resistance and resistance to sunlight and action of microorganisms. This lack of chemical reactivity is due to the strong hydrogen bonding that exists between the electronegative nitrogen atoms and the hydrogen atoms on the polymer backbone and also the strong dipole-dipole interaction between the –CN groups of adjacent chains (see Figure 5.8).

5.4.5.1 Acids and Alkalis. Acrylic fibres have high resistance to mineral acids and to weak alkalis, but strong alkalis cause rapid degradation, especially at elevated temperatures. For example immersion of acrylic fibres in 5% sodium hydroxide at 80 °C for 1 day will cause complete degradation, whilst immersion for 1 day at 75 °C in 60% sulfuric acid causes only slight degradation.

5.4.5.2 Solvents. Most solvents are incapable of diffusing into the fibre matrix and causing swelling. The only exception to this is the highly polar organic solvents of small molecular size, such as dimethylformamide, dimethylsulfoxide and dimethylacetamide, used for wet spinning the fibres.

5.4.5.3 Oxidation. Oxidising agents have little effect on acrylic fibres, especially in comparison with other fibre types. For example it has been shown that exposure of the acrylic fibre Orlon to bleaching fluid for up to 9 hours reduces its tenacity by about 9%, whilst under similar conditions the tenacity of nylon is reduced by 50%, and cotton is almost totally destroyed.

5.4.5.4   Action of Heat.   The resistance of acrylic fibres to heat is also very good. Even exposure for two days at 150 °C causes no loss in strength and the fibres even retain their colour and stability when subjected (for shorter periods) to temperatures up to 230 °C. More prolonged exposure at these higher temperatures will cause yellowing and browning, then eventually, after some sixty hours, the fibres will become black. Acrylic fibres are used as the precursor for the production of carbon fibres, by a process that involves firstly oxidation at about 300 °C and then pyrolysis at about 2000 °C. This process is described fully in Section 6.6.2.

5.4.6   Physical Properties of Acrylic Fibres

There are so many variants of acrylic fibres that it is impossible to give meaningful overall representative values for physical properties. In general, however, it can be said that the popularity of acrylic fibres over the years has arisen from their balance of satisfactory performance in terms of appearance, ease of dyeing to heavy depths, ease of care and durability. Whilst physical properties are only modest they are sufficient for applications in home furnishings and apparel.

The fibres are most commonly produced in staple form, with staple lengths between 25–150 mm, and processed by traditional staple yarn operations. A range of fibre deniers is produced: lower deniers (around 1.5–3) are used for apparel fabrics, whereas at the opposite end of the scale, fibre filaments of about 15-denier are used for carpet yarns, where stiffness and wear resistance are the prime requirements of performance. The tenacity of fibre filaments ranges between 20–35 cN tex^{-1} amongst the variants, and elongation at break ranges between 30–65%. What might be regarded as a 'standard' acrylic fibre, Orlon 42 which is a copolymer of acrylonitrile and vinyl acetate, has a tenacity of 26 cN tex^{-1} and an elongation at break of 33%. Generally, the strength of acrylic fibres is adequate for most textile purposes but not exceptional, and acrylics are fairly stiff when compared with nylon and polyester. The moisture regain properties of acrylic fibres is in the region 1.0–4.5%, depending on the variant. Usually they are less prone to the build-up of static electricity than polyester or nylon, so acrylic fabrics do not 'cling' to the body as much. The general properties of acrylic fibres are given in Table 5.3.

The main disadvantage of acrylic fibres is their behaviour under hot, wet conditions. At temperatures of over 55 °C the strength of acrylic fibres under wet conditions decreases from that under dry conditions at the same temperature. This behaviour is accentuated as the temperature

Table 5.3 Properties of acrylic fibres.

Specific gravity	1.14–1.19. Modacrylics tend to be denser at around 1.3.
Tenacity	20–35 cN tex^{-1}, and about 15% weaker when wet.
Elongation at break	33–60%.
Elastic recovery	Fairly inelastic. Only 45% recovery from a 5% stretch.
Resilience	Good, and abrasion resistance is also good.
Moisture regain	1–4.5%.
Reaction to heat	Acrylics fibres do not show a melting point, but soften at about 450 °C.
Stability to light	Excellent.
Resistance to chemicals	Good resistance to acids and alkalis. Excellent resistance to bacterial action, mould.
Launderability	Acrylic garments should be hand washed or dry cleaned. They can be tumble dried but only on a cool setting.

Table 5.4 Physical characteristics of acrylic fibres measured in dry and aqueous media.

		Dry	*Wet*
Tenacity at break (cN tex^{-1})	25 °C	21.0	18.5
	95 °C	9.0	4.0
Elongation at break (%)	25 °C	33	35
	95 °C	58	90

approaches the boil and under wet conditions the fibres become very extensible and have low tenacity at break (see Table 5.4). The adverse influence of hot, wet conditions on the dimensional stability of acrylic fabrics has implications for laundering. Hand washing in luke-warm water, with minimal agitation is necessary. Otherwise, acrylic fibres are unaffected by dry cleaning solvents, so dry cleaning is a suitable alternative, and sometimes preferred if water soluble softeners have been applied.

5.4.7 Uses of Acrylic Fibres

When acrylic fibres reached their height of popularity their main uses were in apparel, home furnishings and industrial applications. Acrylic fibres competed head-on with the areas in which wool was traditionally the dominant fibre, because they had the same characteristics of bulkiness, warmth and handle, and indeed were superior to wool in terms of resistance to chemical and microbiological attack, and in physical properties, such as higher tenacity and work of rupture.

Thus acrylics became widely used in apparel, for knitwear, fleeces and the flame-resistant versions for children's sleepwear, for which flame resistance in the UK is a legal requirement. Acrylic fibres possess good wicking properties and can be used in garments for sportswear. They have typically been used in blends with wool or with cotton for socks, where their bulk gives added warmth, but even here blends with nylon are often preferred. Socks for trekking now contain nylon, polyester and polypropylene fibres that have superior wicking properties and have overtaken the use of acrylics for this market. In home furnishings acrylic fibres were used extensively for carpets and upholstery. However, whilst carpets made from acrylic fibres could be dyed to bright, attractive shades, they possessed inferior durability and their low resilience meant that pile height reduced fairly quickly in areas of most traffic. By this measure carpets made of wool and wool/nylon blends are much better. The poor hot-wet properties of acrylic fibres also make steam cleaning of acrylic carpets impossible. In addition to their use in making flame-retardant fibres, coarse denier modacrylics are used to make artificial fur fabrics, for use in soft toys and wigs.

During recent years there has been a trend by consumers back towards the natural fibres of wool and cotton, and developments in nylon and polyester (such as microfibre variants and novel cross-sectional shapes) have stimulated their popularity over that of acrylics in blends. Acrylic fibres still retain their importance for industrial applications, but these are in the areas where acrylics have a superior performance that justifies their higher cost. Typically the main applications are those for which long-term light and weather fastness and resistance to mould is essential, such as outdoor fabrics for awnings, boat covers, sun umbrellas, car tops and garden furniture. A number of variants have been developed over the years, an important one being an antimicrobial version in which the internal void structure of a highly microporous variant is treated with a bactericide. These types of fibres have been developed for the sportswear market. Other variants include antistatic fibres for specialised carpets, and anti-soiling fibres.

5.5 POLYOLEFIN FIBRES

5.5.1 Introduction

Polyolefin fibres include all those fibres whose polymer chains are essentially saturated, aliphatic hydrocarbons. Commercially, they comprise manufactured fibres whose polymer chains consist of $\geq 85\%$ by mass of olefin units. By far the most important commercial polyolefin fibres are polypropylene (PP) fibres, although polyethylene (PE) fibres also have some commercial significance. Some other polyolefin fibres have been produced for industrial application, but commercial interest in these types of fibre is limited. This section, therefore, focuses on PE and PP fibres.

PE fibres were first melt extruded from low density PE on a commercial scale in the 1930s. Polymer chain branching in low density PE is extensive, and so the mechanical performance of the fibres was generally unsuitable. Although the fibres were cheap to produce, their low melting point and inability to retain dyes also made them unattractive. Their commercial impact was hence almost negligible.

A key development came in 1953, however, when Karl Ziegler, who directed the Max Planck Institute for Coal Research at Mülheim in Germany, discovered a method of synthesising PE of high molar mass with very little chain branching. Ziegler licensed his process for producing PE to Montecatini in Italy. PE fibres melt extruded from this high density PE entered the market in the late 1950s; they possessed much better mechanical properties. More recently, since the 1980s, outstandingly strong PE fibres of very high molar mass have become available, using special extrusion techniques, such as gel-spinning and solid-state extrusion.

In 1954, Guilio Natta and his group at Milan Polytechnic in Italy initiated studies to discover whether Ziegler's approach could be used to synthesise other polyolefins. They soon realised that PP of high molar mass could be synthesised in a similar way, and they speedily filed a patent. Moreover, PP fibres could be readily produced by melt spinning. Since then, commercial production of PP fibres has steadily risen, and fibres with good mechanical properties are now widespread. Indeed, amongst synthetic fibres, PP fibres are second only to PET fibres in terms of commercial production. It is noteworthy too that, unlike most synthetic fibres where production of polymer and fibre is mostly linked within the same companies, the production of PP fibres is mostly spread among much smaller companies, who buy in specific grades of PP and focus on particular sectors of the fibres market. Like PE fibres, PP fibres cannot be dyed unless they have been modified.

The polymer chains in PE and PP fibres possess markedly different configurations. High density PE consists primarily of chains comprising methylene, $-CH_2-$, groups, as the degree of chain branching is small. The individual chains tend to adopt a zig-zag form, as shown schematically in Figure 5.9. In PP fibres, by contrast, the polymer chains can adopt three different configurations, also shown schematically in fully

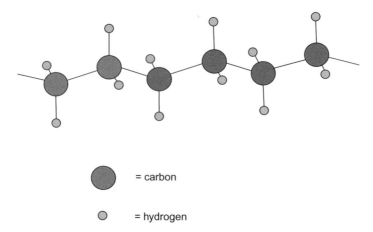

= carbon

= hydrogen

Figure 5.9   Zig-zag form of PE chains.

extended planar projections in Figure 5.10. In planar projections of isotactic PP chains, all the methyl side-groups are situated uniformly on the same side of each chain, although in practice the chains adopt a three-dimensional helical configuration, as shown in Figure 5.11. In syndiotactic PP chains, the methyl groups regularly alternate between the two sides of each chain. Syndiotactic PP chains may also adopt a helical configuration. In the case of atactic PP chains, the methyl groups are arranged randomly on both sides of each chain.

5.5.2   Manufacture of Monomers

Ethene and propene are obtained commercially from the high temperature pyrolysis of hydrocarbons, produced from the distillation of petroleum and fractionation of natural gas. The yields obtained depend on the composition of the hydrocarbon feedstock and the conditions of pyrolysis. Propene can also be obtained as a major by-product in the production of petrol by the catalytic cracking of higher hydrocarbons. Thus, monomer is readily available, without specific synthesis being needed.

5.5.3   Production of Polyolefins

Polyolefins are generally produced by the process devised by Karl Ziegler and Giulio Natta, although, as discussed below, there has been increasing interest in recent years in a method utilising metallocene catalysts. The Ziegler–Natta process utilises solid organometallic

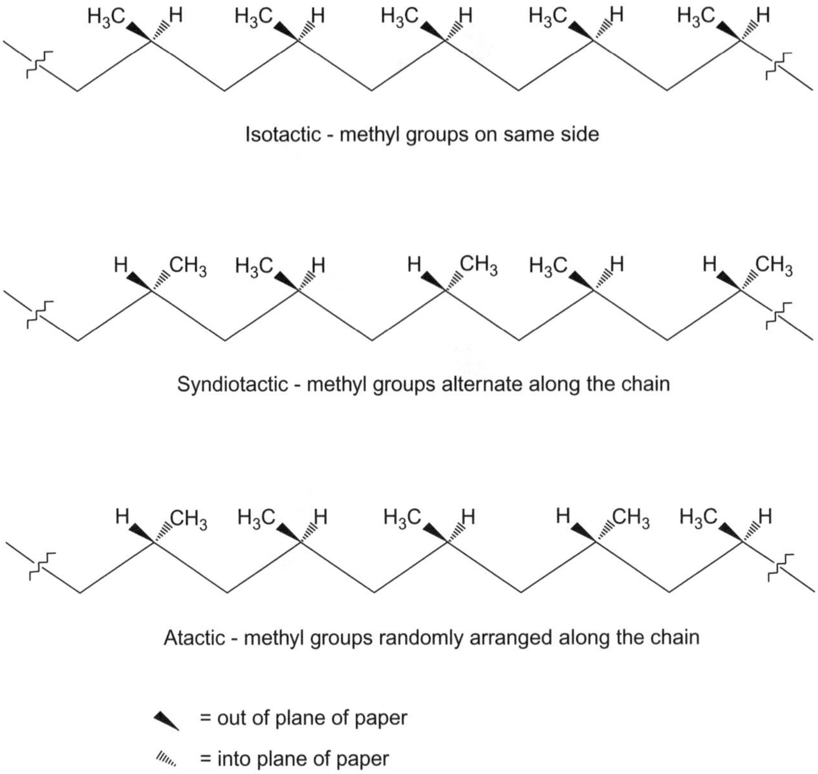

Isotactic - methyl groups on same side

Syndiotactic - methyl groups alternate along the chain

Atactic - methyl groups randomly arranged along the chain

⬥ = out of plane of paper

⁘ = into plane of paper

Figure 5.10 Configurations of PP chains.

catalysts, comprising an alkyl compound of a Group I, II or III metal combined with the halide or ester of a transition metal. The nature of the catalyst is highly influential in determining the degree of stereoregularity in the polyolefin chain produced, and many Ziegler–Natta catalysts have been devised which promote very high stereoregularity. Moreover, catalysts have been designed that favour formation of isotactic PP chains over those of other configurations. PP chains of isotactic content up to 95% can be achieved. Examples of Ziegler–Natta catalysts are given in Table 5.5.

The molar mass dispersities of fibre grade PE produced by the Ziegler–Natta process vary generally between 5 and 20. For PP, the molar mass dispersities generally range between 4 and 8. Thermal cracking, however, can lower molar mass dispersity to 3–6. It may also reduce molar mass to a level that is more suitable for fibre formation.

Metallocenes are organometallic compounds based principally on zirconium. Metallocenes react with methylalumoxane to form an active

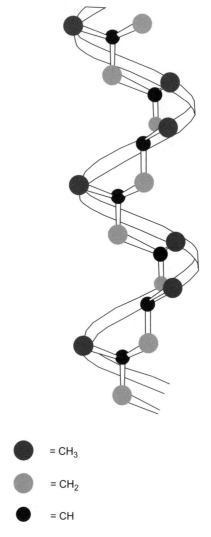

● = CH₃

⬤ = CH₂

● = CH

Figure 5.11   3-D helical configuration of PP chains.

Table 5.5   Examples of Ziegler–Natta catalysts.

Organometallic compound	+	Transition metal compound
$Al(C_2H_5)_3$ or $Al(C_2H_5)_2Cl$ or $Be(C_2H_5)_2$	+	$TiCl_3$
$Al(C_2H_5)_3$ or $Al(C_2H_5)_2Cl$	+	VCl_3 or $TiCl_4$
$Al(C_2H_5)Cl_2$	+	VCl_4

catalyst complex. The structure of methylalumoxane approximates to:

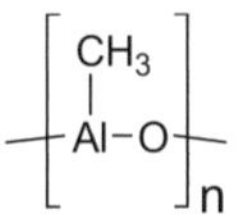

$$\left[\begin{array}{c} CH_3 \\ | \\ Al-O \end{array} \right]_n$$

where $n = 10$–30. The metal in the metallocene molecule is the active centre for the polymerisation process and is sandwiched between two cyclopentadienyl ligands. These cyclopentadienyl ligands are also bridged through either one or more carbon atoms or through a silicon atom. In addition, two other ligands, usually chlorine atoms, are co-ordinated to the metal atom. The structure of a zirconium metallocene can be represented as:

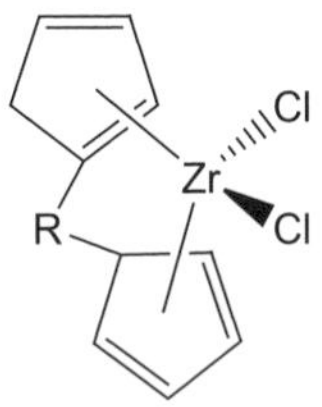

Catalyst systems based on metallocenes are gaining increasing import-ance for the production of PP. These systems are more active than Ziegler–Natta catalysts, and can be designed so that only isotactic PP chains are produced. Moreover, metallocene grades of PP possess small molar mass dispersities, typically *ca.* 2.5. However, some small ir-regularities can appear in the PP chains, where individual propene monomers have been inserted back to front.

5.5.4 Fibre Processing

Polyolefin fibres can be produced by melt extrusion processes, followed by drawing and further treatments such as texturising. Whereas com-mercial PP fibres are routinely produced by melt extrusion, the process is used less commonly for manufacturing PE fibres. Instead, many types of commercial PE fibre are produced using gel spinning and solid-state extrusion techniques. The grades of PE possess very high molar mass ($M_w \sim 10^6$), and the fibres produced from them are used for a number of high-performance applications of the kind discussed in Chapter 6. Due to the high molar mass of PE, melt viscosities are too high for practical melt extrusion.

For gel spinning, PE fibres are formed in the gel state. The PE chains exist essentially as colloidal aggregates within each fibre. In commercial processes (see Figure 5.12), PE is extruded as an approximately 5% solution in tetralin, decalin or paraffin oil (kerosene) into a small air gap, before entering an extraction bath containing water at room temperature. A gel multifilament is formed, in which almost all the solvent still remains. The multifilament is then drawn at 100–150 °C to a draw ratio of 30–100. The solvent is made to leave the filament at this stage.

In solid-state extrusion, PE of very high molar mass is used again, but with no solvent present. The process fundamentally consists of three stages: compression of PE powder under controlled conditions, rolling and 'ultra-drawing' (see Figure 5.13). During compression, the PE powder is converted into a cohesive sheet, which is then fed between a pair of heated, counter-rotating rollers. As the sheet travels through the rollers, its thickness is reduced to <20%, and then slit into tapes of a desired width. The tapes are drawn over long hot-plates to a draw ratio of >10.

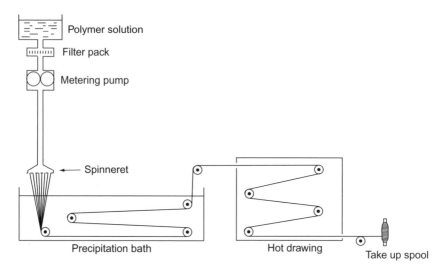

Figure 5.12   Gel spinning of PE fibres.

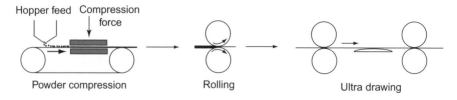

Figure 5.13   Solid-state extrusion of PE fibres.

PP fibres are produced by more variations of the melt spinning process than any other type of commercial melt-spun fibre. The choice of process depends on the type of filament to be produced (*e.g.* monofilament, multifilament or filament from slit film) and the application for which the fibre is destined. Thus, multifilament yarns of superior mechanical performance are produced by long air-quench spinning, in which the length of the cooling chamber is 3–10 m and the extrusion velocity is 300–400 m min^{-1}. Short air-quench melt spinning processes have also been developed, in which the cooling chamber is only *ca.* 1 m long. However, extrusion velocities are limited to 100–150 m min^{-1}, and the apparent quality of the yarn is inferior to that of yarns produced in the long air-quench process. PP monofilaments are produced by cooling in a water bath, and coarser fibres are produced by slitting PP film.

The conditions of drawing are also influenced by the type of filament, and indeed there may be more than one drawing stage. Monofilament, for example, is drawn to ratios up to a maximum of 10, using hot water, steam or hot air at carefully controlled temperatures. Drawing is arguably the most critical stage of processing monofilaments, in order to produce the high strength required. Multifilaments are generally drawn to the same ratios, although higher draw ratios for specific applications can be quite readily achieved. Accurate control of draw temperature is particularly important, as even slight variations can affect the final properties of the filaments. Drawing often incorporates twisting with the stretching operation, as in many instances the final yarn needs to be in the form of twisted multifilaments.

5.5.5 Physical Structures

Amongst melt-spun fibres, polyolefin fibres are notable for their rapid crystallisation after extrusion, primarily because the polymer chains are so flexible. However, despite the close similarities in the molecular structures of polyolefin chains, the crystal structures that are produced are distinctly different. To some extent, this difference is due to the size of the alkyl groups projecting at regular intervals from the chain backbone, a factor that influences how closely the chains may be aligned next to one another. In the case of PE, there are no projecting alkyl groups, and so the PE chains can approach one another more closely than is the case for other polyolefin chains. Consequently there is close packing of the PE chains; they all lie next to one another in a planar zig-zag fashion. PE usually adopts an orthorhombic crystal cell structure, though in some cases such as in gel-spun PE fibres, some pseudo-monoclinic domains are also present.

The close alignment that PE chains can adopt is exploited in the gel spinning process. Under carefully controlled conditions, extremely good alignment of the chains can be achieved, as illustrated schematically in Figure 5.14a. This close interchain alignment gives rise to very strong fibres indeed. In the case of PE fibres produced by solid-state extrusion, the polymer chains are less perfectly aligned. Instead, they adopt a 'shish-kebab' structure as shown in Figure 5.14b, in which a central core of extended PE chains (shish) is surrounded by sections consisting of folded chains (kebab). Shish-kebab structures have also been observed in some PP fibres.

As noted above, PP chains adopt a helical conformation. Three distinct crystal forms are known. The thermodynamically most stable form is the α-monoclinic. This form is the most important in the context of PP fibre technology. Both left-handed and right-handed helical chains are present, and for the most part a given helix lies next to helices of opposite handedness (chirality). Of the other two crystal forms, the β-form structure is trigonal, and the γ-form structure is orthorhombic. There is also a 'paracrystalline' form. This form is not truly crystalline; it appears to consist of a structure that approximates to being liquid-crystalline (see Section 6.2.2.2). In many PP fibre-forming processes, the melt extruded filament is paracrystalline, so that on subsequent drawing, much higher draw ratios can be achieved.

5.5.6 Chemistry of Polyolefin Fibres

5.5.6.1 Introduction. Being hydrocarbons, polyolefins exhibit very high chemical resistance overall. They are impervious to chemical attack by acids and alkalis and most organic liquids. However, polyolefins are swollen by a number of organic liquids, especially at elevated temperatures. Indeed, at sufficiently high temperatures, polyolefins dissolve in decalin, tetralin and a variety of chlorinated aromatic compounds.

Polyolefins are, nevertheless, strongly prone to oxidation by strong oxidising agents, such as hydrogen peroxide and concentrated nitric acid. Oxidation leads to reduced mechanical properties and often to discoloration as well. In addition, the resistance to ultraviolet radiation is poor. In practice, commercial polyolefin fibres contain light-stabilisers.

5.5.6.2 Pyrolysis. In the absence of oxygen, PE is stable to over 300 °C and PP to 350 °C. For both polyolefins, these temperatures are well above normal fibre processing temperatures, and so pyrolysis does not usually interfere with fibre processing.

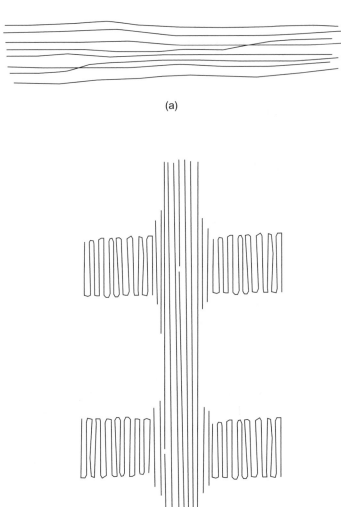

(a)

(b)

Figure 5.14 Alignment of PE fibres: (a) from gel spinning, (b) from solid-state extrusion.

5.5.6.3 Thermal Oxidation. As with polyamide fibres, thermal oxidation occurs through chain reactions involving free radicals. The reactions are initiated through abstraction of hydrogen atoms in the polyolefin chains by excited singlet state oxygen molecules. In the case of polypropylene fibres, those hydrogen atoms bonded to tertiary carbon atoms are the most easily abstracted. A wide range of propagation, chain branching and termination reactions then ensues. Scission of the polyolefin chains also takes place, and the fibres can consequently become severely degraded.

5.5.6.4 *Photodegradation.* Photodegradation is caused mainly by hydroperoxides, and to some extent by ketones, both formed in small quantities during fibre processing. The formation of radicals through the decomposition of hydroperoxides has already been noted (see Section 5.2.7.4).The breakdown of ketones is thought to occur primarily by a process such as:

$$-CH_2-CH_2-CH_2-CO-CH_2- \xrightarrow{h\nu} -CH=CH_2 + CH_3-CO-CH_2-$$

In the case of PP fibres, a further means of initiating photo-oxidation is by the formation of excited charge transfer complexes between PP and oxygen:

$$PolyH + O_2 \rightarrow [Poly-H\cdots\cdots O_2] \xrightarrow{h\nu} [PolyH^+ \cdots\cdots O_2^-]$$
$$\rightarrow Poly^{\cdot} + {}^{\cdot}OOH \rightarrow PolyOOH$$

5.5.6.5 *Stabilisation of Polyolefin Fibres.* The types of stabiliser already described for use with polyamide fibres are also used to stabilise polyolefin fibres. Hindered phenols are widely used as antioxidants, though some of these phenols tend to develop colour in the fibres, due to the quinone structures involved in the stabilisation process. Moreover, reactions with initiator fragments in the polyolefin chains can promote colour development, and the action of pollutants in the atmosphere, such as nitric oxide, can also effect discoloration. The choice of hindered phenol thus requires care, and indeed secondary aromatic amines may be preferred.

The removal of hydroperoxides can also be achieved by thioesters and tertiary phosphites. Reaction of hydroperoxides with thioesters may be summarised as:

$$\overset{O}{\overset{\|}{RSR}} + PolyOOH \rightarrow \overset{O}{\overset{\|}{RSR}} + PolyOH$$

where R is an ester-containing group. Reaction with phosphites may be represented as:

$$P(OX)_3 + PolyOOH \rightarrow \overset{O}{\overset{\|}{P}}(OX)_3 + PolyOH$$

5.5.7 Fibre Properties

The properties of melt-spun PP fibres are shown in Table 5.6 and of gel-spun PE fibres are shown in Table 5.7.

PP fibres have a wide range of applications, partly because the technical performance is remarkably high for a fibre so inexpensive to produce. In addition, the production of PP fibres is comparatively straightforward. PP fibres are used, for example, in carpet backings and carpet facing, sacks and bags, ropes, horticultural netting, and medical and hygiene products. By contrast, the use of PP fibres in apparel is very limited. A major disadvantage is the difficulty in dyeing PP fibres. Nevertheless, they are used in some types of underwear and socks. They are also finding application in sports and leisurewear, such as cycle shorts, swimwear and lightweight outerwear for climbers.

Most types of PE fibre are produced for high-performance applications, although melt-spun PE fibre is now used extensively in artificial sports surfaces. Gel-spun PE fibres are widely used for ballistic protection, such as in helmets and flexible vests for police and military personnel. The fibres also possess good cut and puncture resistance, and so provide good protection against knife attacks and in cut-resistant gloves, chain-saw protective wear and fencing suits. The fibres are also used in nets ropes and sails. PE fibres produced by solid-state extrusion are used for speciality ropes, cord and fishing lines.

Table 5.6 Properties of PP fibres.

Specific gravity	0.90.
Tenacity	30–80 cN tex^{-1} normally, but up to 100 cN tex^{-1} for high tenacity yarns.
Elongation at break	15–35%; $\leq$10% for high tenacity yarns.
Elastic recovery	High; >90% from an 8% stretch.
Resilience	Medium.
Abrasion resistance	High.
Moisture regain	0.04%.
Launderability	PP fibres withstand all chemical conditions commonly encountered in laundering.

Table 5.7 Properties of gel-spun PE fibres.

Specific gravity	0.97.
Tenacity	250–370 cN tex^{-1}.
Elongation at break	3–4%.
Abrasion resistance	High.
Moisture regain	0%.

5.6   ELASTOMERIC FIBRES

5.6.1   Introduction

The interest in producing synthetic fibres that are capable of very high degrees of stretch, but which will snap back to their original length, as an alternative to natural and synthetic rubbers, began during the early 1940s. They are made from a class of materials known as 'elastomers', defined by the Textile Institute as 'any polymer having high extensibility together with rapid and substantially complete elastic recovery'. Much of the early work on the development of elastomers centred around polyamides. One strategy was to replace part of the hexamethylene diamine by a substituted diamine that contained a bulky group (*e.g. iso*butyl group). This bulky group prevented close approach of adjacent polymer chains and caused the molecules to assume a state of high strain when the fibres were stretched, so that they quickly recovered their original state when the tension was released. However, despite intensive research efforts, products based on polyamides never achieved commercial success because they would not withstand repeated stretching cycles. Indeed, many thermoplastic polymers have been developed but most are unsatisfactory for use as textile fibres because they do not recover well from extensions of 2–3 times their original length, have poor tensile strength, are unstable to chemicals, or are difficult to dye to deep shades.

There are different polymeric systems that can be considered to be elastomers (see Table 5.8), of which the most commercially important elastomeric fibres produced today are the 'elastanes', which chemically are polyurethanes. In America these types of fibres are called 'spandex' fibres, where the FTC has a corresponding definition to that of BISFA. In Europe these fibres are marketed by Invista under the brand name Lycra® and by Asahi Kasei Spandex Europe GmbH (formerly Bayer) in Germany as Dorlastan® and Roica®. They first came on to the market in the late 1950s in America (Lycra® and Vyrene) and in the early 1960s in Europe (Bayer's Dorlastan®).

5.6.2   Elastane Fibres

The successful development of elastomeric fibres requires particular structural features of the polymer. Of particular importance is the force required to stretch the fibre and the force with which the fibre snaps back to its original length. The greater these forces are, the less the amount of elastane required in a fabric to produce the desired effect. The molecular chains must be coiled in their stable state and have only weak attractive

Table 5.8 Generic classification of elastomers.

Generic name	Definition	Structure
Elastane	Fibre composed of at least 85% by mass of a segmented polyurethane and which, if stretched to three times its unstretched length, rapidly reverts to substantially to the unstretched length when the tension is removed.	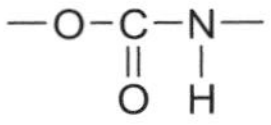 $-O-C-N-$ with $C=O$ and H
Elastodiene	Fibre composed of natural or synthetic polyisoprene, or of one or more dienes polymerized with or without one or more vinyl monomers, and which, if stretched to three times the unstretched length, rapidly reverts substantially to the unstretched length when the tension is removed.	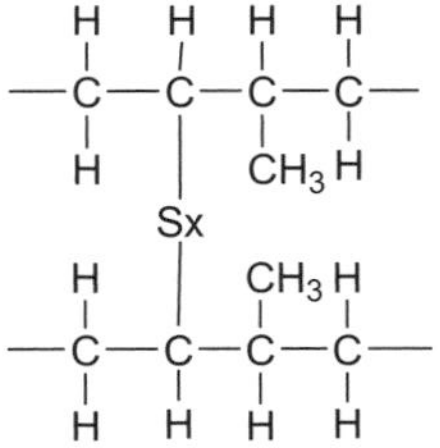 (structure with CH_3, S_x groups)
Elastomultiester	Fibre formed by interaction of two or more chemically distinct linear macromolecules in two or more distinct phases (of which none exceeds 85% by mass) which contains ester groups as dominant functional unit (at least 85%) and which after suitable treatment when stretched to one and a half times its original length and released recovers rapidly and substantially to its initial length.	(structures A and B, repeating unit n, with $-C-O-CH_2CH_2-$ and $-C-O-CH_2CH_2CH_2-$)

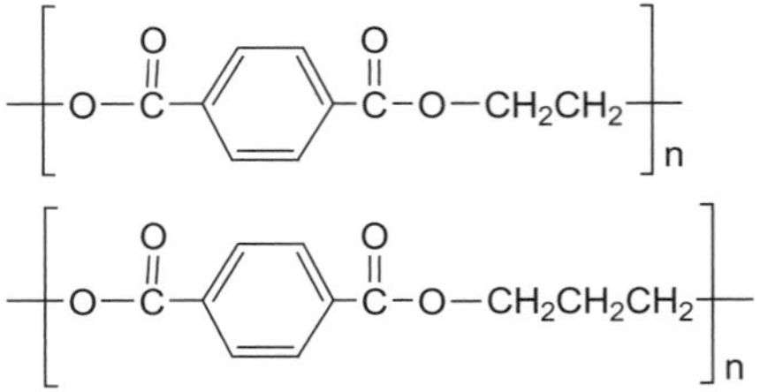

Example of physical arrangement

Table 5.8 (Continued).

Generic name	Definition	Structure
Elastolefin	Fibre composed of at least 95% (by mass) of macro-molecules partially cross-linked, made up from ethylene and at least one other olefin and which, when stretched to one and a half its original length and released, recovers rapidly and substantially to its initial length.	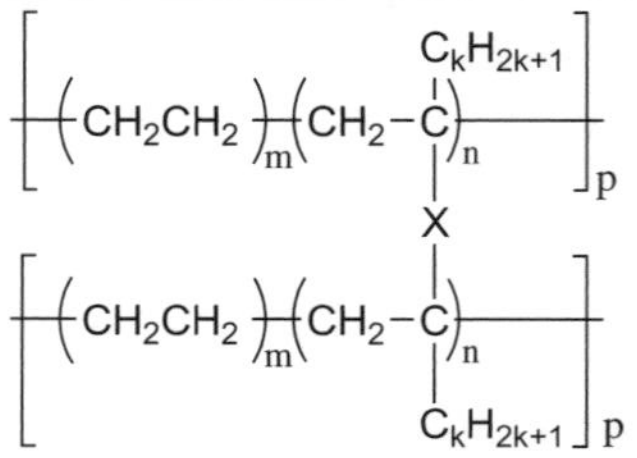 $$\left[\left(CH_2CH_2 \right)_m \left(CH_2 - \underset{C_kH_{2k+1}}{\overset{}{C}} \right)_n \right]_p$$ $$X$$ $$\left[\left(CH_2CH_2 \right)_m \left(CH_2 - \underset{C_kH_{2k+1}}{\overset{}{C}} \right)_n \right]_p$$

(Source: BISFA)

forces between them, so they can be readily extended. However, long-range intermolecular forces of sufficient strength must exist to prevent the molecules slipping over each other under high degrees of stretch. These requirements have been best met with a class of polymeric materials called 'segmented polyurethanes'. These are block copolymers and comprise two alternating segments, one 'soft' and the other 'hard', to form the structure:

$$-A-B-A-B-A-B-A-B-$$

The polymer chain of the 'soft' component has a highly amorphous, randomly coiled structure when the fibre is in the relaxed state, but when the fibres are stretched the coils 'unwind' and the polymer molecules become more aligned to the axis of the fibre. Under tension therefore these 'soft' components, which are either an aliphatic polyether or a copolyester, form crystalline regions. This state is not the most thermodynamically stable and on release of the tension, they very quickly revert to their more stable, coiled state. The 'hard' component is usually an aromatic–aliphatic polyurea. This component of the polymer chain is capable of strong inter-chain bonding, mainly through hydrogen bonding, with the 'hard' components of neighbouring chains, thus providing a network of tie-points through the polymer structure. It is this feature that provides the long-range stability to the structure, preventing molecular slippage. In order to produce fibres which will possess the required degree of stretch, the balance of the 'soft' and 'hard' components is important. If there are too many tie-points, then the stretch will be inhibited; if there are too few, there is the danger of irreversible molecular slippage with a resulting inability of the fibre to recover fully from stretching.

The formation of the segmented polyurethanes is based on isocyanate chemistry and there are many routes possible for their synthesis. All have the same three stages however:

1. Formation of a macroglycol pre-polymer, which is either a polyether or a polyester of molar mass in the range 1000–3000, and which has –OH end groups;
2. Reaction of the macroglycol with a molar excess of an aromatic diisocyanate, so that the product, the pre-polymer (a macrodiisocyanate), has isocyanate end groups; and
3. Formation of the segmented polyurethane by reaction of the macrodiisocyanate with either a diamine or a diol, to form a polyurethane.

In the first stage, the formation of a polyether macroglycol can be achieved by the ring opening of expoxides or a cyclic ether, typically tetrahydrofuran, and polymerisation to form polytetramethyleneether glycol.

$$HO-(CH_2-CH_2-CH_2-CH_2-O)_n-H$$

If a polyester macroglycol is to be formed, a dicarboxylic acid is reacted with a molar excess of a diol, though often a mixture of diols is used in a ratio that gives the desired stretch-recovery characteristics. Typically ethane-1,2-diol and propane-1,2-diol are reacted with 1,6-hexandioic acid (adipic acid) to form a polyethylene–polypropylene adipate:

$$HOCH_2CH_2OOC(CH_2)_4COOCHCH_2OOC(CH_2)_4CO\cdots\cdots OCH_2CH_2OH$$
$$|$$
$$CH_3$$

In each case (both polyether and polyester products) –OH groups exist at the ends of the polymer chains. These structures comprise the 'soft' segment of the final segmented polyurethane.

The second stage of the process, the formation of the pre-polymer, involves heating either the polyether or the polyester macroglycol with twice the molar quantity of a diisocyanate. The particular diisocyanates used are either 4,4'-diphenylmethane-diisocyanate (MDI) or 2,4-toluene-diisocyanate (TDI). If HO–R–OH represents the macroglycol, and OCN–Ar–NCO represents the diisocyanate, the equation for the reaction can be represented as:

$$HO - R–OH + 2\,OCN–Ar–NCO$$
$$\downarrow$$
$$OCN–Ar–NHCOO–R–OCONH–Ar–NCO$$

In practice the reaction is not quite so stoichiometric, and a low molecular weight polymer is formed with a small number (*n*) of repeating units of:

$$-[O–R–OCONH–Ar–NHCO]_n-$$

The pre-polymer molecules now have terminal isocyanate groups and some of the MDI or TDI which remains un-reacted, plays a part in the third stage.

The third stage, the formation of the segmented polyurethane, involves the formation of the 'hard' segment, by a reaction called 'chain extension'. In this reaction the pre-polymer is dissolved in a polar solvent such as dimethylformamide or dimethylacetamide and a bifunctional 'chain extender' is added. This chain extender is usually an aliphatic diamine, such as ethylene diamine, though other diamines, such as propylene diamine or cyclohexylene diamine, can also be used. Where the MDI or the TDI react with the diamine chain extender, the 'hard' component of the polyurethane is formed:

$$-\text{CONH}-\text{Ar}-\text{NHCO}-\text{NH}-\text{CH}_2\text{CH}_2-\text{NH}-$$
$$(\text{CONH}-\text{Ar}-\text{NHCO}-\text{NH}-\text{CH}_2\text{CH}_2-\text{NH})_n-\text{CONH}-\text{Ar}-\text{NHCO}-$$

This 'hard' component alternates with the 'soft' component in the final segmented block copolymer through urethane $-\text{NHCO}-\text{O}-$ linkages. When the required molecular weight of the final block copolymer is reached, a small amount of a monofunctional amine is added to terminate the reaction.

It will be clear from the above, that there is a choice of chemicals available at the various stages and that the molecular sizes of the 'soft' and 'hard' components can be varied. Thus it is impossible to state precisely the chemical formula of any particular elastane. All that can be indicated are the general formulae of the polyether and the polyester type polyurethanes:

A polyether segmented polyurethane

"Hard" segment   "Soft" segment

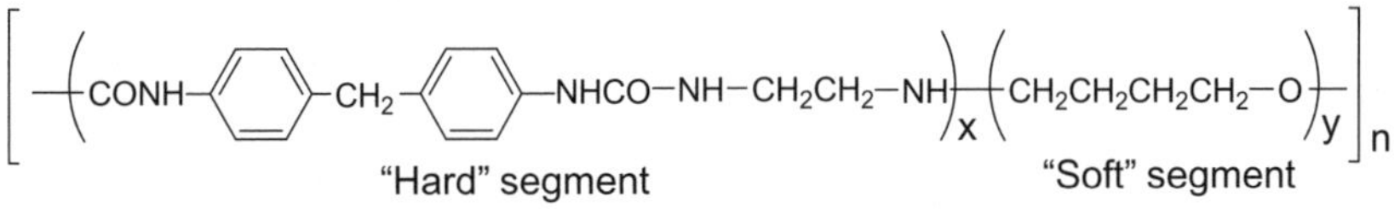

A polyester segmented polyurethane

"Hard" segment   "Soft" segment

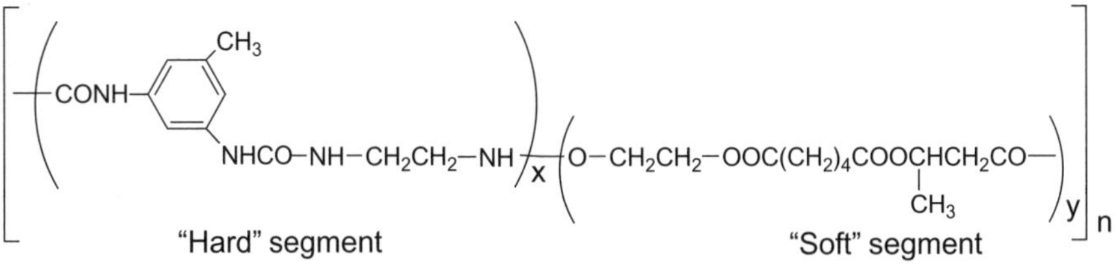

Further variants may include the use of diamine chain extenders that include a tertiary amino group in their structure, the function of which is to enhance substantivity for acid dyes (see 'Dyeing' in appendix). Other additives may include compounds to improve stability to light, or pigments to provide opacity.

Extrusion of the polymer to form fibres can be carried out by wet spinning, dry spinning or melt spinning. Most producers of elastane fibres use the dry spinning method (see Section 5.1.3), since the solvents used (dimethylformamide or dimethylacetamide) evaporate readily in the hot air environment of the spinning chamber. Also faster wind-up speeds (up to $500\,m\,min^{-1}$) and therefore higher production rates are possible, so the economics of manufacture are more favourable. When the individual fibre filaments form as the solvent evaporates, they are sticky and coalesce together if they touch. This is purposely allowed to happen, so that threads of a required linear density can be produced.

Polyurethane fibres are superior to natural rubber because the tie-points that prevent molecular slippage are more uniformly distributed than they are in vulcanised rubber. Where tie-points are close together in the more random configuration of natural rubber, flexibility is limited, and *vice versa* in the regions where the tie-points are fewer. Polyurethanes are approximately twice as strong as rubber fibres (see Figure 5.15), which means that double the force is required to stretch polyurethanes for a given extension. As a result, less polyurethane is required in a fabric blend to give a desired elastic performance than rubber. To put the forces required to stretch elastomeric fibres into context, the forces required to stretch 'conventional' fibres such as cotton or polyester are over 1000 times greater. The force with which these elastomers snap back after stretching to their original length and the elastic recovery are key issues. Polyurethanes do not recover quite as well from stretch as rubber fibres, giving a residual extension, or 'set'. However, if left for a period of time in the relaxed state after stretching, the 'set' tends to diminish.

Care has to be taken in measuring the stress–strain properties of polyurethane fibres because results can vary, depending on the degree of stretch imparted, the duration of time over which the measurements are taken and the number of times the stretching force is applied. Additionally the time allowed for the full recovery from a stretch, to determine the residual 'set', has an influence on the results obtained. Stress–strain diagrams show a marked hysteresis (see Figure 5.16), indicating that the force of retraction of the fibres is less than the force required to extend them.

This behaviour can be interpreted in terms of structural changes taking place during initial stretching, which do not immediately recover during relaxation. If the fibre is stretched again, the force required is much less than for the first stretch and is similar to the relaxation force from the first stretch. This characteristic is known as 'stress-softening' and is thought to be due to dislocation of the chains. Usually after five cycles of stretching and relaxation the hysteresis curves become consistent.

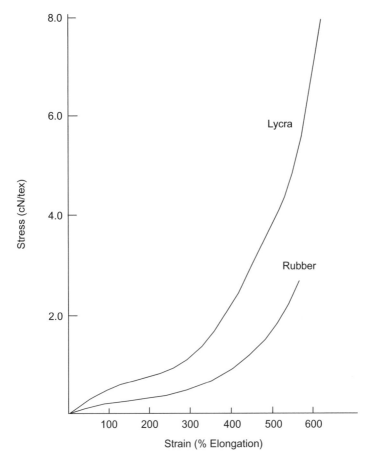

Figure 5.15   Stress–strain curves of Lycra and rubber fibres.

The properties of elastomeric fibres are shown in Table 5.9. In general they are weak fibres, but stronger than rubber and they can be produced as very fine filaments, which creates the sheer handle quality required for ladies' hosiery. The fibres can be used bare, for example in combination with yarns of other fibre types in some knitted fabrics, where the elastane gives stretch characteristics to the fabric. The elastane fibres themselves can be covered, either single-covered where the yarn of another fibre type is wrapped round them, or double-covered where a second layer of the other fibre type is wrapped around the first layer, but in the opposite direction. On stretching, the covering yarns will also be extended, but to a more limited extent than the elastane itself, so covered elastane yarns are used in woven fabrics for under-garments where a controlled degree of stretch is required. Another variant is core-spun yarn, in which a

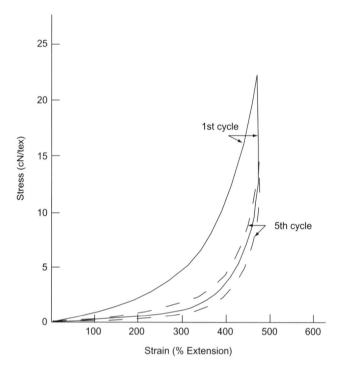

Figure 5.16 Hysteresis curves of an elastane fibre after 1st and 5th extension.

staple fibre of a 'normal' fibre type such as cotton or polyester, is spun around the slightly stretched elastane fibre.

The main uses of elastane fibres are in knitted or woven fabrics where stretch is required. Only a small amount of elastane is required in the overall fabric construction (around 2–4% by weight), but the stretch characteristics induced add much benefit to the comfort and fit of garments made from them. The types of garments that incorporate elastane fibres are extensive, from intimate apparel, through hosiery and socks, to sportswear. Garments for casual wear, such as denim jeans, shirts and tops, often contain small amounts of elastane as well.

5.6.3 Elastodiene Fibres

Elastodienes are composed of natural or synthetic polyisoprene, so as a group include rubber fibres, which are composed of *cis*polyisoprene:

Table 5.9 Properties of elastane fibres.

Fibre length	Continuous filament.
Fineness	2–600 tex. A very wide range, to suit applications in different types of garments.
Specific gravity	1.21.
Tenacity	5–14 cN tex^{-1}, based on the unstretched fibre dimensions. When the fibres are stretched they become thinner, so these values increase by a factor of about 7.
Elongation at break	400–650%.
Elastic recovery	Near perfect up to 500% extension. Some permanent extension after the first few extensions, but not normally progressive.
Resilience	Low, but abrasion resistance is good.
Moisture regain	1.3%.
Reaction to heat	Elastanes are thermoplastic and melt at around 250 °C. They become sticky at about 150 °C.
Reaction to chemicals	Good resistance generally to chemicals, but elastanes can be degraded by repeated exposure to chlorine. The polyether types have better resistance to chlorine and are more suitable for swimwear, for example.
Stability to sunlight	Generally excellent, though they discolour on prolonged exposure.

Whilst the molar mass of raw rubber is in range 100 000 to 500 000, processing causes degradation and the molar mass of the fibre form is much less. The *trans* configuration of polyisoprene enables the chain molecules to align more closely, so gutta percha and balata, which have this structure, are much less elastic than rubber. The chains of the *cis* isomer are able to coil up in their relaxed state, but uncoil under tension. If stretched too much, however, molecular slippage occurs and a degree of permanent extension is induced. Since rubber is thermoplastic its elasticity increases with temperature but the potential for molecule slippage to occur is greater. The class of elastodienes also includes synthetic rubbers such as styrene–butadiene rubber, though these have no applications in textiles.

To limit the extent to which molecular slippage can occur, rubber is vulcanised by treating it with sulfur, though other vulcanising agents, such as peroxides can be used. The sulfur reacts at the carbon atom next to the carbon atom of the double bond of the isoprene units, to form cross links between adjacent polymer chains. The cross links may involve more than one atom of sulfur, the number of sulfur atoms involved depending on the vulcanising agents and the conditions used.

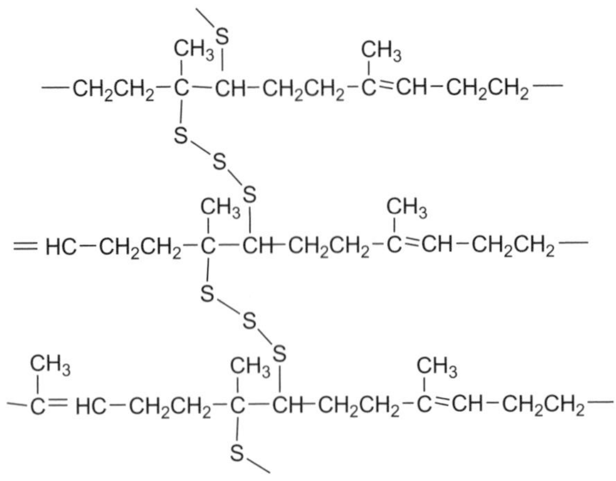

The degree of vulcanisation influences the degree of stretch and the initial modulus of the fibres, so rubbers over a wide range of these properties can be produced. The introduction of the elastanes and more recently the elastomultiester and elastolefins, all of which have better performance characteristics for introducing elasticity into fabrics, has largely replaced the use of rubber fibres for this purpose.

5.6.4 Elastomultiester Fibres

These fibres are a relatively recent innovation and the name 'elasto-multiester' was formally adopted for use in garment labelling by the European Commission in 2006 at the request of the fibre manufacturer Invista. In the USA, the FTC has approved the generic name Elasterell-p (as a sub-class of polyester) for the elastomultiester fibre T-400® produced by Invista. Like elastane, elastomultiester is added in small quantities to woven and knitted fabrics to improve the stretch characteristics and it is increasingly being used in garment manufacture. The T-400® fibre, which has a mushroom-shaped cross section, is promoted as having stretch and recovery properties in between those of textured polyester and elastane fibres, but having better resilience over time and resistance to chlorine bleach than elastane. Typically these fibres can be stretched to one and a half times their original length, compared to elastane fibres which are capable of stretches to five times their original length.

The T-400® elastomultiester fibre is a side-by-side bicomponent fibre (see Table 5.7), in which the two components are 3GT polyester and a modified polyester. On heating, each of the polyester variants shrinks to a different degree, giving the yarn a smooth helical crimp. As the yarn is made of thermoplastic polymers and the crimp is induced by heat rather than by mechanical means, the stretch and recovery of T-400® is very durable. The hysteresis curve measured in the relaxation mode is closer to that measured in the stretching mode than with elastane fibres. These fibres are finding applications in denims, workwear, shirts, socks and other lightweight knitted goods.

5.6.5 Elastolefin Fibres

Elastolefins are a sub-class of olefin fibres (named 'Lastol' by the FTC in the USA in 2003) based on metallocene chemistry. They are comprised substantially (>95% by weight) of ethene and at least one other type of olefin unit. The fibres undergo high-intensity electron-beam radiation to partially cross-link the polymer chains, thereby imparting a constrained geometry to the molecular architecture of the polymer, so they have a low but significant crystallinity. These fibres have the capacity to stretch by one and a half times their original length and are claimed to have high heat resistance and excellent resistance to chemicals, especially chlorine. The leading example of this fibre class is Dow XLA™, manufactured by the Dow Chemical Company. The fibres can be core spun, with a sheath of cotton fibres for example, to give the look and handle of cotton. Apart from the high heat resistance and resistance to chemicals, elastolefins seem to possess similar attributes to the elastomultiester type fibres.

SUGGESTED FURTHER READING

1. J. E. MacIntyre, *Synthetic Fibres: Nylon, Polyester, Acrylic, Polyolefin*, Woodhead Publishing Ltd., Cambridge, UK, 2005.
2. B. L. Deopura, *Polyesters and Polyamides*, ed., B. L. Deopura, R. Alagirusamy, M. Joshi and B. Gupta, Woodhead Publishing Ltd., Cambridge, UK, 2008.
3. S. C. O. Ugbolue, *Polyolefin Fibres: Industrial and Medical Applications*, ed., S. C. O. Ugbolue, Woodhead Publishing Ltd., Cambridge, UK, 2009.
4. B. G. Frushour and R. S. Knorr, *Handbook of Fiber Chemistry*, ed., M. Lewin, Taylor & Francis Group, Boca Raton, USA, 3rd edn, 2007, ch. 12.

CHAPTER 6

High-performance Fibres

6.1 INTRODUCTION

High-performance fibres are so named because they possess at least one major property that is markedly superior to that of the more conventional fibres already discussed in this book. Most high-performance fibres are distinguished by their high mechanical performance. Many of them are also notable for their thermal resistance; they do not catch fire, but instead form layers of carbonaceous char, which can still provide some heat protection. Thus, fabrics constructed from them can be used where protection against fire is paramount. The use of the limiting oxygen index (LOI) has already been discussed in Section 1.6.2, and Table 6.1 compares the limiting oxygen indices of high-performance textiles with those of conventional textiles. In addition, many high-performance fibres exhibit resistance to hazardous chemicals.

Fibres with improved properties have been actively sought over the last century or so, and those discussed in Chapter 5 are testimony to the successes achieved in the middle decades of the twentieth century. Since that time, much progress has been made in the quest to find fibres for specialist industrial, engineering and biomedical applications, although many of these fibres come with a much higher price tag. As the fibres have low extensibility, they are generally unsuitable for use in clothing. A major exception is specialist protective clothing, such as is worn by the military and emergency services personnel.

Among the first high-performance fibres were Nylon 6T fibres, which appeared in the 1960s. The polymer was produced by the condensation of 1,6-diaminohexane with terephthalic acid. Nylon 6T was, therefore, a

The Chemistry of Textile Fibres
By Robert R Mather and Roger H Wardman
© Robert R Mather and Roger H Wardman 2011
Published by the Royal Society of Chemistry, www.rsc.org

Table 6.1 Limiting oxygen index (LOI) values of fibres.

Fibre	LOI (%)
Acrylic	18.2
Cotton	18.4
Polypropylene	18.6
Viscose	18.9
Polyamide	20–22
PET	20–21
Wool	25
para-Aramid (*e.g.* Kevlar)	29
*meta-Aramid (*e.g.* Nomex)	29–30
Aromatic polyester	30
*Polybenzimidazole (PBI)	41
*Polyphenylene-2,6-benzobisoxazole (PBO)	68

*flame-resistant fibres.

polyamide, which was strengthened by virtue of the aromatic rings in the polymer chains, with extra reinforcement being derived from interchain hydrogen bonding. Aramid fibres were first synthesised not long afterwards, however, and they quickly superseded Nylon 6T fibres. Indeed, aramid fibres still enjoy extensive use. Many other high-performance fibres are now also commercially available, and it would be over-ambitious to discuss them all in this chapter. The range of fibres included in this chapter is, therefore, necessarily selective.

Fibres with high mechanical performance can be broadly divided into three groups: organic polymer fibres, carbon fibres and inorganic fibres. At the molecular level, they differ in the dimensionality of their bonding, but they are similar in that there is strong bonding in the direction of the fibre axis. In high-performance polymer fibres, as in nearly all polymer fibres, the chains are virtually one-dimensional, being highly oriented along the fibre axis. Indeed, the high orientation of the chains in gel-spun PE fibres has already been noted (Section 5.5.5), and these fibres certainly merit the label of 'high performance'. In other high-performance polymer fibres, the one-dimensional character of the chains is still very clear, even though the chains contain bulkier aromatic rings and side groups.

Carbon fibres traditionally consist of large sheets, very similar to those in graphite. Suitable high temperature processing under tension gives rise to strong fibres, in which the sheets are more oriented in the direction of the fibre axis. Another type of carbon fibre, currently being developed, is based on carbon nanotubes.

Three-dimensional networks of polymers, so-called 'thermoset resins', can be produced in fibre form. The fibres are not as strong as many other high-performance fibres, but possess high thermal resistance. Three-dimensional inorganic networks, such as those constituting ceramic and glass fibres, also possess high thermal resistance.

In this chapter, the three groups of high-performance fibres are treated in turn. Amongst the polymeric fibres considered are aramid fibres, which are aromatic polyamides, and thermotropic liquid crystalline polymers, which are for the most part aromatic polyesters. Also included are polybenzimidazole fibres and fluoropolymer fibres. Carbon fibres are then discussed, and finally thermoset and ceramic fibres.

6.2 ARAMID FIBRES

6.2.1 Introduction

The term, aramid, denotes aromatic polyamide. In commercial terms, aramid chains must possess amide groups, of which at least 85% are joined directly to two aromatic rings. Fibres containing certain polyamide–imide structures are also often classed as aramid fibres. However, these fibres are less prominent in commercial terms and are not discussed further in this chapter.

The first commercial aramid fibres appeared in the 1960s. Their polymer chains contained primarily *m*-disubstituted benzene rings, and they were notable for their thermal stability and resistance to combustion. However, only a few years later, aramid fibres with chains containing *p*-disubstituted benzene rings appeared. In addition to good thermal stability, these fibres also possess outstanding mechanical properties.

6.2.2 Production of Aramid Fibres

6.2.2.1 m-Aramid Fibres. *m*-Aramid polymers are produced from the polymerisation of 1,3-diaminobenzene (*m*-phenylene diamine) with isophthaloyl chloride at low temperature:

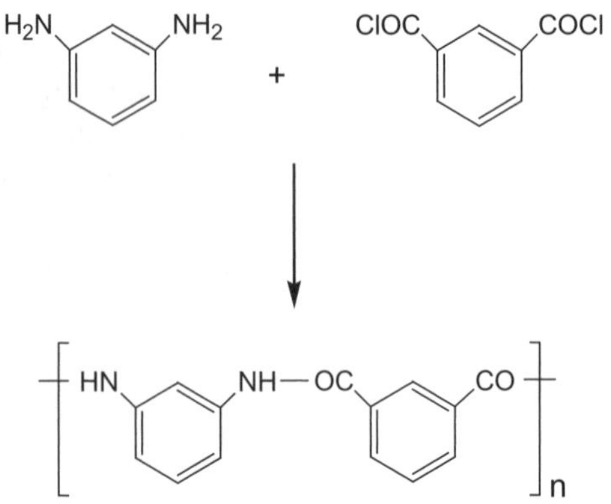

The polymer produced is poly(*m*-phenylene isophthalamide), MPIA. In the process devised by DuPont to produce the MPIA (Nomex®), the preferred solvent for the polymerisation is dimethyl acetamide (or sometimes dimethyl formamide). A solution of MPIA in dimethyl acetamide containing some lithium chloride or calcium chloride and a very small quantity of water is then dry spun. The filaments are washed in water to extract the inorganic salt, and drawn at 90 °C to a draw ratio of 4–5.

Another fibre based on *m*-aramid is Conex®, produced by Teijin. The polymerisation process consists of two stages. In the first stage, the two comonomers react with each other in tetrahydrofuran, and a slurry of an oligomer is formed. After the slurry has been contacted with sodium carbonate, polymer is formed, and is then dissolved in hot *N*-methyl pyrrolidone. Filaments are wet spun or dry-jet wet spun from the solution, and the *N*-methyl pyrrolidone is extracted with dichloromethane (methylene chloride). After being washed, the filaments are then drawn in boiling water to a draw ratio of *ca.* 3, dried, drawn further to a draw ratio of *ca.* 1.4 and heat-set.

6.2.2.2 p-Aramid Fibres. Although *m*-aramid fibres possess good thermal stability, their mechanical properties are quite poor, principally because their polymer chains cannot closely pack together. On the other hand, closer alignment of the chains constituting *p*-aramid fibres is much better. By the same token, however, *p*-aramids are more difficult to dissolve, for the purpose of fibre processing.

An early *p*-aramid fibre from DuPont (1970) was poly(*p*-benzamide):

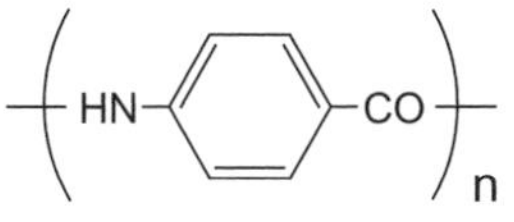

The fibre was marketed as 'Fibre B'. The polymer can be produced by the polymerisation of either *p*-thioaminobenzoyl chloride or of *p*-aminobenzoyl chloride:

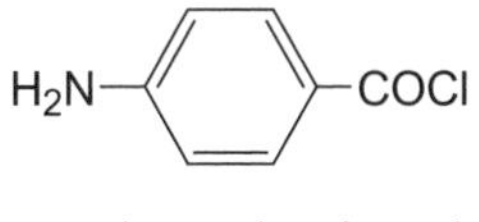

The solvent used for the polymerisation is dimethyl acetamide, or sometimes tetramethyl urea. In both cases, lithium chloride is present to increase the solubility of the polymer. The lithium cation is considered to associate with dimethyl acetamide.

Under the correct conditions, the polymer solution formed can be
solution spun to produce filaments. Unlike corresponding aliphatic polyamides, such as Nylon 6, poly(*p*-benzamide) decomposes before and
during melting, and so melt spinning is precluded. However, provided the
molar mass and concentration of the polymer, and also the concentration
of lithium chloride, are sufficiently high, the solution adopts a liquid
crystalline state rather than the normal isotropic state. The polymer
chains consist essentially of rods in liquid crystalline arrays, instead of
the usual random coil structures, as illustrated schematically in
Figure 6.1. When the liquid crystalline solution is pumped through the
holes in a spinneret, the arrays readily orient in the direction of flow of

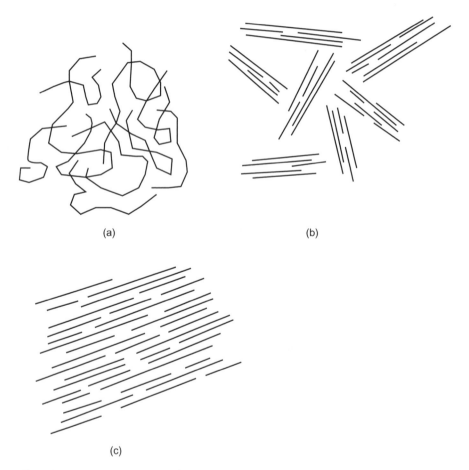

Figure 6.1   Schematic representation of polymer states in solution: (a) random coils,
(b) rods in liquid crystalline arrays, (c) rods in orientated liquid crystalline
arrays.

the solution, as also shown in Figure 6.1. The distance between adjacent arrays, however, is still large enough to prevent severe aggregation, and hence precipitation of the polymer. An important consequence of this orientation is that the viscosity of the solution, as it passes through the spinneret holes, is much lower than it would be in an isotropic state. Thus, the spinning process is considerably facilitated.

Filaments can be formed by either wet or dry spinning, and they are then washed and dried. As the polymer arrays are all oriented in the direction of the filaments, drawing is not carried out. Heat treatment in an atmosphere of nitrogen improves the filaments' tensile properties.

In 1973, DuPont replaced Fibre B with poly(*p*-phenylene tereph-thalamide) (PPTA), under the name Kevlar®. Not only did PPTA fibres exhibit outstanding mechanical performance, but they also addressed a key problem in the production of poly(*p*-benzamide) – the high cost of the monomer. PPTA is produced from the condensation of 1,4-diami-nobenzene and terephthaloyl chloride:

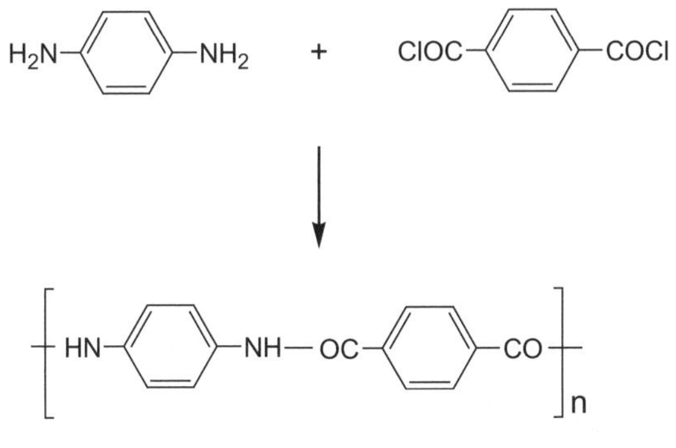

The polymerisation is normally carried out at 10–20 °C. The solvent used in the process is a mixture of hexamethylphosphoramide (HMPA) and *N*-methyl pyrrolidone, though HMPA has been found to be a potential carcinogen. The stoichiometries of the solvent and reactant mixtures are important influences on the molar mass of PPTA obtained. The highest molar mass is achieved where the volume ratio of HMPA to *N*-methyl pyrrolidone is *ca.* 2:1, and the optimum concentration for both reactants is *ca.* 0.25 mol l^{-1}. To counteract the problem of using HMPA as a solvent component, a competitive product, Twaron®, was produced by Akzo (now Teijin). The solvent was *N*-methyl pyrrolidone containing 10–20% (w/w) calcium chloride. Other suitable polymer-isation mixtures have also been identified.

The PPTA formed is washed and dried, and then added to concen-trated sulfuric acid, to make a 20% solution prior to extrusion at 80 °C.

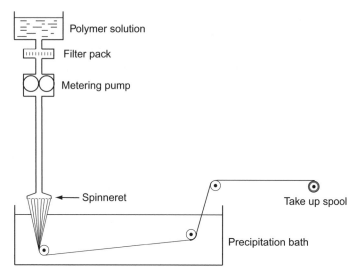

Figure 6.2 Schematic diagram of the dry-jet wet spinning process.

The solution is in a liquid crystalline state, and so fibre extrusion is considerably facilitated. The spinning process adopted, dry-jet wet spinning, is similar to the process used for producing gel-spun PE filaments (see Section 5.5.4), and is illustrated schematically in Figure 6.2. After egress from the spinneret holes, the filaments of dissolved PPTA pass through an air gap of 0.5–1.0 cm into water maintained at 0–5 °C in a coagulation bath. While travelling through the air gap, the hot filaments are still in a liquid crystalline state, and the stretching they undergo gives rise to a very high orientation of PPTA chains in the direction of the filament axis. The cold water in the coagulation bath removes the sulfuric acid from the PPTA filaments, which are then further washed in a separate bath. Heat treatment of the filaments for a few seconds under tension at a temperature of *ca.* 550 °C enhances the orientation of the polymer chains still further. The extent to which the filaments stretch under tension is thus considerably reduced.

6.2.3 Physical Structure

PPTA fibres are composed of highly oriented, extended polymer chains, and are almost completely crystalline. The chains form rigid sheets and are linked through hydrogen bonding. The fibres of the heat-treated *p*-aramid, Kevlar 49, contain highly ordered fibrillar columns,

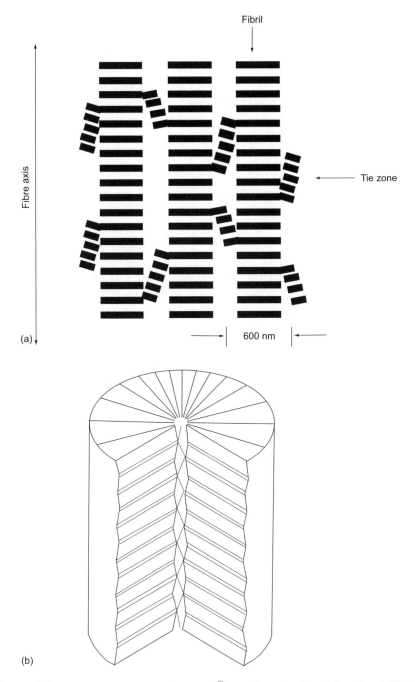

Figure 6.3 Schematic diagram of Kevlar® 49 fibre showing: (a) ordered fibrillar columns linked at various points by tie zones, (b) radially pleated structure.

ca. 600 nm wide and up to several cm in length. The columns are linked at various points by tie zones (see Figure 6.3a). When viewed under an optical microscope, the fibres of Kevlar 49 are observed to have a series of transverse bands. The spacing of the bands is 500–600 nm. To explain the observation, a radial pleated structure has been proposed, in which the pleated sheets are *ca.* 500 nm long (see Figure 6.3b). To form the pleats, the alternating bands in each sheet are arranged at approximately equal but opposite angles of *ca.* 170°.

6.2.4 Chemistry of Aramid Fibres

6.2.4.1 Introduction. Aramid fibres possess very high thermal stability. The glass transition temperature, T_g, of *m*-aramids is *ca.* 275 °C, and thermal degradation starts at *ca.* 375 °C. These high values are attributable to the rigid structure of the *m*-aramid chains compared with those of the synthetic fibres discussed in Chapter 5, and to the high level of interchain bonding. The extreme order of the polymer chains in *p*-aramid fibres confers even greater structural rigidity. T_g is raised to 340 °C and the polymer only begins to degrade at 550 °C.

Aramid fibres are also resistant to nearly all chemical treatments. Most organic liquids have little effect on them. Although resistant to acids and bases under most application conditions, aramid fibres are susceptible to hydrolytic attack by strong acids and bases at elevated temperatures. Aramid fibres are also prone to photodegradation.

6.2.4.2 Hydrolysis. As with polyamide fibres, the amide groups in aramid fibre chains can be hydrolysed. Chain scission occurs:

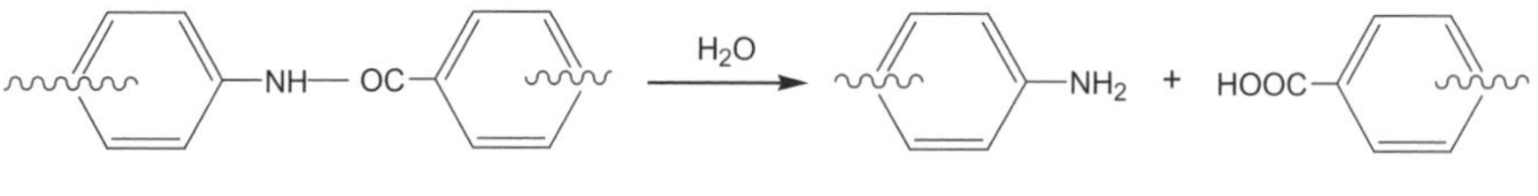

and the fibres are consequently weakened. Hydrolysis can be caused by strong acids during prolonged exposure at ambient temperatures, and to strong acids and bases at elevated temperatures.

6.2.4.3 Photodegradation. Photodegradation is often a serious problem in the applications of aramid fibres, unless means have been undertaken to stabilise them. The first step is photo-induced scission of the amide bonds. In the absence of oxygen, this step is followed by a number of reactions:

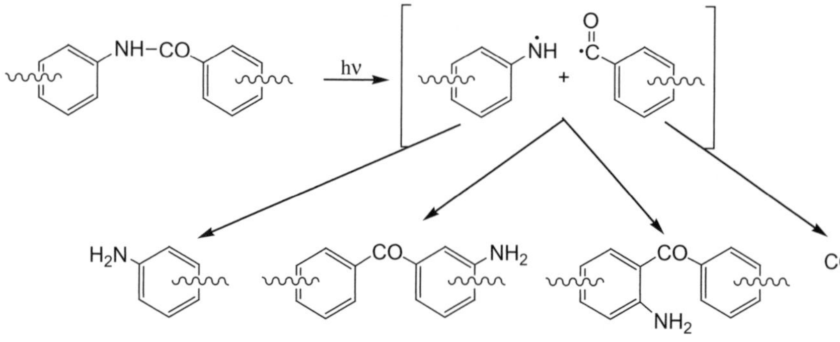

There is also scope for some crosslinking between chains. In air, the radicals formed as a result of chain scission react with oxygen to form peroxides, which then decompose to form carboxylic acid and possibly nitroso groups:

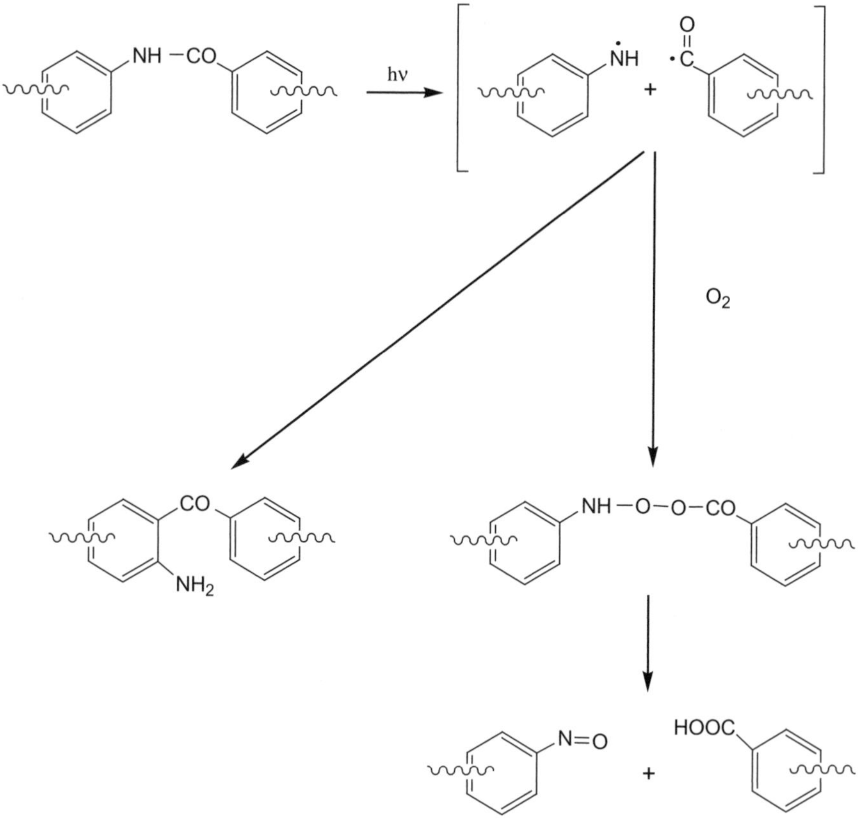

Despite their susceptibility to photodegradation, however, aramid fibres are to some extent self-screening. For example, filaments lying within a

Table 6.2 Some properties of aramid fibres.

	m-Aramid fibres	*p-Aramid fibres*
Specific gravity	1.38	1.44–1.47
Tenacity	40–50 cN tex^{-1}	190–240 cN tex^{-1}
Elongation at break	20%	1–4%
Moisture regain	5%	1–7.0%, depending on grade
Limiting oxygen index	28–30%	29%

multifilament yarn are highly protected, whilst those at the yarn surface are degraded. For *m*-aramid fibres, the incorporation of a photostabiliser into the wet spinning solution can confer resistance to photodegradation. Moreover, a well established means of protecting *p*-aramid rope yarn is to enclose it in a polyurethane or polyethylene sheath containing carbon black.

6.2.5 Fibre Properties

Some properties of aramid fibres are shown in Table 6.2.

Whilst *m*-aramid fibres possess very good thermal resistance, their mechanical properties are similar to those of PET fibres. By contrast, *p*-aramid fibres possess both good thermal and mechanical properties. Nevertheless, *m*-aramid fibres are widely produced. They are used in protective clothing against hostile environments and in industrial filters. Due to their high flame resistance, they are used extensively in furnishings as well. Fabrics blended from wool and *m*-aramid fibres are used as fire- and smoke-resistant materials – for example, for seating fabrics in trains and aircraft.

A variety of grades of *p*-aramid fibres are available. Some are used as cord in heavy duty tyres for aircraft and lorries. Other grades are used for composite applications in aircraft, ropes and cables. A particularly important application is in ballistic protection, such as in bullet proof vests, helmets and vehicle armour. Some protective gloves contain *p*-aramid fibres too.

6.3 AROMATIC POLYESTER FIBRES

6.3.1 Introduction

Aromatic polyesters consist of aromatic comonomeric units joined by ester groups along each polymeric chain. These polymers are, therefore, analogues of the polyesters discussed in Section 5.3. Numerous aromatic polyesters have been synthesised, but few have been processed commercially into fibre form. Unlike the polyesters already discussed, most aromatic polyesters are unsuitable for melt processing into fibres because of their tendency to degrade before melting. To reduce the melt temperature sufficiently, several modifications can be made to the polyester chains, in order to disturb the ordered structure of the polymer chains: the addition of bulkier pendant groups, non-linear monomers and flexible spacer units. These polymers can exist in a liquid crystalline state. In contrast to aramids, however, the liquid crystalline state exists within a particular temperature range, sandwiched between the solid state and the truly (isotropic) liquid state. These aromatic polyesters are, therefore, thermotropic liquid crystalline polymers (TLCPs).

TLCPs first aroused commercial interest in the 1970s. A whole host were synthesised in commercial and academic laboratories, and a number of them were commercialised. However, the price of these aromatic polyesters is high, and they therefore tend to be found only in niche market products. The most significant aromatic polyester fibre in commercial terms is Vectran®, developed originally by Hoechst Celanese and now marketed by Kuraray. The fibre is melt spun from the parent polymer, Vectra®. Others have included fibres extruded from copolymers of PET and *p*-hydroxybenzoic acid (HBA), such as Rodrun® developed by Unitika, and from Ekonol®, developed originally by Carborundum and marketed by Suitomo Chemical and Nippon Exlan.

6.3.2 Production of Aromatic Polyester Fibres

Vectra® polymer is synthesised by an ester exchange reaction between 6-acetoxy-2-naphthoic acid and *p*-acetoxybenzoic acid:

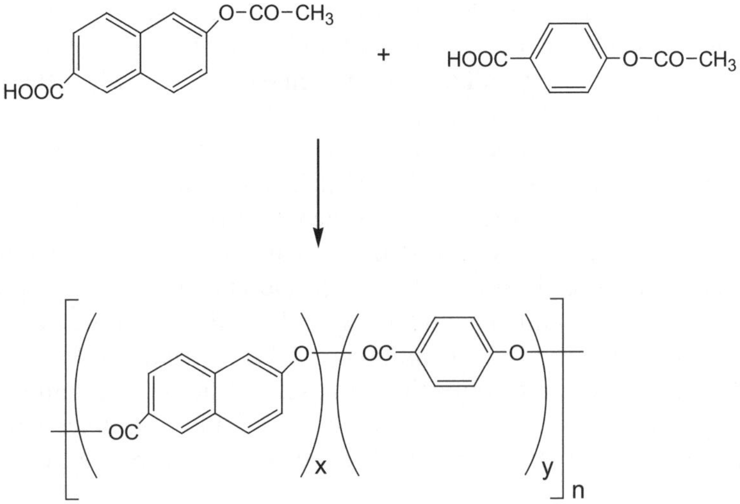

The two comonomers are the acetylated derivatives of 6-hydroxy-2-naphthoic acid (HNA) and HBA, respectively. The temperature required for the synthesis is above the melt temperature of the Vectra® product, and the high viscosity of the isotropic melt can restrict the polymerisation process. One approach to overcoming this problem is to prepare a low molar mass prepolymer instead and then to polymerise the prepolymer in the solid state. A polymerisation process based on non-aqueous dispersion of the polymer product has also been successfully developed.

HBA/PET polymers are synthesised in a corresponding manner:

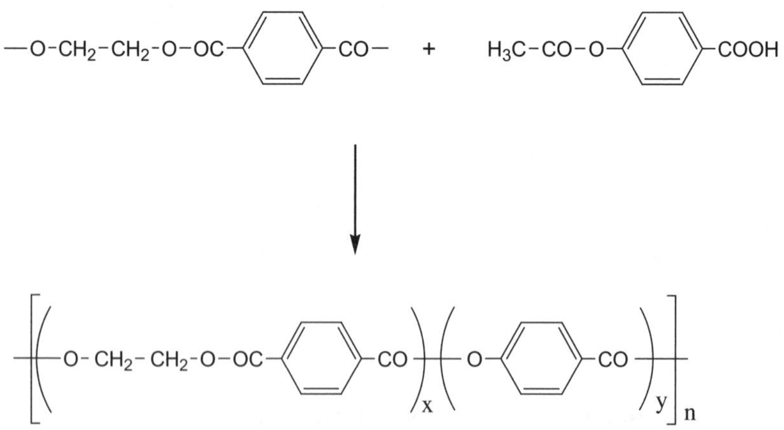

At 275 °C, *p*-acetoxybenzoic acid reacts both with itself and with PET to produce short polymer segments, terminated by carboxylic acid groups and acetoxy ($CH_3 - CO - O -$) groups. Heating under pressure, forces the reaction between the segments to yield the HBA/PET polymer. A thermotropic liquid crystalline state is achievable only when the molar fraction of HBA in the polymer exceeds 30%.

Ekonol® is synthesised from a variety of precursors that include *p*-acetoxybenzoic acid, terephthalic acid, isophthalic acid and 4,4′-diacetoxybiphenyl (see the reaction scheme on p. 213).

Aromatic polyesters are extruded in a manner similar to conventional PET. They are directly spun from the liquid crystalline state. Spinning temperatures are usually *ca.* 300 °C or slightly higher, though Ekonol® fibres are spun between 360 °C and 400 °C. The extruded fibres are often heat treated in an inert atmosphere for several hours at a temperature just below that at which the solid becomes liquid crystalline. This heat treatment considerably improves fibre mechanical properties, especially fibre tenacity.

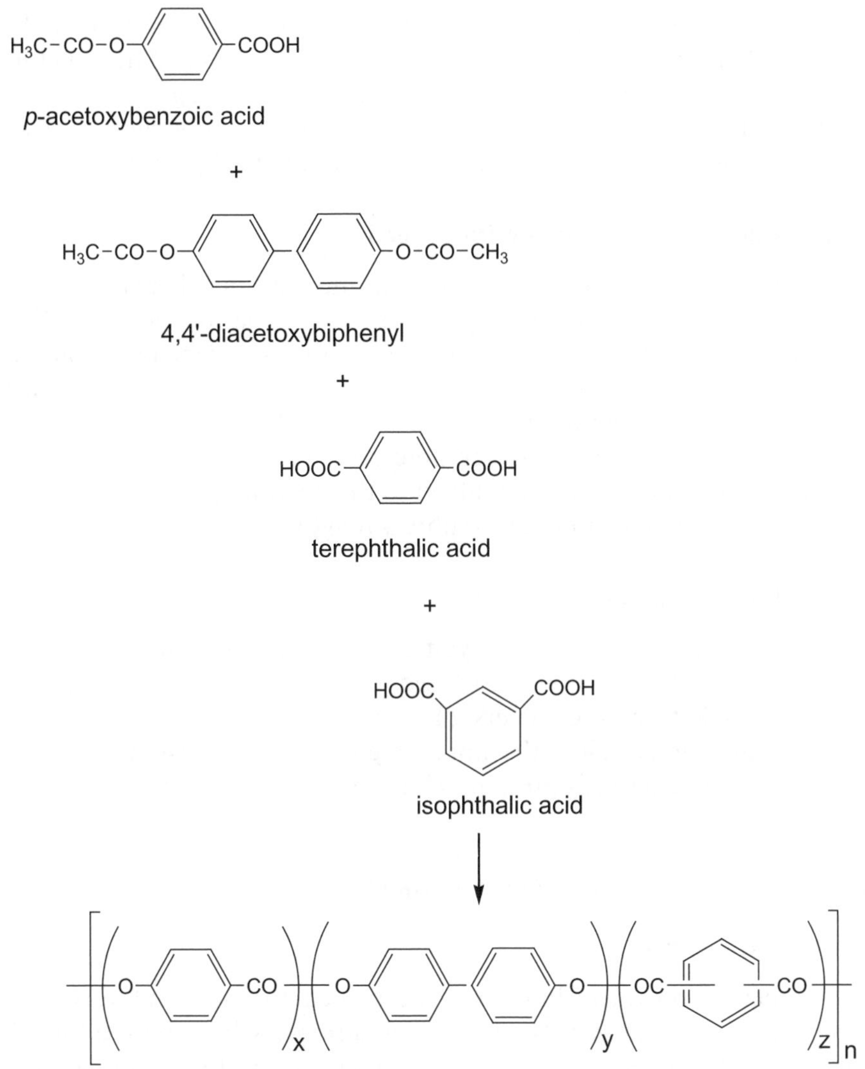

H₃C–CO–O–⬡–COOH

p-acetoxybenzoic acid

+

H₃C–CO–O–⬡–⬡–O–CO–CH₃

4,4'-diacetoxybiphenyl

+

HOOC–⬡–COOH

terephthalic acid

+

HOOC–⬡–COOH

isophthalic acid

↓

$$\left[\left(O\!-\!\bigcirc\!-\!CO \right)_x \left(O\!-\!\bigcirc\!-\!\bigcirc\!-\!O \right)_y \left(OC\!-\!\bigcirc\!-\!CO \right)_z \right]_n$$

6.3.3 Fibre Structure

Vectran® fibres have been developed with a number of HBA/HNA compositions, ranging from 25:75 to 75:25. The comonomers are randomly distributed along each polymer chain. In all these fibre grades, there is very high orientation of the chains in the direction of the fibre axis. It is observed too that Vectran® filaments are composed of fibrils of *ca.* 500 nm in width and that there are transverse banded structures, features similar to those in *p*-aramid filaments.

In HBA/HNA polyester chains, the comonomers are not randomly arranged. There are well ordered regions in each chain that are rich in polymerised HBA and disordered regions rich in PET. The HBA polymeric sections exist in ordered domains of size 30–40 µm. The structure of Ekonol® fibres is not well reported. It is clear, though, that the biphenyl units, which are absent in the other aromatic polyesters, have an important influence on fibre structure and mechanical properties.

6.3.4 Chemistry of Aromatic Polyester Fibres

Aromatic polyester fibres are resistant to most chemical environments. They are resistant to organic liquids and are stable to acids at <90% concentration and bases at <30% concentration depending on the duration and temperature of exposure. Their resistance to UV radiation is, however, quite poor and protection is required. The protection can be a jacket over aromatic polyester fibre rope or a protective film over fabric. Photostabilisers are not added to the polymer prior to extrusion, as they often significantly reduce fibre strength.

6.3.5 Fibre Properties

Some properties of aromatic polyester fibres are shown in Table 6.3.

Aromatic polyester fibres are used in yacht ropes, sailcloth, bow strings, bicycle frames, catheters and control cables in surgical devices. These applications reflect the fibres' high abrasion resistance, cut resistance and impact resistance, in addition to their high strength.

6.4 POLYBENZIMIDAZOLE FIBRES

6.4.1 Introduction

A number of polymers have been developed to withstand very high temperatures, and a variety of chemical strategies have been adopted. One of these is the incorporation of thermally unreactive aromatic rings in the polymer chains, and the success of this strategy has already been noted for aramids (see Section 6.2). The presence of resonance stabilised entities and of protective side groups is also useful for conferring thermal stability. Very high stability can be achieved if single bonds can be

Table 6.3 Some properties of aromatic polyester fibres.

Specific gravity	1.4
Tenacity	200–250 cN tex^{-1}
Elongation to break	2–4%
Moisture regain	<0.1%
Limiting oxygen index	30%

largely eliminated, as these bonds tend to be susceptible to thermal degradation, giving rise to chain scission. A ladder structure can also be useful, as illustrated by polyimidazopyrolone. This polymer only starts to decompose at *ca.* 600 °C.

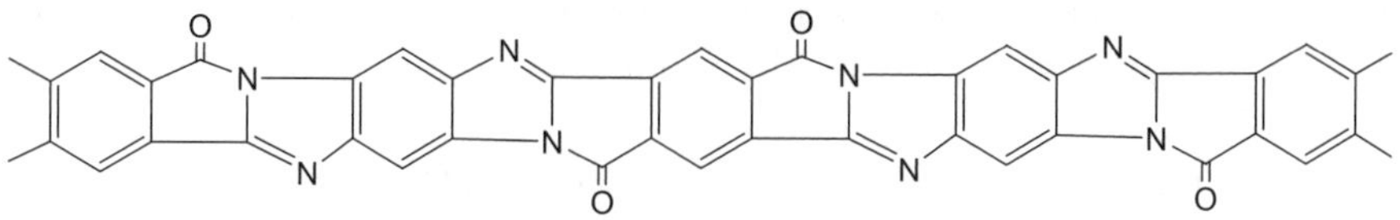

However, the structures devised to provide high thermal stability also render fibre processing more challenging. The polymers are degraded at temperatures below their melting points, so melt spinning is precluded. In addition, solvents may be difficult to find for solution spinning. Nevertheless, several types of thermally resistant fibres are available, and many have been very successful commercially. They do not catch fire readily and so are useful fibres for fire resistant clothing. Some of these fibres are listed in Table 6.4, together with the solvents used for solution spinning and values of maximum temperature for practical use. One of the most successful thermally resistant fibres is polybenzimidazole (PBI). Its limiting oxygen index is > 41. When exposed to a hot flame, it forms a char rather than catching fire. This fibre is now discussed, as illustrative of thermally resistant fibres.

6.4.2 Production of PBI Fibres

PBI is synthesised by reaction between diphenyl isophthalate and 3,3′,4,4′-tetraaminobiphenyl. Diphenyl isophthalate is prepared from phenol and isophthaloyl chloride. 3,3′,4,4′-tetraaminobiphenyl, however, is prepared by a long sequence of reactions starting from 2-nitrochlorobenzene:

Table 6.4 Thermally resistant fibres.

Polymer	Solution spinning solvent	Maximum temperature for practical use (°C)
Polybenzimidazole (PBI)	Dimethyl acetamide + lithium chloride	420–450
PBO	Polyphosphoric acid	450–500
p-aramid	Dimethyl acetamide + lithium chloride	*ca.* 400
PIPD	Polyphosphoric acid	*ca.* 450

PBO = Poly(*p*-phenylene-benzobisoxazole).
PIPD = Poly{2,6-diimidazo[4,5-b:4′,5′-e]pyridinylene-1,4-(2,5-dihydroxy)phenylene}.

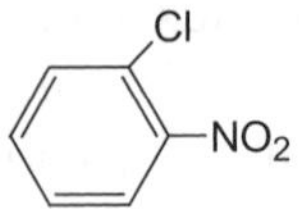

The product from this sequence contains many impurities. These impurities are removed by dissolving the product in boiling water, filtering and recrystallising.

The polymerisation process entails two stages. In the first stage, equimolar portions of the comonomers undergo a melt polymerisation reaction in an inert atmosphere. Water and phenol are evolved, and their evolution causes foaming of the prepolymer formed. After the foam is heated for 1 hr or so, it is allowed to cool and is ground into a fine powder. In the second stage, the prepolymer is heated under nitrogen for 2–3 hr at *ca.* 400 °C, and the polymerisation process is completed. The reaction scheme can be summarised as seen on p. 217.

To prepare the **PBI** for processing into fibres, a 20–25% solution of it is prepared in dimethyl acetamide containing lithium chloride. The polymer solution is dry spun into a column of hot nitrogen. Absence of oxygen is essential to avoid the formation of a gel through oxidative crosslinking of the polymer chains. The filaments are formed as the dimethyl acetamide is evaporated. They are then thoroughly washed, in order to remove lithium chloride and any remaining dimethyl acetamide, and dried. After washing and drying, the fibre is drawn in an atmosphere of nitrogen at *ca.* 400 °C. The fibre is still prone to shrinkage at elevated temperatures, however. To overcome shrinkage, the fibre is treated with dilute sulfuric acid, and heated. Sulfonation occurs to yield:

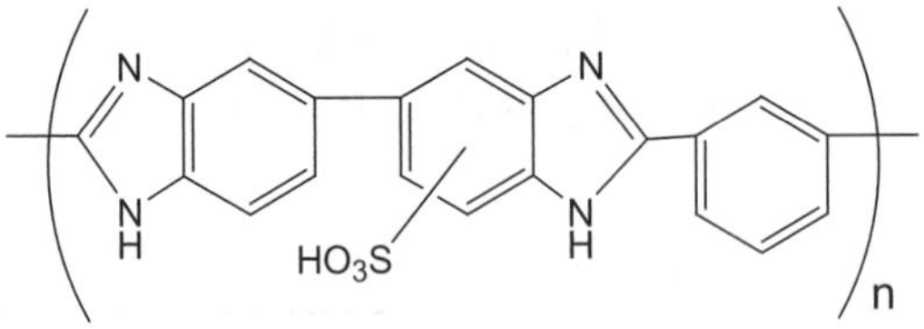

and the fibre is stabilised.

6.4.3   Fibre Properties

Some fibre properties are given in Table 6.5. **PBI** fibres possess mechanical properties no better than those of many conventional fibres, yet the thermal resistance of **PBI** fibres is outstanding. Applications include

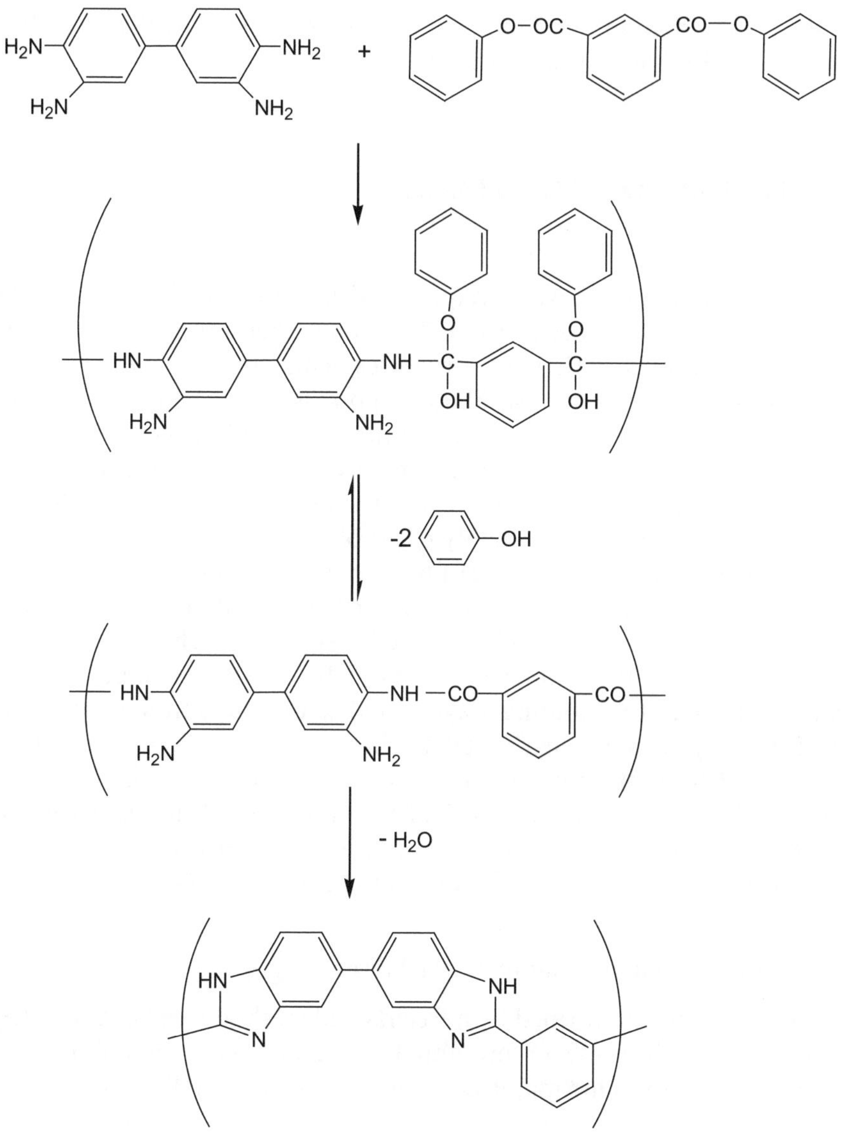

Table 6.5 Some properties of polybenzimidazole fibres.

Specific gravity	1.43
Tenacity	240–270 cN tex^{-1}
Elongation to break	28–30%
Moisture regain	15%
Limiting oxygen index	>41%

uniforms for firefighters, fireblocking fabrics in aircraft, flue gas filters and fabrics for thermal insulation.

6.5 FLUOROPOLYMER FIBRES

6.5.1 Introduction

Some applications of textiles require resistance to hostile chemical environments as a key specification. The fibres, therefore, have to consist of polymer chains with inert structures. Due to their inert chemical structures, the fibres are also likely to possess good thermal resistance, and in particular a good flame resistance. A hostile chemical environment may also be a high temperature environment (as in a factory or warehouse fire), however, so a combination of chemical and thermal resistance in a fibre can be highly desirable, and even essential.

One chemically inert group of fibres consists of fluoropolymer fibres. Although they are expensive, they nevertheless find commercial application because of their stability in the most extreme chemical environments. Several of these fibres are listed in Table 6.6. It can be noted, in particular, that the limiting oxygen index of polytetrafluoroethylene (PTFE) fibres is very nearly 100%; they are almost completely flame resistant. The limiting oxygen indices of the other fibres in Table 6.6, whilst much lower than that of PTFE fibres, are still markedly higher than those listed in Table 6.1. These good properties arise from the chemical inertness of the C–F bonds along the polymer chains.

6.5.2 Production of Fluoropolymer Fibres

Fluoropolymers are formed by a free radical method (see Section 5.4.3). Conversion of the polymer into fibre form can be achieved in most cases by melt extrusion. One major exception, however, is PTFE. Its melting

Table 6.6   Some properties of fluoropolymer fibres.

Fluoropolymer	Structure	Melt temperature (°C)	Limiting oxygen index (%)	Tenacity (cN tex^{-1})	Elongation to break (%)
PTFE	$-(CF_2-CF_2)_n-$	327–340	95	8–14	20–30
PVF	$-(CH_2-CHF)_n-$	ca. 170		20–40	15–30
PVDF	$-(CH_2-CF_2)_n-$	160–170	44	40–45	10–40
ETFE	$-(CHF-CF_2)_n-$	270–275		ca. 30	25
ECTFE	$-(CFCl-CF_2)_n-$	240–245	48	ca. 30	25

point is high: 320–340 °C. The PTFE melt is too viscous for melt spinning, and it also starts to degrade. In addition, PTFE is highly insoluble. Fibre processing by standard methods is, therefore, precluded. In a process developed by DuPont for producing filaments of Teflon®, finely divided particles of PTFE are blended in a solubilised form of cellulose. This mixture is wet spun into a bath containing acid, whereupon the insoluble form of cellulose is regenerated. A heat treatment then strips the cellulose from the PTFE, and the PTFE particles are at the same time fused into continuous filaments. The residue of burnt-off cellulose, however, gives the filaments a dark brown colour, which can, if desired, be removed by a bleaching process.

In the paste-extrusion process used by companies such as W. L. Gore, fine PTFE powder is mixed with a hydrocarbon lubricant and compressed into preforms. Each preform is extruded into a film or rod, and the lubricant is removed. The film or rod is slit, heated sufficiently to fuse the PTFE particles and stretched, to form high tenacity filaments.

A split–peel process has been developed by Lenzing. A cylindrical PTFE block is mounted on a high-precision lathe, from which a continuous film of PTFE is turned. The film is split, and the resulting filaments are stretched and tempered.

6.5.3 Fibre Properties

In addition to the chemical and thermal properties already discussed, values for fibre tenacity and elongation to break are also listed in Table 6.6. It will be noted that despite their excellent thermal properties, the tenacity of PTFE fibres is much the lowest listed. Thus, PTFE fibres are generally applied in such products as filters, braiding and gaskets, where high strength is not routinely required. They are invaluable in many chemical engineering applications, where resistance at high temperatures to corrosive chemicals is essential. Exposure may be required for several months.

The other fibres listed in Table 6.6, especially PVF and PVDF, are more flammable than PTFE fibres, though still more flame resistant than other types of commercial fibre. They are also stronger than PTFE fibres, so they are normally applied where a combination of strength and chemical resistance is required. For example, PVF and PVDF fibres are used in a number of filtration products. ETFE and ECTFE fibres are used in filtration products, as well as gaskets and conveyor belts.

6.6 CARBON FIBRES

6.6.1 Introduction

Carbon fibres essentially comprise two-dimensional sheets, which are almost identical to the graphene sheets constituting graphite (see Figure 6.4). As described below, these sheets are oriented during processing of carbon fibres into fibrous forms.

Although fibres carbonised from cotton yarns were exploited by both Sir Joseph Swan and Thomas Edison as electrical lamp filaments in the late 1870s, it was not until *ca.* 1960 that there was serious commercial interest in producing carbon fibres of high strength. Initially, these fibres were produced from viscose precursor fibres and isotropic pitch. Pitch is a by-product from refining petroleum and coal coking. Nowadays viscose is rarely used as a precursor, however, because of technical difficulties with the graphitisation step of the process. In addition, the oxidation stage in the conversion of isotropic pitch is too long to render

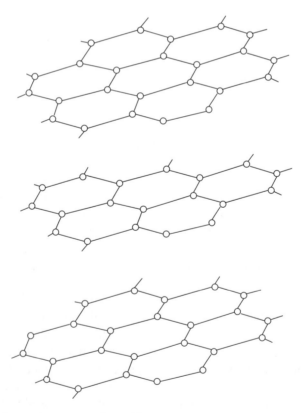

Figure 6.4   Layered structure of graphite.

the process commercially attractive. Instead, high strength carbon fibres are produced very largely from polyacrylonitrile (PAN) fibres and from mesophase pitch (MP) precursors.

6.6.2 Carbon Fibres from PAN Fibres

Conversion of PAN fibres to carbon fibres involves three principal stages: oxidative stabilisation, carbonisation and graphitisation. The first stage, oxidative stabilisation, involves the production of a ladder polymer by means of the cyclisation of the pendant nitrile groups in the PAN chains, as shown in Figure 6.5. The fibres are held under tension at 200–250 °C. The ladder polymer eventually formed contains rows of ketone groups, and is oriented in the direction of the fibre axis. In the second stage, carbonisation, the black oxidised fibre is slowly heated under tension to 1000–1500 °C in an inert atmosphere. Crosslinking now occurs, an example of which is shown in Figure 6.5. A number of gases, such as ammonia, steam, nitrogen and hydrogen cyanide, are evolved in this stage, depending on temperature and duration. The fibre now contains mostly carbon atoms and a few nitrogen atoms in aromatic ring structures. The fibre is still not graphitic, however, in that the aromatic sheets are not stacked in a regular manner one on top of another. Moreover, mechanical performance is still insufficient. In the third stage, graphitisation, the carbonised fibre is heated under tension to temperatures >2000 °C. In this stage, the structure becomes more ordered approaching a true graphitic form, and the mechanical properties of the fibre are considerably enhanced. The loss of fibre mass in the overall process is *ca.* 50%. Further improvement in mechanical performance can be achieved by drawing at >2000 °C, often to *ca.* 30% stretch.

6.6.3 Carbon Fibres from Mesophase Pitch

Pitch consists of aromatic ring structures of low molar mass. Some examples are shown on p. 222.

Heating pitch over an extended period at 400–450 °C converts it to 'mesophase pitch', as it is called. In this process, the molecules are polymerised into long sheet-like structures, which contain a high proportion of liquid crystalline material.

The mesophase pitch is melt extruded. The aromatic sheets become highly oriented in the direction of the axis of the extruded filaments. The filaments are then oxidised in air for *ca.* 2 hr at *ca.* 300 °C, and extensive crosslinking occurs between the aromatic structures. Finally, the filaments are first carbonised and then graphitised in processes similar to those used on oxidised PAN fibre.

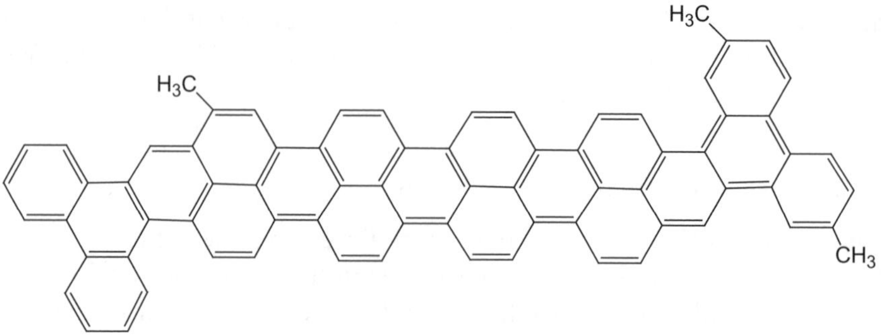

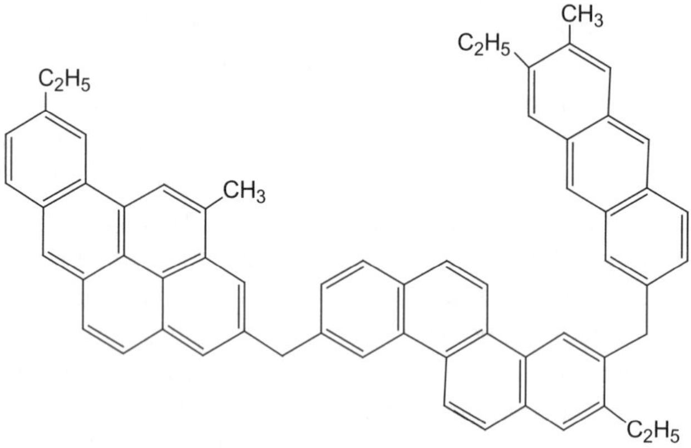

6.6.4   Fibre Structure

The dimensions of the sheets formed during the graphitisation stage vary widely between different grades of carbon fibre, and depend on the heat treatment undergone during processing. The sheets are generally in the form of ribbons, whose width lies between 5 and 100 nm. The ribbons are often undulating rather than flat.

The organisation of the sheets within a PAN-based carbon fibre is often complex. In many types of these fibres, the sheets near the fibre surface are all oriented parallel to the direction of the fibre axis. In the core of the fibre, the sheets are much more wrinkled and even folded in on themselves. This arrangement gives rise too to an extensive network of void space within each fibre.

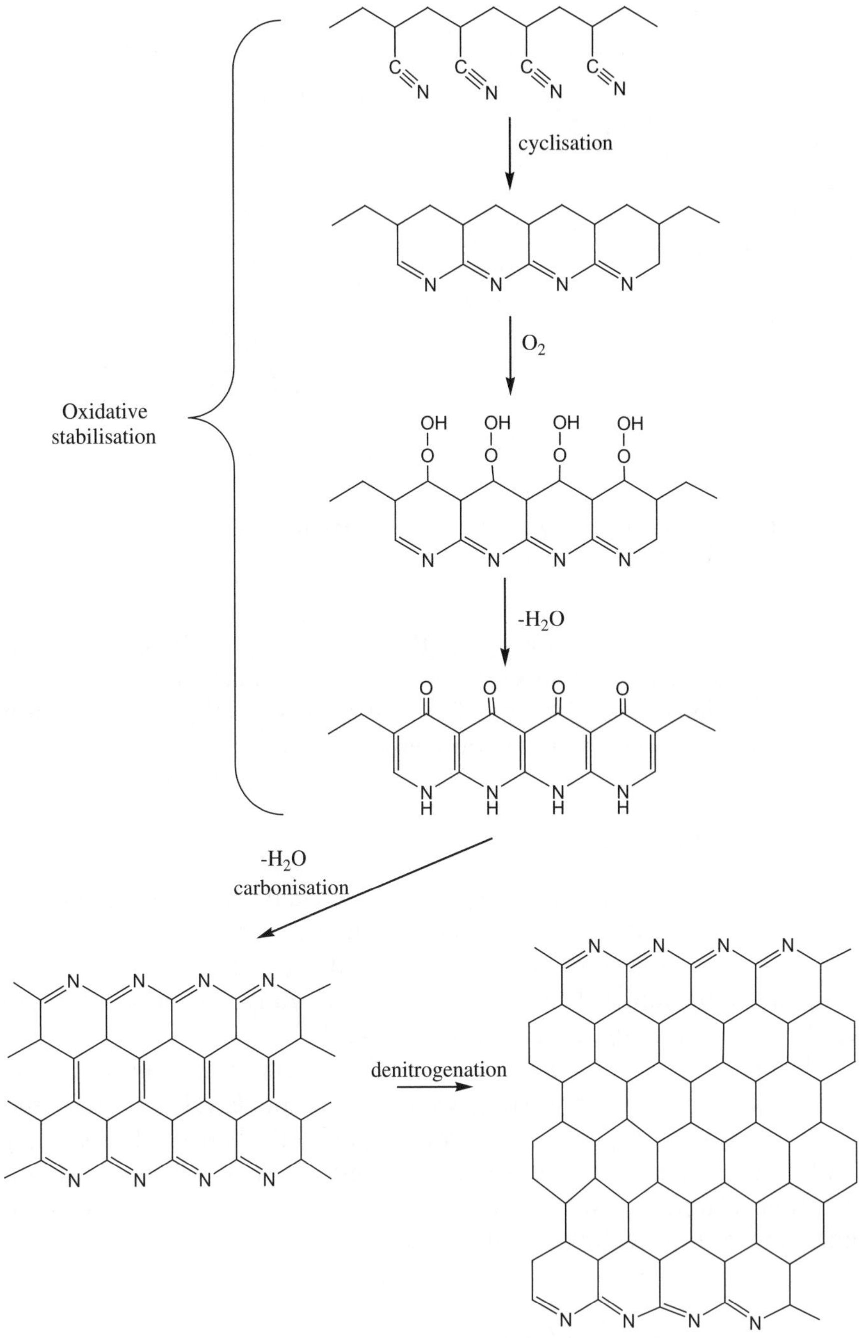

Figure 6.5 Chemical changes occurring during oxidative stabilisation, crosslinking and graphitisation of PAN fibres.

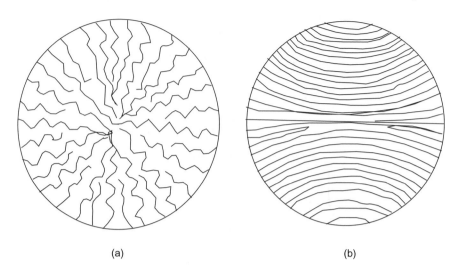

(a)   (b)

Figure 6.6 Organisation of graphitic sheets in MP-based carbon fibres: (a) radial folded, (b) flat layer.

The organisation of the graphitic sheets in MP-based fibres is different, although each sheet is still oriented parallel to the fibre axis. Figure 6.6 illustrates schematically in fibre cross-section two common arrangements of the sheets in commercial fibres.

6.6.5 Fibre Properties

A wide variety of carbon fibres are commercially available, depending on the precursor and processing conditions, so it is not surprising that the properties of carbon fibres are also quite varied. Specific gravity is generally in the range of 1.7–2.2. Fibre tenacity ranges from *ca.* 50 to 130 cN tex^{-1}, or even higher. Extension to break is 0.3–2.0%.

The major application of carbon fibres is as reinforcement in composite materials. PAN fibres are used in composite components for aircraft, and in sports equipment, including racquets, golf club shafts and skis. They are also used in parts for industrial machinery. They even find application in surgical implants, such as replacement hips and heart valves. Pitch fibres are used to reinforce cement and in parts of off-shore oil rigs.

6.6.6 Carbon Nanotube Fibres

The discovery of carbon nanotubes in 1991 by Sumio Iijima followed the important discovery of fullerenes in the previous decade. Fullerenes are composed of carbon atoms joined together covalently in clusters to form hollow structures. Many different types of fullerene exist, and among these are carbon nanotubes.

Single walled nanotubes (SWNTs) consist of hollow cylinders, whose walls are formed from rolled up single graphene sheets. At the ends of each cylinder are hemispherical caps. Examples of single walled nanotubes are shown schematically in Figure 6.7. The diameter of a single walled nanotube can be as low as 1.5 nm, with a length of >1 μm. Multiwalled nanotubes (MWNTs) are, in essence, a collection of concentric single walled nanotubes with differing diameters. The overall

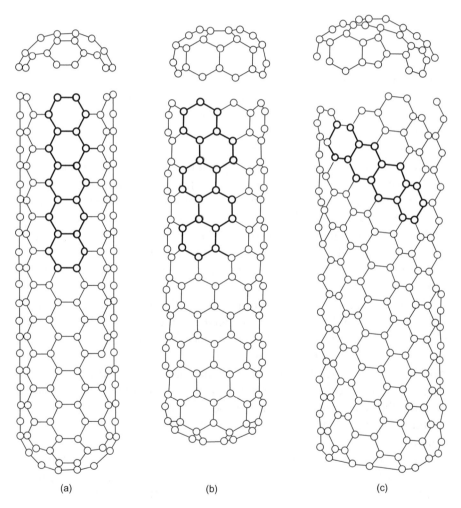

Figure 6.7 Single walled carbon nanotube structures: (a) armchair, (b) zig-zag, (c) chiral.

diameter of a MWNT ranges between 4 and 50 nm, and its length is often several μm. Double walled nanotubes of lengths in the order of 1 mm have also been reported.

Carbon nanotubes can be produced by three different techniques. In one technique, a vapour is created by an arc discharge between two carbon rods placed end to end and separated by *ca.* 1 mm, in a space filled with an inert gas at low pressure. The arc discharge vaporises one of the rods, and a deposit of carbon nanotube is formed on the other rod. In the absence of catalyst, MWNTs are formed, but if a transition metal, such as nickel or cobalt, is present, SWNTs are generally formed.

In the laser ablation technique, a pulsed or continuous laser beam vaporises a graphite target at *ca.* 1200 °C in an atmosphere of inert gas. A plume of hot vapour is formed. On cooling of the vapour, a soot is formed, some of which is in the form of nanotubes. Transition metal catalysts can assist formation of the nanotubes, or even promote formation of SWNTs.

Neither of these two methods can be readily scaled up to commercial production levels. The third technique, utilising chemical vapour deposition, is more attractive in this respect. In chemical vapour deposition, a material is produced in the vapour phase from one or more precursors that are caused to react by heating or by electrical discharge. The material is often deposited as a coating on a substrate, such as silicon or glass. Carbon nanotubes can be produced by 'cracking' carbon monoxide, methane or ethyne (acetylene). The heated substrate is coated with a transition metal, often iron, cobalt or nickel. The type of nanotube produced depends very much on the processing conditions. SWNTs of high purity, for example, have been produced from carbon monoxide.

Carbon nanotubes possess a number of special properties. They exhibit exceptional mechanical properties, and they are electrical conductors or semiconductors. Over the past decade or so, there has been great activity in the production of continuous filaments from carbon nanotubes. The principal approaches that have been adopted are spinning from liquid crystalline suspensions of carbon nanotubes, spinning directly from a 'smoke' of nanotubes as they are being formed by chemical vapour deposition, and spinning from MWNTs grown as 'forests' on a silicon substrate coated with a film of iron. Although none of these processes is yet available on an industrial scale, the intense research and development being undertaken is very likely to render commercial production possible in the near future. Currently, carbon nanotube fibres of tenacity 100–200 cN tex^{-1} are achievable, and indeed tenacities as high as 600 cN tex^{-1} and higher have been measured in some samples. The discrepancy between the two sets of values reveals that the mechanical performance of continuous filaments is hampered by defects within them. Successful commercial production will no doubt be achieved as these defects are eliminated.

6.7 THERMOSET POLYMER FIBRES

6.7.1 Introduction

Thermoset polymers are changed irreversibly into permanently hardened structures when given particular treatments. A change may be triggered by heat, a chemical reaction between two species at ambient temperatures (as with some epoxy resins) or irradiation, and almost invariably arises from the onset of crosslinking of polymer chains into three-dimensional structures. Yet, despite this crosslinking, some thermoset polymers can be converted by melt spinning into textile fibres. These fibres are melamine–formaldehyde fibres and phenol–formaldehyde fibres. In both these cases, the thermosetting results from heat treatments.

Thermoset polymer fibres are designed so that when they are heated, polymerisation and crosslinking continue further. In melamine–formaldehyde fibres, crosslinking is rapid, whereas in phenol–formaldehyde fibres, crosslinking occurs far more slowly. On severe heating, both types of fibre char, an indication of high levels of fire resistance. Indeed, under controlled conditions, phenol–formaldehyde fibres can be converted to carbon fibres, although these carbon fibres are not as strong as those derived from PAN or mesophase pitch. Nevertheless, these carbon fibres are used in composites for the exit nozzles of rocket motors, and in products for the elimination of static electricity in copying machines.

6.7.2 Melamine–Formaldehyde Fibres

The only grade of melamine–formaldehyde fibre produced commercially is Basofil®, produced by BASF. The chemical structure of melamine is:

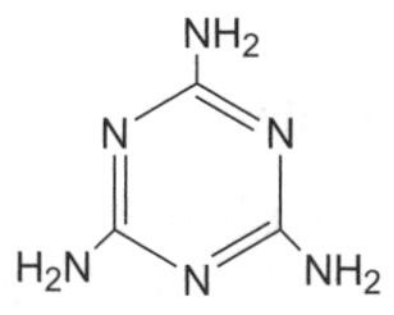

On reaction with methanal (formaldehyde), each primary group is converted to an alcohol which, at acidic pH, can undergo further reaction with the loss of water:

$$R-NH_2 \;+\; HCHO \;\rightleftharpoons\; R-NH-CH_2OH$$

$$R-NH-CH_2OH \;+\; H_2N-R \;\rightleftharpoons\; R-NH-CH_2-NH-R \;+\; H_2O$$

$$2\,R-NH-CH_2OH \;\rightleftharpoons\; R-NH-CH_2-O-CH_2-NH-R \;+\; H_2O$$

In this reaction sequence, R represents a melamine residue. Since each molecule of melamine possesses three primary amine groups, it can be seen that three-dimensional networks are readily developed. Neighbouring melamine units are linked either by methylene ($-CH_2-$) groups or by dimethylene ether ($-CH_2-O-CH_2-$) bridges. The ratio of the two types of linkage depends on pH and temperature.

The production of melamine–formaldehyde fibres involves the polymerisation process and subsequent spinning of the resin formed. During the melamine–formaldehyde polymerisation reaction, the viscosity of the resin progressively increases. When the viscosity has reached a suitable level, the resin can be extruded into filaments, but the resulting filaments are very brittle. Substituted melamines are therefore added, which reduce the degree of crosslinking and increase elongation to break.

Some of the properties of melamine–formaldehyde fibres are shown in Table 6.7. The strength of the fibres is not high, even in comparison with the strength of conventional fibres. The fibres do, however, exhibit good resistance to most common organic liquids. They are also resistant to hydrolysis by alkalis, but show less resistance to acids if exposed to them over extended periods.

Melamine–formaldehyde fibres find applications where good thermal and chemical resistance are required, without the need for high mechanical performance. They are often used in conjunction with other types of fibre. In felts used as high temperature filters and in fire-fighting clothing, they are present as blends with *m*-aramid fibres. Blends of melamine–formaldehyde and other types of fibre are used in fire blocks.

6.7.3 Phenol–Formaldehyde Fibres

The only commercial grade of phenol–formaldehyde fibre is Kynol®, produced by the Gun-ei Chemical Industry Company, Japan. These fibres are known as 'novoloid' fibres. They are produced from novolak resin, a mixture of polynuclear phenols of various chain lengths formed from the acid-catalysed reaction of phenol with methanal (formaldehyde). Novolak molecules possess on average 5–6 benzene rings, though with a range of 2–13 rings. An example of a novolak molecule is:

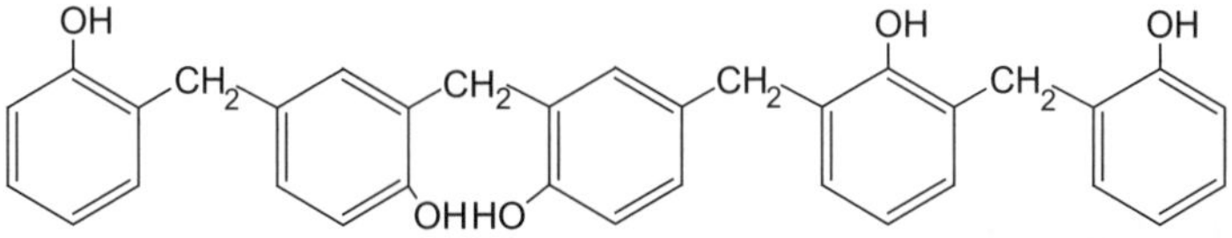

After melt spinning of the novolak resin, the fibres produced are strengthened by a curing process, in which a three-dimensional cross-linked polymeric structure is formed. The polymeric structure in Kynol® fibres, for example, is given as:

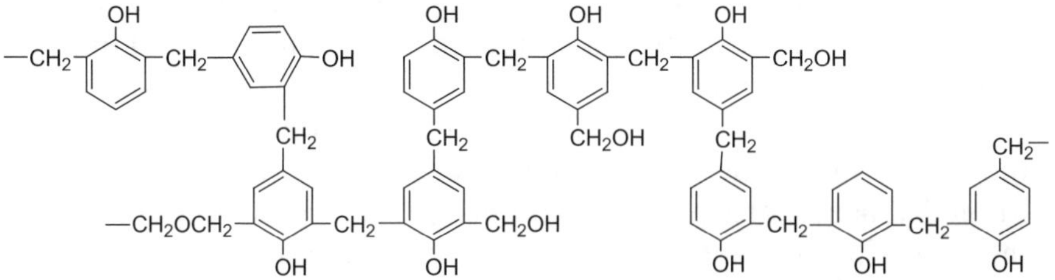

Some important properties of phenol–formaldehyde fibres are listed in Table 6.7. Above 150 °C in air, the fibres become reactive. However, they char rather than catch fire on exposure to hot flames. Their limiting oxygen index is similar to that of melamine–formaldehyde fibres. They display good resistance to most common organic liquids, but are attacked by strong alkali and by sulfuric and nitric acids.

Phenol–formaldehyde fibres are used in a variety of applications. They are used, for example, in protective clothing for firefighters, welders, foundry workers and racing car drivers, and have replaced

Table 6.7 Some properties of thermoset polymer fibres.

	Melamine–formaldehyde	*Phenol–formaldehyde*
Specific gravity	1.4	1.3
Tenacity (cN tex^{-1})	18–20	12–16
Elongation to break (%)	15–18	30–50
Moisture regain (%)	5	6
Limiting oxygen index (%)	32	30–34

asbestos in roof insulation and in curtains designed to protect from fire, intense heat and hot metal splash. They are used extensively in smoke barriers and heat insulating materials in aircraft, trains and passenger ferries. Phenol–formaldehyde fibres are also used in some composite materials.

6.8   CERAMIC FIBRES

6.8.1   Introduction

Developments in engineering technologies, especially in the aerospace and nuclear reactor industries, have required materials of high strength, whose stability in air is maintained at temperatures up to 600–700 °C, and even up to 1000 °C or higher. Ceramic fibres very largely meet these stringent requirements and have found extensive use as reinforcements in metal and ceramic matrix composite materials. The commonest ceramic fibres are silicon carbide and alumina (aluminium oxide) fibres.

Two standard commercial processes are utilised for producing ceramic fibres: chemical vapour deposition and spinning. In chemical vapour deposition, the ceramic is deposited onto a heated tungsten or carbon filament, through the introduction of a gas, or gas mixture, containing silicon atoms. In the spinning technique, a precursor polymer is spun into filaments, whereupon the filaments are heated in order to convert them to ceramic fibre.

6.8.2   Production of Silicon Carbide Fibres

Silicon carbide monofilaments of diameter >0.1 mm can be produced by chemical vapour deposition onto a tungsten or carbon filament. The process is carried out in a vertical tubular reactor, normally *ca.* 2 m in length. The gas, or gas mixture, is introduced at injection points along the reactor, which is maintained at 1400–1500 °C. A common reactant is methyl trichlorosilane, which is converted to silicon carbide with loss of hydrogen chloride:

$$CH_3SiCl_3 \rightarrow SiC + 3HCl$$

The silicon carbide fibres are formed on the filament core. Alternatively, a carbon filament is exposed to a mixture of silane (SiH_4) and hydrogen gases in the reactor. The gases react with the carbon surface to produce silicon carbide vapour, which is deposited on the filament.

Silicon carbide multifilaments, such as Nicalon® fibres (commercialised by Nippon Carbon), can be produced from a melt spinning process. A suitable polymeric precursor, *e.g.* polydimethylsilane:

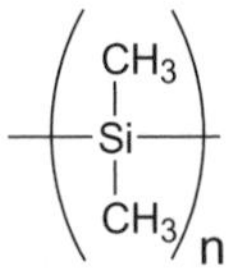

$$\left(\begin{array}{c} CH_3 \\ | \\ -Si- \\ | \\ CH_3 \end{array}\right)_n$$

is melt spun at 280 °C in a nitrogen atmosphere. The filaments formed are heated in air for 30 min at 190 °C, so that oxygen can crosslink the polydimethylsilane chains. The filaments are then heated to > 1100 °C in a nitrogen atmosphere, with release of methane and hydrogen. The silicon carbide filaments formed are predominantly crystalline, and also contain some carbon and silica (SiO_2).

6.8.3 Production of Alumina Fibres

Alumina fibres are produced by solution spinning and subsequent heat treatment. The precursor is a basic aluminium salt, whose general formula is $Al(OH)_{3-n}X_n$, where X is an inorganic ligand (commonly Cl) or an organic ligand. A viscous aqueous solution of the aluminium salt is prepared, before spinning to produce a gel fibre, which is then dried and heat treated. Initially, the aluminium salt is decomposed by the heat treatment, and aluminium hydroxides are precipitated out of solution. As the temperature is raised higher, alumina is formed in the fibres. If, however, these fibres are exposed to temperatures > 1100 °C, there is a danger that the fibres can become extremely brittle. This problem can be substantially overcome by the presence of 3% (w/w) of silica in the fibres.

6.8.4 Fibre Properties

Some fibre properties are given in Table 6.8.

Table 6.8 Some properties of ceramic fibres.

	Silicon carbide (chemical vapour deposition)	Silicon carbide (spun)	Alumina
Specific gravity	3–3.1	2.4–2.8	3.6–3.9
Tenacity (cN tex^{-1})	80–100	70–120	30–50
Elongation to break (%)	0.6–0.8	1.0–1.5	0.3–0.5

Due to their outstanding resistance to very high temperatures, ceramic fibres are primarily used in materials that are required to withstand these temperatures. Applications of these materials include gas turbines in aircraft, the walls of nuclear reactors and heat exchangers. Fabrics made from ceramic fibres are used for the filtration of gases at high temperatures.

SUGGESTED FURTHER READING

1. J. W. S. Hearle, ed., *High-Performance Fibres*, Woodhead Publishing Limited, Cambridge, UK, 2001.

CHAPTER 7
Other Speciality Fibres

7.1 NANOFIBRES

7.1.1 Introduction

Nanofibres are not formally defined by the Textile Institute but are generally understood to be fibres whose diameters are in the range 1–100 nm (1×10^{-9} to 1×10^{-7} m), though there is a divergence of opinion, with some workers regarding diameters of up to 500 nm as nanofibres. Nevertheless, nanofibres are approximately 40 times thinner than microfibres and about 1000 times thinner than human hair (see Figure 7.1). Individual molecular chains have diameters smaller than 1 nm (1×10^{-9} m) so it is clear that nanofibres are extremely fine and have a thickness in the order of just a few molecular chains.

Interest in nanofibres has centred on the novel properties they possess, which arises from their very high surface area in relation to their mass. Nanofibres can have a surface area of about $40\,\mathrm{m^2\,g^{-1}}$, some 1000 times greater than standard grade fibres, one consequence of which is significantly greater moisture absorbency. In absolute terms the tensile strength of individual nanofibres is very small, but when their fineness is taken into account the values for their tenacities are very large. One explanation for the high tenacities is that the molecular chains are more highly aligned and oriented in the nanofibres, and, since there are relatively few molecular chains across the width of a nanofibre, a high degree of order exists. In this more ordered configuration, the molecular chains are less likely to slip over each other when the nanofibre is under

The Chemistry of Textile Fibres
By Robert R Mather and Roger H Wardman
© Robert R Mather and Roger H Wardman 2011
Published by the Royal Society of Chemistry, www.rsc.org

Figure 7.1   Comparison of size of nanofibres and human hair. (Photograph courtesy
of J. McVee, Heriot-Watt University.)

tension than they are in standard fibres of greater diameter. This feature
of the structure gives nanofibres tremendous potential in composite
materials with high mechanical performance.

Nanofibres are so different in their physical properties from standard
grade, or even microfibres, that they have a wide range of applications in
high added value products. Such applications include consumer products (*e.g.* garments, wipes), medical products (biodegradable tissue
scaffolds, drug release membranes) and industrial uses such as filters,
separators for batteries and fuel cells. Nanofibres have considerable
potential to extend the applications of textile fibres well beyond their
traditional uses in the future.

7.1.2   Methods for Making Nanofibres

There are numerous methods for producing nanofibres:

- Phase separation
- Melt fibrillation
- Electrospinning
- 'Islands-in-the-sea'

- Gas jet
- Nanolithography
- Self assembly
- Melt blowing

The two main methods for the production of nanofibres are electro-spinning and the 'islands-in-the-sea' method and only these two will be described in detail. The technique of melt blowing is a widely used method for producing fibres of very small diameter, so it is also included.

7.1.2.1 Electrospinning. Electrospinning is not an especially new technique and much of the basis for the concept was established at the beginning of the twentieth century. The technique, as a method for producing fibres, was really established in a patent by an American called Formhals in 1934 but no commercialisation resulted, and it was not until the 1980s that an American company called Donaldson began to produce electrospun nanofibres for its filtration products.

In the electrospinning method, a solution of the polymer is pumped at a constant rate through a very fine orifice at the tip of a syringe (see Figure 7.2) towards a conductive collector plate, made of copper or aluminium, at an optimal distance (between 15–30 cm) away. A high voltage, of about 15–25 kV, is applied between the tip of the syringe and the collector plate. The collector plate, which may be directly below the syringe, or at right angles to it, is electrically earthed.

If no voltage is applied, drops of the polymer solution simply fall under gravity to the collector plate, but if the voltage, and hence the intensity of the electric field, is increased an electric bi-layer forms in the spinning solution (see Figure 7.3). The thickness, D, shown in Figure 7.3 is called the Debye length and is related to the thickness of the ionic accumulation in the region towards the solution surface. This region is only a few nanometres thick so the external electrostatic environment only influences the molecules at or near to the surface of the polymer solution. As the voltage is further increased, the charge distribution in the body of the droplets at the tip of the needle causes them to assume a conical shape, known as a 'Taylor cone'. If the voltage exceeds a threshold value the electrostatic repulsion overcomes the surface tension that causes the droplets to form and a continuous charged jet of the fluid emerges from the syringe. The charged jet travels in a straight line for just a short distance after emerging from the syringe and then, as the solvent evaporates, the charge migrates to the fibre surface. The jet then experiences a whipping action due to the electrostatic repulsion at small

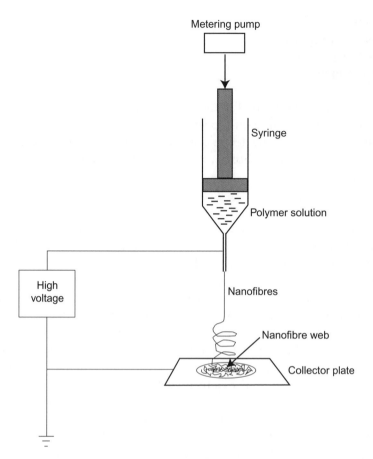

Figure 7.2 Schematic diagram of electrospinning apparatus.

bends in the fibre causing it to spiral toward the collector plate. During the passage from the syringe to the collector plate, the solvent evaporates leaving the polymer in nanofibre form. If the collector plate is a flat stationary plate, the spiralling of the jet towards it causes the nanofibres to accumulate in a randomly oriented web. An alternative is to use a cylindrical rotating plate to capture the nanofibres, in which case the fibres align in a more parallel fashion to each other. The process produces nanofibres with diameters over a broad a range from about 50 to 2000 nm.

In order to produce nanofibres successfully the following parameters have to be optimised:

- The viscosity and surface tension of the polymer solution, which depends on the concentration of the solution, the values of M_w and

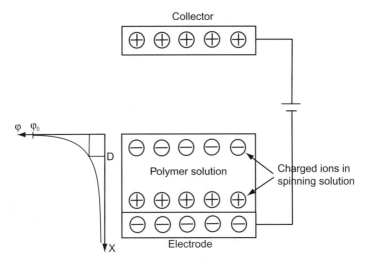

Figure 7.3 Electric bi-layer in a polymer solution under the influence of an electric field.

M_n of the polymer, and the solvent used. Surface tension is the primary force that opposes coulomb repulsion so it is critical in determining electrospinnability. It is also essential that the solution has an electrical conductivity, because solutions of zero conductivity cannot be electrospun. Conductivity can be increased by additions of a small quantity of an inorganic salt.

- The electric field intensity as determined by the applied voltage.
- The flow rate of the polymer solution from the syringe.
- The distance of the collector plate from the tip of the syringe.
- The volatility of the solvent. Ideally all of the solvent should have evaporated by the time the nanofibres reach the collector plate and it is amazing that the vast majority does. Solvent volatility has considerable influence on the shape and porosity of the nanofibres formed. For example, if the solvent evaporates only slowly, the nanofibres have a flatter ribbon-like structure, rather than the preferred round structure. If a solvent of high volatility is used, a more porous structure, with a regular array of pores is obtained. In this respect electrospinning is similar to dry spinning (see Section 5.1.3).
- The dielectric constant of the solvent. A high dielectric constant leads to a better dispersion of charge density on the surface of the liquid and hence better electrospinnability.
- The temperature and humidity of the environment, which affects the evaporation of the solvent.

Table 7.1 Polymer–solvent systems for electrospinning.

Polymer	Solvents
Nylon 6 and nylon 6,6	Methanoic acid
Polyacrylonitrile	Dimethylformamide
Polyester	Trifluoroacetic acid or dimethyl chloride
Polyvinylalcohol	Water
Polystyrene	Dimethylformamide or toluene
Polybenzimidazole	Dimethylacetamide
Polyimides	Phenol

These parameters depend very much on the polymer–solvent system employed. Typical systems are shown in Table 7.1. It will be apparent from Table 7.1 that a wide range of polymers can be electrospun. The polymers indicated in Table 7.1 are all thermoplastic. It is possible to electrospin cellulosic nanofibres from solution in *N*-methylmorpholine-*N*-oxide (NMMO), though the process is more complicated than for thermoplastic fibres and there are considerable practical difficulties. Cellulosic derivatives, such as methyl cellulose (MC), carboxymethyl cellulose (CMC), hydroxyl propoxy methyl cellulose (HPMC) and methyl ethyl hydroxyl ethyl cellulose (MEHEC), however, have all been successfully electrospun. It is not always essential that the polymer is capable of being dissolved and some polymers, such as polyethylene, polypropylene and nylon 12, can be electrospun from the molten polymer. The Japanese company, Toray Industries Inc., is producing nanofibres of nylon, polyester and polyphenylene sulfide from the molten polymer on a commercial scale.

One problem with electrospinning is the low production speed. Typical speeds are less than $10\,g\,h^{-1}$, which is not a viable rate. The problem can be overcome by using a multi-nozzle array, but this is technically complex and clogging of the nozzles can create additional difficulties. Also, as can be seen in Table 7.1, the solvents that have to be used to dissolve the polymers are hazardous chemicals and so effective solvent recovery from the atmosphere around the electrospinning unit is essential.

A company in Liberec in the Czech Republic, Elmarco, is producing nanofibres with diameters of between 200–500 nm, using a needleless electrospinning technology called Nanospider™, which was developed at the Technical University of Liberec. The electrospinning originates from the surface of the polymer solution so the process does not require a nozzle. A cylindrical spinning electrode immersed in the polymer solution is connected to a very high voltage source (up to 100 kV) and the nanofibres are deposited on the grounded collector electrode. It is claimed that production rates of up to $30\,\mathrm{m\,min^{-1}}$ of nanofibre webs over 1 m wide are possible. The research group at Liberec have also success-fully used the needleless method to electrospin polycaprolactone and polypropylene nanofibres, though the presence of a surfactant additive such as dodecyl trimethyl ammonium bromide in the melt significantly improves the consistency of the fibre diameter and morphology.

An interesting development of the standard electrospinning process is 'core-shell' electrospinning, which is sometimes called 'co-axial' electro-spinning. This technique requires the use of a cleverly designed needle structure on the syringe (the spinneret), which enables two polymer solutions to be spun, one on the outside of the other (see Figure 7.4). Essentially, the spinneret comprises two small capillary tubes, one set inside the other, through which the two polymer solutions are pumped at carefully controlled rates. To produce core-shell nanofibres, the outer shell component must be made of an electrospinnable polymer solution, though it is not necessary that the core material is polymeric, or is even an electrospinnable fluid. This is because the electrospinnability of the core-shell fluid depends on the solution at the surface, which is where the

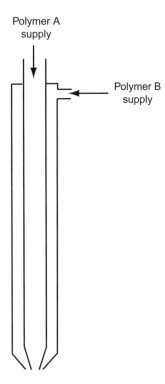

Figure 7.4 Diagram of syringe for electrospinning core-shell nanofibres.

electrical charge develops. The main requirement is that the whipping motion of the nanofibre as it progresses to the collector plate does not rupture the core-shell structure. Core-shell electrospinning can be used to incorporate functional chemicals, or nanoparticles, such as magnetic particles, into the core of the shell. If the shell is made of a biodegradable polymer, the slow release of the active agent within is possible if that is a desired feature. Another possibility is to use a liquid as the core, which can be removed after electrospinning, leaving hollow nanofibres.

One other variation that has been explored, though has not reached commercial production, is a dual syringe system (see Figure 7.5) in which the two nozzles have opposite charges. The electrospun nano-fibres with opposite charges discharge in a direction between the two nozzles and attract each other. They can be pulled out and wound onto a roller, in the form of high bulk webs of interlaced nanofibres which, it is claimed, is useful for fillers and thermal insulation webs.

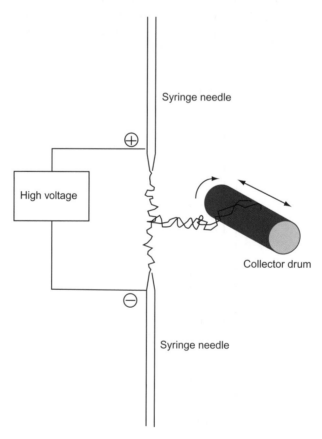

Figure 7.5   Diagram of dual syringe system for electrospinning bulked nanofibres.

7.1.2.2 Melt Blown Nanofibres. The melt blown spinning process is a variation of melt spinning (see Section 5.1.2) that produces a web of very fine randomly arranged fibres. The molten polymer is extruded from a two-dimensional array of spinnerets. As the fibres form, they are delivered by a stream of a hot, rapidly moving air current in such a way that their motion becomes random, causing them to intertwine, towards a forming screen. A polymer of low viscosity is required. The entangling of the fibres occurs before they have fully cooled to a rigid form, so they tend to stick at their points of contact and adhesion between them is therefore promoted. The spinnerets vary in size, so that a mixture of sub-micron and greater than sub-micron fibres are formed, the purpose of the larger fibres being to trap and immobilise the smaller diameter fibres in the formation of the web. In general the fibres formed have diameters approximately four-fold larger than those of electrospun fibres. The webs formed are very light, about $5 \, g \, m^{-2}$, but still about $40 \times$ heavier than the webs that can be made with electrospun nanofibres. Melt blown webs have a pore size range which is useful for many filtration products, and they are also widely used for medical drapes and disposable hygienic garments.

7.1.2.3 'Islands-in-the-Sea'. The 'islands-in-the-sea' method involves the melt spinning of fibres made from two different polymer types, in such a way that the extruded fibre filament comprises one of the polymers present as a number of sub-filaments (the 'islands'), surrounded by a 'sea' of the other polymer. When the 'sea' polymer is dissolved away the individual, separate 'island' filaments, of nanofibre diameter, remain. Extrusion of the 'islands-in-the-sea' fibres in the first place requires some complex engineering of the spinneret. A number of two-component polymer flow systems that give either side-by-side or sheath-core fibre filaments are bundled together to pass through the spinneret.

The whole process is essentially one of classical melt spinning, though the need for the multiple bicomponent flow systems of the two molten polymers to the spinneret adds more complexity. The number of these flow systems determines the number of 'islands' in the 'sea' and an American company, Hills Inc. has successfully developed the technology to produce over 1000 islands. Extrusion speeds of $5 \, kg \, h^{-1}$, and wind-up rates of $2500 \, m \, min^{-1}$ are claimed, which are commercially viable production rates. An example of the 'islands-in-the-sea fibre' is shown in Figure 7.6. By adjusting the relative rates of flow of the 'island' polymer and the 'sea' polymer it is possible to change the amount of the 'sea' polymer around the 'island' polymer. After extrusion and drawing, the

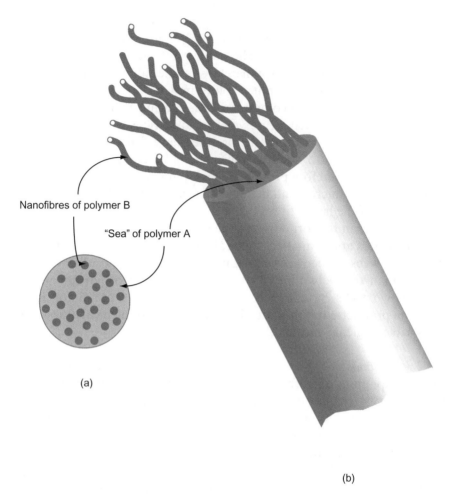

Nanofibres of polymer B

"Sea" of polymer A

(a)

(b)

Figure 7.6 Schematic diagram of 'islands-in-the-sea' method of nanofibre preparation: (a) cross section of fibres, (b) longitudinal view.

yarns are woven or knitted into fabric form, the 'sea' polymer is then dissolved away. As long as the solvent used to dissolve the 'sea' polymer is not a solvent for the 'island' polymer, very fine fibres in the form of either a loosely woven or knitted structure are left.

Nanofibres of polypropylene, polyester and polyamide have been successfully extruded as the 'islands' in a 'sea' of polystyrene. The concept can be extended to extruding two different 'island' polymers in the 'sea', so that what remains after the 'sea' polymer is dissolved away is a bicomponent nanofibre woven or knitted structure. Another variation on this method is to minimise the amount of the 'sea' polymer during extrusion, when the 'island' fibres can then be slit away by mechanical action.

7.1.3 Uses of Nanofibres

The difficulties in making nanofibres in large quantities have rather restricted their use in the production of composites for large scale items. Applications have therefore been more widespread in the medical field where relatively small amounts are required. Typical uses are in wound dressings, medical implants and tissue scaffolds. The most important feature of these scaffolds which is exploited in biomedical applications is their ability to mimic the extracellular matrix in natural tissue. Several biocompatible polymers are used in various biomedical applications, including polycaprolactone, polylactic acid, polyvinyl alcohol, chitosan, polyhydroxybuyrates and their copolymers. Biodegradable polymers are preferred, since no post-operative surgery is required after implantation. Biodegradable polyester nanofibres are widely used and biodegradable block copolymers, for example of polyethylene glycol (the 'soft' component) and polycaprolactone (the 'hard' component) are also being investigated. Another medical application of nanofibre webs is that of slow release drug delivery, where the micropore structure makes them suitable temporary hosts for chemicals such as drugs.

A major use of nanofibres is in filtration products. The very tiny pores of nanofibre webs are much smaller than those of melt blown webs. This pore size is tunable, by adjusting the process parameters, so that a variety of filters for different applications can be made. The main advantage of nanofibre webs, however, is their ability to filter submicron particles from air or water due to their high surface area and highly porous structure. The thin nanowebs have poor mechanical properties, in that they are not especially rigid structures, so they are laid on a substrate of much coarser fibres (see Figure 7.7). In this form they are used to make HEPA (high efficiency particulate air) filters for clean rooms, hospitals, laboratories, and also filters for automotive uses, such as oil and fuel filters, and in the food industry for beverages.

Nanofibres also have potential for use in catalysis. They can be used as a surface on which enzymes, capable of breaking down toxic chemicals for example, can be immobilised. The fibre manufacturer DuPont produces nanofibre sheets based on nylon 6,6 which are a uniform web of randomly oriented nanofibers used for separators in energy storage devices such as:

- Electrochemical double layer capacitors
- Aluminium electrolytic capacitors
- Lithium ion and alkaline batteries

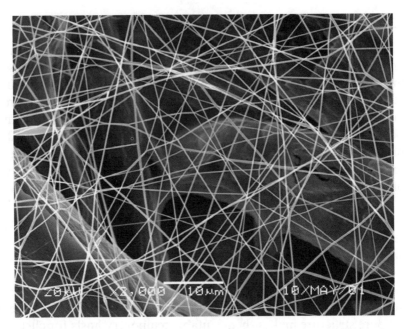

Figure 7.7   SEM photograph of nanofibres laid on to a cellulose substrate for an air
filter. (Photograph courtesy of Donaldson Company Inc., Minneapolis,
MN, USA.)

Teijin Fibres Limited produces Nanofront™, a nanofibre spun by an
'islands-in-the-sea' composite spinning technology giving nanofibres of
700 nm. Their high surface area is claimed to give enhanced water absorption, absorbability of particles and anti-translucency. Applications
of these fibres are in functional sportswear and inner wear. They are also
used for wipes, where their small pores are very effective in collecting oils
and small particles from surfaces.

7.2   ELECTRICALLY CONDUCTING FIBRES

7.2.1   Introduction

The textile fibres normally used for apparel do not conduct electricity. In
fact, they generally have a very high electrical resistance, so high that
in some cases the extent to which they accumulate electrical charge
and develop static electricity can be uncomfortable in a garment (see
Section 1.6.3). There are applications, however, where it is desirable to
have fibres that will conduct electricity, particularly in the developments
that are currently taking place in the field of smart fabrics and interactive

textiles, the acronym known as SFIT. Wearable electronics for the healthcare sector are a good example of the developments taking place. In this application, textile-based sensors capable of monitoring the physiological condition of the wearer (heart rate, breathing rate, temperature, *etc.*) need to be incorporated into close-fitting garments and it is also necessary to transmit the electrical signals from these sensors through the garment to some form of display or transmitter. For this purpose of linking sensors with output devices or actuators, it is desirable to use textile fibres to conduct these electrical signals. An interesting area of development is in the development of artificial muscles. More generally, the possibility exists for the development of textile based chemical sensors that respond to pH, dangerous gases and biological molecules, for use in safety clothing. Other uses are for fabrics which provide electromagnetic shielding, protection against interference of sensitive electronic apparatus or which can be used in heated car seats.

7.2.2 Methods for Making Electrically Conductive Fibres

There are four main methods for making fibres that will conduct electricity and each will be discussed in turn.

1. Very fine metal wires
2. Inherently conductive fibres
3. Coating textile fibres with an electrically conductive layer
4. Addition of conductive particles into a fibrous polymer matrix

None of these methods is ideal in every respect; each has a disadvantage for one reason or another.

7.2.2.1 Very Fine Metal Wires. On the face of it, this approach would seem the most sensible way to conduct electricity around a garment. Indeed the electrical conductivity of metals is by far the highest of any other material (see Figure 7.8), but metal wires, even very fine ones, have a number of significant disadvantages when incorporated into textile fabrics and garments:

- They are expensive. The metals with highest conductivity are copper and silver, both of which are expensive metals.
- They can be up to five times heavier than textile fibres
- They are brittle, especially copper, which can break easily.

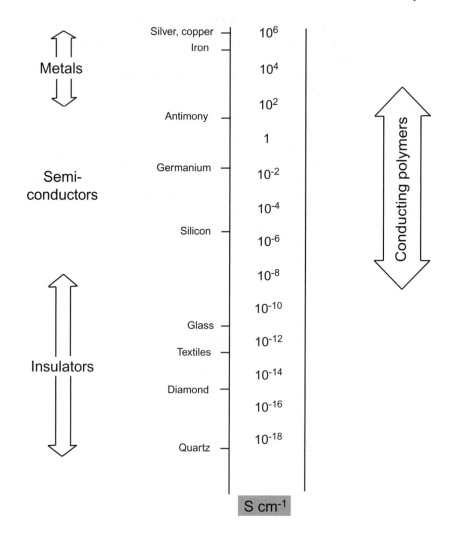

Figure 7.8   Conductivities of some materials.

- They can abrade textile manufacturing equipment during yarn manufacture or weaving.
- They are not as flexible as textile fibres and can adversely affect the drape and handle of fabrics.

Where metallic wires are incorporated in a fabric, the electrical resistance depends on the diameter of the wire and the density of the wires in the fabric structure. Also if the fabric has to be cut into the shapes required for garment construction, the conductivity will be reduced as the interlaced wires will be broken. However, there are applications where the use of metallic wires is quite satisfactory, such as in carpet backings to give antistatic performance and in fabrics for electromagnetic interference (EMI) shielding.

7.2.2.2 Inherently Conductive Polymers (ICPs). As will be apparent from Figure 7.8, textile fibres are very poor conductors of electricity and are more properly regarded as insulators. A considerable amount of research has been carried out over the last thirty years or so to develop polymers that are electrically conducting and can be extruded as fibres. The potential advantage of fibrous conductive polymers is their ease of processing on conventional textile machinery and of blending with other (non-conducting) textile fibres. A problem that exists with the most useful ICPs, however, is that of extruding them in fibre form and of a quality that meets the usual mechanical performance criteria required of textile fibres.

There are three types of systems that provide for electrical conductivity in polymers. The first type is ionically conducting polymers, in which the electrical conductivity is due to the movement of ions through the polymer, which is typically a polyelectrolyte such as polyacrylic acid. The conductivity of these systems depends on the size of the ions and their mobility through the polymer matrix, and smaller, more highly mobile ions such as lithium ions give higher conductivities. Nevertheless, the conductivities of these systems are generally lower than those of the other two types of ICPs.

The second type of system is that of redox polymers, in which electron transfer takes place between immobilised redox centres by a 'tunnelling' mechanism across an insulating barrier. In order to possess a reasonable conductivity, there needs to be a high concentration of redox centres in the polymer matrix to increase the probability of tunnelling.

The third type of system is that based upon the movement of electrons in a conjugated double bond structure, which provides a source of highly mobile π-electrons. It is this third type of ICP that has been the subject of most interest and development. The movement of electrons through a fibrous polymer can occur in three ways, as illustrated in Figure 7.9. At the molecular level, electrons move along the conjugated polymer chain. Conductivity is therefore enhanced in polymers of high molar mass, though if the molar mass is too high distortions in chain symmetry may cause a reduction in conductivity. For each type of polymer, there is an optimum molar mass at which the highest conductivity can be achieved. However electrical conductivity also depends on the ability of

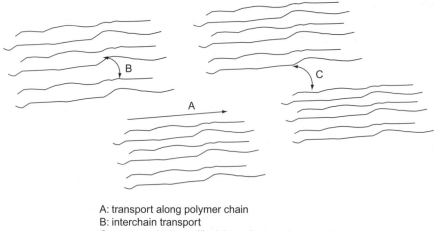

A: transport along polymer chain
B: interchain transport
C: transport by 'tunnelling' through amorphous regions

Figure 7.9   Electron transfer mechanisms in conductive polymers.

the π-electrons to hop from one polymer chain to another, and this charge hopping is facilitated by the close approach of adjacent polymer chains, which in turn is promoted by low inter-chain distances and a high degree of molecular order. A fairly rigid, planar structure is therefore required, and a common feature of ICPs is a repeating unit that contains an aromatic ring. The individual regions of high molecular order (the crystalline regions) can be regarded as being metallic in character, but they are surrounded by amorphous regions where conductivity is markedly lower. The third transfer route of electrons is that of tunnelling through the amorphous regions from one 'metallic' region to another and this route is also important in influencing the overall conductivity.

The intrinsic conductivity of all ICPs is low, typically in the range 10^{-12}–$10^{-5}\,S\,cm^{-1}$, but it was soon discovered that the presence of a dopant, which promotes the formation of protonated forms of the polymer, caused an increase of several orders of magnitude. The dopant also serves another important function which is to improve the processibility of the polymer. The incorporation of dopant molecules into polymeric structures confers solubility in polar solvents and hence enables extrusion into fibre form by wet spinning.

A wide range of conjugated ICPs has been developed over the years, the first of which was polyacetylene (PA), though the ICPs that have become most commonly used in textile applications are polyaniline (PANI), polypyrrole (PPy) and, to a lesser extent, polythiophenes (PT).

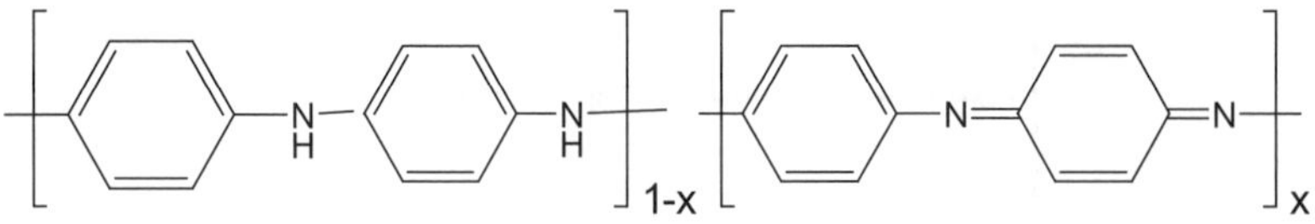

Polyaniline (PANI)

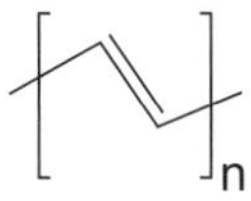

Polyacetylene (PA) 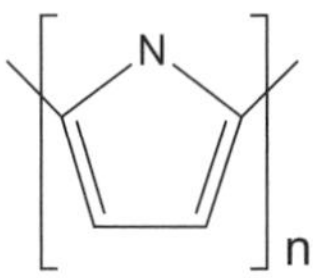   Polypyrrole (PPy) 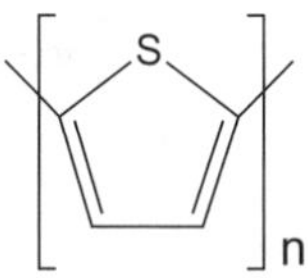   Polythiophene (PT)

In terms of conductivity, doped PA compares favourably with metallic copper (see Figure 7.10), but it is difficult to process into fibres and worse, it is not very stable. Whilst doped PANI, PPy and PT have lower conductivities, they are easier to process and have better stability (see Table 7.2). Although the stability of PT is generally good, its conductivity decreases over time, a property that has limited its industrial application. A general difficulty in extruding ICPs in continuous fibre form, including PANI and PPy, is their lack of solubility in common solvents making wet or dry spinning difficult, and their infusibility which makes melt spinning impossible.

Polyaniline (PANI). Of all the ICPs mentioned, polyaniline (PANI) has probably generated the most interest due to its good stability, relatively high conductivity and ease of preparation. It is typically prepared by the oxidative polymerisation of aniline in an acidic aqueous medium and is formed as a precipitate. In its doped form, it is soluble in highly polar solvents such as *N*-methyl-2-pyrrolidone, *m*-cresol, 1,4-diaminocyclohexane and dimethyl propylene urea. PANI can exist in three redox states, known as the leucoemeraldine (fully reduced), the emeraldine salt (half oxidised) and the pernigraniline (fully oxidised) forms. Of these, only the half oxidised, protonated and anion-doped emeraldine form is conductive, the other two forms being mainly insulating in character.

When the emeraldine form is protonated by doping, a polaron structure is formed, which is highly conducting. A mechanism to account for the conductivity envisages electron 'holes' which are nitrogen cation radicals, alternating along the polymer chain with neutral nitrogen atoms. An electron from the neutral nitrogen jumps to the hole at

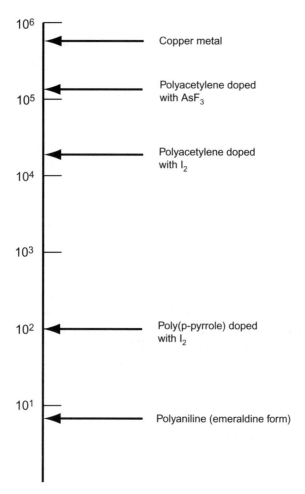

Figure 7.10 Electrical conductivities of some inherently conducting polymers used in textiles S cm^{-1}.

Table 7.2 Properties of common ICPs (undoped).

Polymer	Conductivity ($S\ cm^{-1}$)	Stability	Processability
Polyacetylene	10^3–10^5	Poor	Limited
Polyphenylene	1000	Poor	Limited
Poly(phenylene vinylene)	1000	Poor	Limited
Poly(phenylene sulfide)	100	Poor	Excellent
Polypyrroles	100	Good	Good
Polythiophenes	100	Good	Excellent
Polyanilines	10	Good	Good

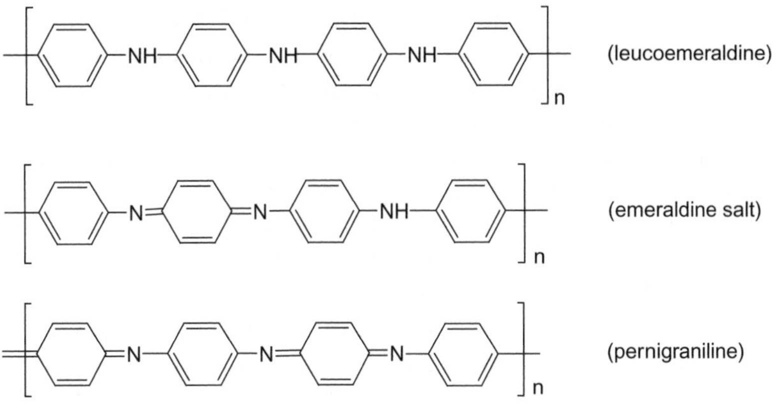

(leucoemeraldine)

(emeraldine salt)

(pernigraniline)

Redox states of polyaniline

the adjacent nitrogen, which in turn becomes neutral. The nitrogen from which the electron jumped then becomes a 'hole' itself. In this way the electron hops along the polymer chain (or between adjacent chains), giving the conductivity. If a bipolaron structure forms, electron hopping is not possible, because holes alternate with neutral nitrogen atoms.

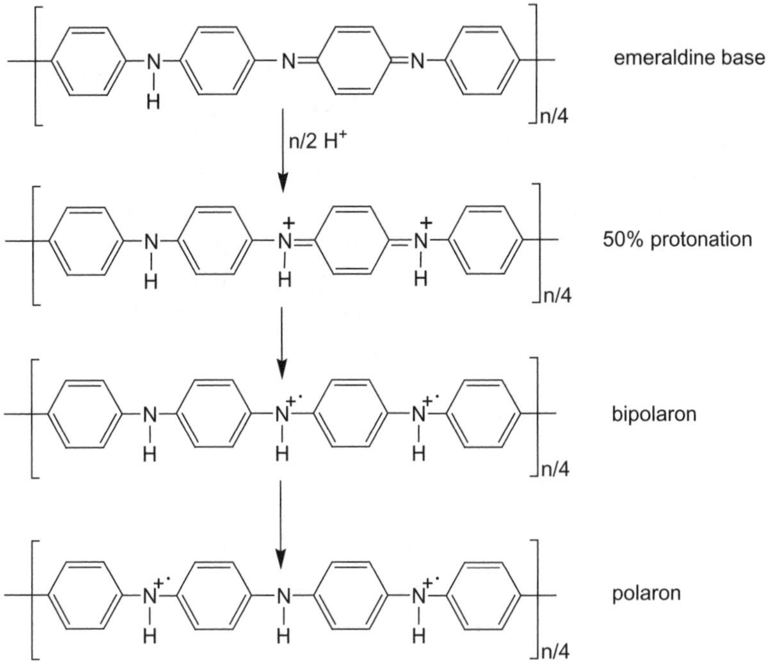

emeraldine base

$n/2\ H^+$

50% protonation

bipolaron

polaron

Doping of emeraldine base by an acid

Similarly, in the case of the leucoemeraldine and pernigraniline forms, protonation by the dopant may occur at any point along the polymer chain, not necessarily at alternating nitrogen atoms, so there is less chance of creating the polaron structure required for conductivity.

The compounds used as dopants are usually strong acids, such as methane sulfonic acid, 2-acrylamido-2-methylpropane-1-sulfonic acid, camphor sulfonic acid and dodecylbenzenesulfonic acid. The level of dopant added to PANI is important and for the perfect emeraldine base form of PANI, that is when there exists an alternate arrangement of quinoid and benzenoid rings in the polymer chains, doping at a level of 50% gives the highest possible conductivity. At a loading greater than 50% dopant, conductivity reduces, possibly due to bipolaron formation.

PANI has a wide range of uses when cast as a film, but it is also produced in fibre form (*e.g.* Panion™) for SFIT applications. The fibre is prepared from ultra high molar mass PANI by wet spinning from solution in a mixture of dichloroacetic acid and 2-acrylamido-2-methyl-propane-1-sulfonic acid. The extruded as-spun fibre can be stretched (drawn) by up to five times to induce orientation and has very good conductivity of approximately $1000 \, S \, cm^{-1}$. The applications of PANI fibres are in the fields of antistatic textiles, sensors and actuators, all important requirements for SFIT products.

Polypyrrole (PPy). Polypyrrole (PPy) has a high electrical conductivity and reasonably good environmental stability, and is therefore a polymer of much interest. As with PANI, the synthesis and subsequent processing of PPy has presented considerable difficulties in forming films or fibres, due to the strong interchain interactions, but methods have recently been reported for synthesising PPy with a good solubility in many organic solvents. The use of ammonium persulfate as the oxidant and bulky dopant molecules such as dodecylbenzene sulfonic acid or di(2-ethylhexyl) sulfosuccinate is the key to producing PPy that is soluble in solvents such as *N*-methyl pyrrolidone, dimethyl sulfoxide or dimethyl formamide.

Continuous polypyrrole fibres, doped with di(2-ethylhexyl) sulfo-succinate, have been produced by wet spinning from solution in dichloroacetic acid into a coagulation bath containing 40% dimethyl-formamide in water. It was necessary to keep the fibres in the coagulation bath for 30 minutes to enable them to solidify completely and then in a bath of water for twenty four hours to further reduce the solvent content. Clearly the fibre formation process is a long one and the fibres formed tend to be brittle having an elongation at break of only 2–3%, which limits their application. However, by reducing the temperature of the polymerisation to around −15 °C instead of 0 °C and increasing the concentration of the oxidant (ammonium peroxydisulfate), PPy of higher molar mass can be obtained and with a molecular structure of higher linearity. On subsequent extrusion, fibres can be obtained with a conductivity of about 30 S cm^{-1}. The fibre properties are also improved, the tensile strength being some five times higher and elongation at break almost three times greater.

The successful electrospinning of PPy fibres has also been reported. A solvent soluble form of PPy, made using dodecylbenzene sulfonic acid dopant, has been electrospun from solution in chloroform, though the fibre diameters, at 3000 nm, were outside the range considered to be nanofibres. The non-woven fibre web formed had a conductivity of 0.5 S cm^{-1}, however, which it is claimed is higher than that achievable with cast films. The webs could have application as an electrode material in rechargeable batteries.

Polythiophene (PT). The conductivity of polythiophene (PT) decreases over time and as a consequence this has limited its use, certainly in fibre form. Substituted thiophenes are more stable, especially in their anion-doped state, and of these the 3,4-(crown ether) thiophene, PEDOT, has been most studied. PEDOT is poly(3,4-ethylenedioxythiophene) and has an excellent conductivity in the doped state (up to 550 S cm^{-1}), as well as a high stability.

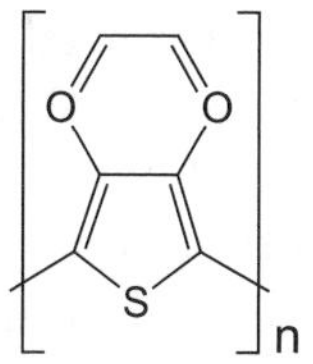

Polyethylenedioxythiophene (PEDOT)

As with many conductive polymers, PEDOT is relatively insoluble, but a very effective dopant for it is poly(styrene sulfonate) (PSS) and PEDOT-PSS is widely used in films where high conductivity is required. It is not extruded as a textile fibre, but PEDOT has been applied as a coating to textile fibres to impart conductivity (see Section 7.2.2.3).

7.2.2.3 Coating Textile Fibres with an Electrically Conductive Layer. An effective and well established method of making electrically conducting fibres is to coat the fibres with a very thin layer of a conductive material. The technique has the advantage that 'ordinary' textile fibres, such as cotton, wool, nylon and polyester, can all be used so that the resulting conductive fibre possesses all of the inherent mechanical properties of these fibres.

One method to achieve electrical conductivity is to coat a fibre such as polyester with a metal layer, for example silver or copper. A problem of coating fibres with metals is the low adhesion between them and the method is only really applicable if a very thin coating of metal is required. A technique used to deposit thin layers is low-pressure plasma sputtering, which gives a metallic layer of 100–200 nm thickness. Another way to introduce metals into fibres is by electroless deposition. The fibres are treated in an aqueous solution of the metal salt, and after the solution has absorbed into them, a chemical reduction is carried out to deposit the free metal. Whilst this method does not influence the inherent mechanical properties of the fibres, it is technically complex and only low conductivities are achieved. A variation on this method that has been developed for fibres which are wet spun (acrylics, viscose and Lyocell) is to pass the fibres in the 'never-dried' state, that is the fibres leaving the coagulation bath still in their wet, gel-like state, into a solution of the metal salt and a reducing agent. Due to the fact that the fibres are still in a swollen condition, diffusion of the metal salt solution into them is much enhanced. Another method for making electrically conducting nylon or polyester fibres is to incorporate conductive copper sulfide into them.

Coating fibres with a conductive polymer (an ICP) is a method that is the subject of much interest amongst researchers. The attraction is that reasonable conductivity can be achieved without detrimental effect on the mechanical properties of the original textile fibre. This technique has been used to apply a coating of PPy to the regenerated cellulosic fibres viscose and Lyocell. The method involves the wet impregnation of pyrrole into the cellulose fibre, followed by oxidative polymerisation using $FeCl_3$. Whilst the PPy becomes firmly bound into the surface of the cellulosic fibres and has a high washfastness, the conductivity deteriorates with time, due to air oxidation. A PPy coating has also been applied to wool, and whilst initial conductivity was good, it was markedly reduced after laundering, probably due to the alkaline conditions causing dedoping. Atmospheric oxidation of the PPy also causes the conductivity to decay over time. Clearly if this approach is to be successful, a protective coating, for example of a silicone oil, is required, though the durability of such aftertreatments may limit their effectiveness in normal textile usage. The application of a PPy coating to nylon 6 fibres has shown that a reasonably linear relationship exists between the fractional increase in electrical resistance and applied strain, indicating a possible sensing application.

Another method for coating fibres is by a vapour-phase treatment of fibres (pre-treated with oxidant, such as $FeCl_3$) with pyrrole, to form a coating of PPy. This vapour-phase oxidative polymerisation route has been reported to give conducting cotton–PPy, aramid–PPy, poly-acylonitrile–PPy and poly(p-phenylene terephthalamide)–PPy fibres. PEDOT has also been coated on to polyester fabric, though from a solution of EDOT in methanol sprayed onto the fabric which was then placed in an oven at 70 °C to allow for the completion of the polymerisation to PEDOT.

A variation on this method, recently reported by researchers at the University of Gent, Belgium, is the electroless deposition of firstly polypyrrole, then copper, on to a polyaramid surface. For the electroless deposition of copper a solution of $SnCl_2$ and $PdCl_2$ was used to activate the polypyrrole-coated polyaramid surface. The conductivity of the copper/PPy coated surface was found to be comparable with metallic conductivity.

Electrically conducting PPy-coated fibres retain their inherent mechanical properties but also have the electrical and microwave properties of PPy. In addition to their use in clothing for conferring attributes such as antistatic properties, their ability to absorb a portion of the microwave region of the spectrum may mean that they have application in military camouflage, where electromagnetic interference shielding by absorption, rather than reflection, is more important. Elastomeric fibres have also been coated with PPy, creating opportunities for application as a strain sensor in cases of high deformation.

7.2.2.4 Incorporating Conductive Particles into Fibres. In this method, electrically conductive fillers, such as carbon black or metal particles, are added to either the polymer melt (in the case of melt spinning) or to the polymer solution (in the case of wet or dry spinning) just before extrusion. This method relies on adequate contact between the filler particles in the extruded fibre, and to achieve this, high loadings of the filler are usually required. The loading required depends on the filler and its particle size and shape. Fillers that are spherical in shape (*e.g.* metal fillers) and isolated from each other in the polymer matrix are required in greater amount than structured fillers, which have a high tendency to aggregate (*e.g.* carbon black). Nevertheless, typically 15–20% of carbon black is required and thorough mixing of the filler in the polymer matrix is necessary. At such high loadings the fibres become difficult to process and the mechanical properties of the fibre can be adversely affected. In the case of single walled carbon nanotubes (see Section 6.6.6) only about 2.5% is required however, but adequate dispersability is often difficult to achieve. Another rather obvious disadvantage is the colour – carbon black is black (!) and so it is not possible to produce conductive fibres with other colours by this method. If black fibres are undesirable they have to be hidden in a garment. Since the fibres are also highly opaque they cannot be used for optoelectronic applications. For technical applications, such as antistatic properties, the colour and opacity is not usually an issue. Fibres containing carbon black can cause contamination and are therefore unsuitable for use in cleanrooms. Metal fillers have a number of disadvantages which limit their usefulness. They make the fibres heavier, give them a poor surface finish and make the polymer matrix susceptible to oxidative degradation. As with carbon fillers, they also lower the mechanical properties of the host fibrous material.

The addition of carbon fibres to synthetic fibres such as polyester or nylon is used by many garment manufacturers who make antistatic garments for workers in, for example, electronics and computing companies, and environments where a spark could cause an explosion. For such applications, the polyester or nylon contains about 1–2% carbon in the form of carbon fibre.

An interesting development of incorporating conductive particles into fibres is to incorporate carbon nanotubes (CNTs) into a conductive fibre such as PANI. PANI fibres doped with 2-acrylamido-2-methyl-1-propane sulfonic acid have excellent conductivity, but only modest mechanical properties, whilst CNTs have both excellent mechanical properties and form highly conductive fibres. It has been found that composite fibres, formed by wet spinning PANI fibres containing CNTs as a reinforcing material, show significant improvement in both conductivity and mechanical strength. Although there are considerable difficulties in successfully dispersing CNTs in the PANI and then wet spinning, the incorporation of just 2% of CNT gives an improvement of 150% in tensile stress, a 110% increase in Young's modulus and a thirty-fold enhancement of conductivity. The chemistry of carbon nanotubes is covered in Section 6.6.6.

7.2.2.5 Polymer Blending. The difficulty of extruding ICPs into fibre form has stimulated research into producing blends of ICPs with normal (non-conducting) textile fibres. The advantage of blending the two polymer types, rather than incorporating carbon black or metallic particles, is that the conducting polymer becomes oriented along the fibre axis during extrusion and drawing, giving longer conducting channels. Attempts of melt spinning blends of PANI and polypropylene have not been successful in terms of producing conductive fibres, however, blends such as PANI/polyester, PANI/polyamide-11 and PANI/polyacrylonitrile (the PANI doped with usually dodecylbenzene sulfonic acid) have very good fibre mechanical properties and improved electrical conductivities of approximately 10^{-3}–10^{-1} S cm^{-1}.

7.2.3 Inherently Conducting Nanofibres

A considerable amount of research work is being carried out into the formation of ICPs and blends of ICPs with other (non-conducting) polymers by electrospinning. The electrospinning of PANI fibres was first reported in 1995, and in 2000 the electrospinning of a blend of polyethylene oxide (PEO) and PANI doped with camphorsulfonic acid from a solution in chloroform was reported. The PEO/PANI fibres formed had diameters in the range of 950–2100 nm, though in later work diameters of < 100 nm were produced. A blend ratio of 50 : 50, gave a conductivity of approximately 0.1 S cm^{-1}. Similar conductivities have been achieved with blends of PANI with polystyrene and with poly-acrylonitrile. Electrospun blends of PEDOT/PSS and polyacrylonitrile have also been reported, where it was found that conductivity is greater in the fibres of smaller diameter.

7.3 OPTICAL FIBRES

7.3.1 Introduction

Optical fibre technology is usually associated with communications networks and for the transmission of data over long distances. The prime requirement of optical fibres is a high transparency because this property governs the distance over which a signal can be transmitted effectively. The deterioration of the signal through an optical fibre is called the attenuation, measured as the rate of decrease in decibels per kilometre (dB km^{-1}), and clearly the requirement is for a fibre to have as low an attenuation rate as possible.

The intensity of a light beam, I, decreases exponentially with distance, d, along a straight optical fibre, according to the expression:

$$I(x) = I(0)10^{-\alpha d/10} \qquad (7.1)$$

where $\alpha = $ *attenuation coefficient* of the optical fibre, and gives the attenuation of a beam in dB km^{-1}. The value of α varies with wave-length and with the nature of the fibre. Glass fibres for optical data transmission have attenuation rates of less than 1 dB km^{-1} at their optimum wavelength (1500 nm), but for the polymer optical fibre PMMA the minimum is about $100\,dB\,km^{-1}$ and occurs in the region 500–600 nm.

In general, attenuation increases as core diameter decreases, because cores of small diameter have a greater number of imperfections that cause Rayleigh scattering. Also light strikes the core-cladding interface more frequently in a small diameter core, each interaction leading to light loss. Attenuation of the light beam can also occur through absorption, and the highest quality optical fibres are achieved with quartz and glass, which are appropriate for transmission over large distances, since they have the high transparency required. Glass fibres made of these materials, however, are expensive and require high accuracy in fibre connections. At the point of connection to the light source, usually a laser diode, a complex lens system is required for high coupling efficiency, all of which adds to the cost of these systems. An increasingly important class of optical fibres is polymer optical fibres (POFs) which, whilst having inferior transparency to quartz fibres, are suitable for transmissions over short distances (less than 200 m), such as in homes or cars, and are much cheaper to install.

The basis of optical fibres being able to transmit light along them is the physical principle of total internal reflection, where any light travelling down the core that strikes the interface with the cladding will be reflected back into the core, as long as the angle of incidence is less than the critical angle (see Figure 7.11). If light strikes the core from outside the acceptance cone, it is lost in the cladding.

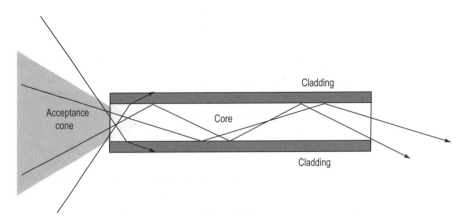

Figure 7.11   Reflection within an optical fibre.

The sine of the angle of aperture of the acceptance cone is called the *numerical aperture* (*NA*), the value of which depends on the refractive indices of the core and cladding:

$$NA = \sin\theta_A = \sqrt{n_{co}^2 - n_{cl}^2} \qquad (7.2)$$

Within the core the light will not escape from the fibre if it strikes the core-cladding interface at an angle less than the critical angle, θ_C, which is given by:

$$\theta_C = \sin^{-1}(n_{cladding}/n_{core}) \qquad (7.3)$$

where n_{core} and $n_{cladding}$ are the refractive indices of the core and the cladding, respectively. For a POF having a PMMA core (refractive index = 1.492) and a fluoropolymer cladding (refractive index = 1.417), $\theta_C = 72°$. For a glass fibre $\theta_C = 16°$, so a PMMA POF is capable of transmitting light through bends more effectively than glass optical fibres.

If an optical fibre has a large core, of around 100 microns in diameter, some of the light rays that make up the digital pulse may travel a direct route along the fibre, whereas others may zigzag along it as they bounce off the cladding. These alternative pathways cause different groupings of light rays, referred to as modes, to arrive separately at the receiving point at the other end of the fibre. The pulse, an aggregate of the different modes, begins to spread out, losing its well-defined shape. The need to leave spacing between pulses to prevent overlapping limits bandwidth, that is, the amount of information that can be sent. Consequently, this type of fibre, known as a 'multi-mode' fibre, is best suited for transmission over very short distances, in an endoscope, for instance. A much simpler situation exists when light travels along a very thin fibre, because then there are virtually no wall reflections. There is very little dispersion of the pulse of light and such fibres are referred to as 'single-mode' fibres. The radius of the core in single-mode fibres is typically only a few microns in diameter.

7.3.2 Glass Fibres

Glass is a hard, transparent inorganic material, formed when white sand (silica; SiO_2) is fused with metal oxides or carbonates to give a mixture of silicates. The structure of glass is somewhat different from most other materials, and certainly different from other types of textile fibres, in that it is completely amorphous, existing as a super-cooled liquid, as though it has formed by cooling the melt so quickly that a crystalline structure has not had time to form. Fused silica forms an excellent glass, but its melting point is over 1800 °C, which is a very high temperature and the process is costly in terms of the energy required to maintain this temperature. Pure silica glass is produced, but only for applications where the high cost is justified, for example in optical applications where a high transparency is required, particularly in the ultraviolet region of the spectrum. Otherwise, it is usual practice to add other compounds to the glass to reduce the temperature of melting and to facilitate extrusion into fibre form, and these compounds can also confer other beneficial attributes to the fibres.

Glass fibres are manufactured by a melt extrusion process, on the basis of a 'high heat/quick cool' strategy to ensure an amorphous product. Molten glass, at a temperature of about 1200 °C, exits under gravity from a 'bushing' (the equivalent of the spinneret in conventional fibre melt extrusion (see Section 5.1.2) containing between 200–8000 holes, each 1–2 mm in diameter. The bushings are made of a platinum–rhodium alloy to resist erosion by the molten glass continually passing through them. After emerging from the bushing, the fibres are mechanically drawn at high speed, with simultaneous cooling by spraying with water, so the final diameter of the fibres is between 4–34 μm. Before winding the web of parallel glass fibre filaments onto a cone, the fibres are coated with a 'size', which is a chemical coating. The size is composed of a lubricant, binder and/or coupling agent. The lubricant reduces friction between the fibre filaments and prevents abrasion during winding and subsequent weaving operations. If the glass fibres are to be used as reinforcing elements in plastic composites, the coupling agent is added give the glass affinity for the resin to be reinforced. Consequently, the coupling agent strengthens the bonding at the interface between the glass fibre and the resin matrix. Since the resins used are very varied, the formulation of the size is critical to the structural integrity of the final composite. Sizes for textile glass fibres typically comprise:

- One or more film formers, such as polyvinyl acetates, or poly-urethane resins;
- A lubricant, such as an oil or wax;
- A wetting agent, usually a fatty acid-based polyamide; and
- One or more coupling agents, usually organo-functional silanes, of general formula $R^1Si(OR^2)_4$, (where R^1 = an amine group, $R^2 = H$ or a small alkyl group), such as γ-aminopropyltriethoxysilane.

The silanes are mostly hydrolyzed to silanols before they are added to the size. The silanols react with the glass surface, forming a coupling agent layer of a thickness of approximately 5 nm, covering the fibre surface like a protective film. In the beginning, this protective film is still soluble as an oligomer, but then condenses to cross-linked structures later on, resulting in a siloxane.

The need to operate the manufacturing process at commercially viable temperatures by incorporating compounds into the silica has led to the development of a range of types of glasses, according to the compounds added. The original glass used for fibre manufacture was soda-lime glass, called A-glass. This is formed using silica (SiO_2), limestone ($CaCO_3$) and sodium carbonate (Na_2CO_3), but it has a lack of resistance to alkali. The corrosion process can be represented as:

$$-Si-O-Si + OH^- \rightarrow -Si-O^- + SiOH$$

with a consequent deterioration in the physical properties of the fibre.

A better quality of glass, E-glass (the 'E' standing for 'electrical' because of its high electrical resistance), was developed which contains about 50% silica and a range of other compounds, including the oxides of boron, aluminium and calcium, together with fluorspar, boric acid and clay. It can be regarded as an alumina–borosilicate glass and it is the type most commonly used to make fibreglass for composites. Whilst E-glass has good resistance to alkali, its resistance to acid is less good and it is attacked by chloride ions. Also it can be eroded by a leaching action when placed in hot water for 24 hours. Another type, C-glass or E-CR glass (corrosion-resistant), is an alumina–lime silicate with <1% alkali oxides, developed to be resistant to attack by acids. A high strength glass, known as R-glass, has been developed. R-glass contains higher levels of silica oxide, aluminium oxide and magnesium oxide than E-glass and is 40–70% stronger than E-glass. D-glass is a borosilicate glass with a high dielectric constant.

The uses of glass fibres are principally in reinforcement, in the manufacture of composites for boat hulls, surfboards and other structures. They are also used for thermal and electrical insulation, for sound absorption and for heat-resistant fabrics. A major use of glass fibres, however, is for communication and the capacity of glass fibres for data transmission is huge. Even a single silica glass fibre can carry hundreds of thousands of telephone channels, and the attenuation rate of modern single-mode fibres (see Section 7.3.3 for an explanation of single- and multi-mode fibres) is so small (about $0.2\,dB\,km^{-1}$) that signals can be transmitted for many tens of kilometres. If transmission over much longer distances is required, amplification is necessary. Although it might be expected that light in the visible region of the spectrum is transmitted along the optical fibres, the wavelength most often used is in the infrared region at 1.5 μm, because at this wavelength losses in silica are minimal. To amplify signals, erbium-doped fibre amplifiers (EDFAs) are used. An EDFA comprises a short length of erbium-doped glass fibre (usually a single-mode fibre) connected through dichroic pump couplers to two laser diodes (see Figure 7.12). Light of 980 nm is pumped from the two laser diodes in opposite directions, causing excitation of the erbium and stimulated emission at 1.5 μm, thereby enhancing the incoming signal. By incorporating a number of such EDFAs along an optical fibre, data can be transmitted over considerable distances.

The main advantages glass optical fibres have over electrical cables are their greater data carrying capacity, lighter weight and insensitivity to electromagnetic interference.

7.3.3 Polymer Optical Fibres (POFs)

Polymer optical fibres are lightweight, flexible and cheap and allow for faster data transmission rates than copper cables. They do not suffer from electromagnetic interference either. The polymer that is used most commonly for the manufacture of optical fibres is polymethylmethacrylate (PMMA), with polycarbonate (PC) also much used, and to a lesser extent polystyrene (PS).

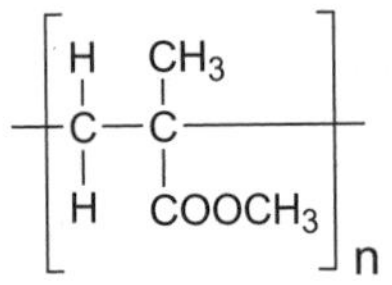

Polymethylmethacrylate
(PMMA)

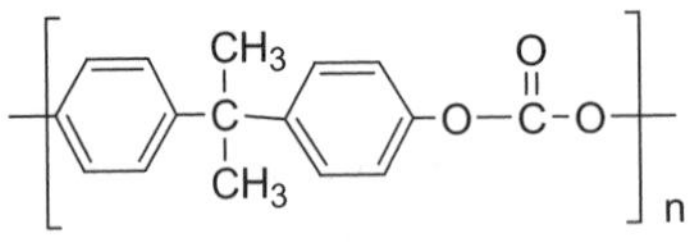

Polycarbonate
(PC)

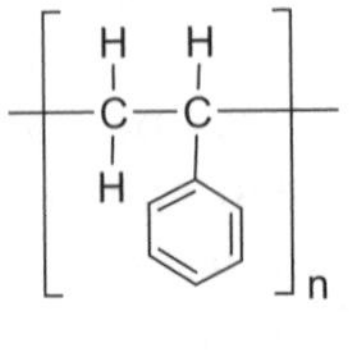

Polystyrene
(PS)

With PMMA core optical fibres, transmission speeds of 500 Mb s^{-1} over 50 m can be reached with an attenuation rate of about 130 dB km^{-1}.

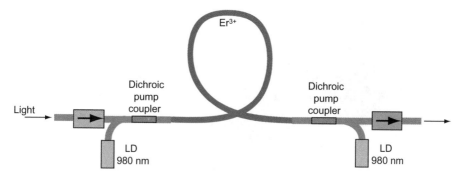

Figure 7.12 Erbium-doped fibre amplifier. (Diagram copied from Encyclopedia of Laser Physics and Technology of RP Photonics Consulting GmbH)

The wavelength of light used for optical transceiver modules is 650 nm, which is a wavelength at which the absorption by PMMA is quite low. However PMMA degrades rapidly in humid, high-temperature environments (over 85 °C) and its useful life is then reduced to a few thousand hours. PC is more stable at higher temperatures and humidities, but its attenuation rate is higher and can approach 300 dB km^{-1}. Over the last ten years, another POF has been introduced by Asahi Glass, a fluorinated polymer called CYTOP, which is poly(perfluorobutenylvinyl ether). This polymer is totally amorphous and the absence of crystalline domains prevents light scattering.

CYTOP

The transmission is improved to such an extent in CYTOP that the attenuation rate is only about 20 dB km^{-1}, so signals can be transmitted over distances of up to 1 km.

POFs are comprised of three layers (see Figure 7.13);

- A core, made of PMMA, PC or PS, which is circular in cross-section;
- The cladding, made of a fluoropolymer, which has a lower refractive index than the core; and
- A protective cover, called the jacket, made of polyethylene, polyvinylchloride or chlorinated polyethylene.

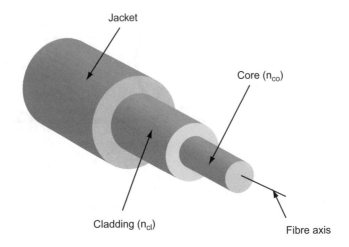

Figure 7.13 Structure of a polymer optical fibre.

In comparison with fibres for textile uses, the fibres for communication applications are very large in diameter, of about 500 µm, with some diameters of 1000 µm. Of this diameter, the cladding is only a thin coating, the core comprising some 96% of the fibre.

Usually the core has either a uniform (step-index; SI) or a graded-index (GI) refractive index profile, but multi-step structures are also produced (see Figure 7.14).

A step-index POF has a large core, up to 100 microns in diameter. It is a multi-mode fibre, best suited for transmission over very short distances, in an endoscope, for instance.

A graded-index POF contains a core in which the refractive index diminishes gradually with distance, r, from the centre axis out toward the cladding, according to the formula:

$$n(r) = n_{co}\sqrt{1 - 2\Delta(r/a)^g} \qquad r \le a \qquad (7.4)$$

$$n(r) = n_{co}\sqrt{1 - 2\Delta} \qquad r > a \qquad (7.5)$$

where a = radius of the core, r = distance from the fibre axis, n_{co} = refractive index of the core, n_{cl} = refractive index of the cladding and Δ = relative difference between the indices:

$$\Delta = \frac{n_{co}^2 - n_{cl}^2}{2n_{co}^2} \qquad (7.6)$$

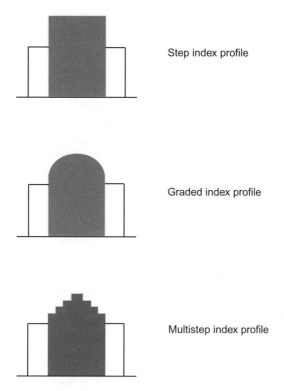

Figure 7.14 Different types of refractive index profiles.

The factor *g* is called the profile exponent. A parabolic profile is obtained when $g = 2$, and a step-index profile is obtained in the limit $g \to \infty$. The higher refractive index at the centre of a graded-index core makes the light rays moving down the axis advance more slowly than those near the cladding. Also, rather than zigzagging off the cladding, light in the core curves helically because of the graded index, reducing its travel distance. The shortened path and the higher speed allow light at the periphery to arrive at a receiver at about the same time as the slow but straight rays in the core axis. As a result, a digital pulse suffers less dispersion and these fibres act as single-mode fibres.

POFs are made in two stages. In the first stage, a 'preform' is made, which is a solid cylindrical rod, between 0.5–1.0 m in length and several centimetres in diameter, and then in the second stage this preform is drawn to a length of several kilometres. To make a step-index POF, the preform is made by a melt spinning process. To make a graded-index preform a polymer hollow fibre of the required polymer (*e.g.* PMMA) is made, which acts as the cladding in the final POF. The hollow fibre cylinder is then filled with a mixture of:

- The monomer used to make the same polymer as that of the hollow cylinder;
- A dopant whose refractive index (RI) is higher than that of the cladding polymer, the higher RI being required to cause total internal reflection. Dopants used can be bromobenzene (RI = 1.60), benzyl *n*-butyl phthalate (RI = 1.54) or benzyl benzoate (RI = 1.568);
- An initiator; and
- A chain transfer agent.

Polymerisation of this mixture is then carried out by heating the cylinder (at about 95 °C) and simultaneously rotating the cylinder about its longitudinal axis for about 24 hours. A gel phase forms on the inner wall of the cylinder. The rate of polymerisation is faster here, so polymer is formed more slowly towards the middle of the cylinder, until finally the hollow cylinder is filled with a polymer having a higher refractive index than the cylinder wall. The monomer molecules have a lower molecular volume than the dopant molecules, so are able to diffuse more rapidly to the gel phase, consequently the distribution of the dopant molecules in the final polymer is such that they are more concentrated in the central region of the core, giving this region a higher RI than the peripheral regions near the wall.

In the second stage the preform is drawn at about 200 °C to the desired diameter, which can be some 300-fold smaller. During this extension, the relative structural dimensions of the core and the cladding of the original preform are maintained. Directly after drawing, the protective jacket (a plastic coating) is applied.

7.3.4 Uses of Optical Fibres in Textiles

POFs are being employed in technical applications such as the development of sensors for 'smart' textiles. 'Smart' textiles are textiles that are capable of monitoring their environment and providing a response that protects the wearer, and sensors that can monitor the physiological condition of the wearer in real-time are an integral requirement. They are easily incorporated into textiles and their ability to measure colour, opacity or turbidity can be used to monitor the blood or even detect the location and extent of bleeding, for example in the case of injury. Wearable optical sensors can also be used for the non-invasive measurement of body fluids such as the pH of sweat or urine. These sensors detect the colour change that takes place in a pH sensitive sol-gel layer of about 0.5–5 μm applied to the fibre core.

POF sensors can be embedded into woven composites that are used to construct strong lightweight structures, where they can monitor stresses and strains and the overall structural integrity. These types of sensors are highly resistant to corrosion.

Another technical use of optical fibres is for making flexible display screens for clothes, or for exhibiting information or designs in cars or buildings. Display screens made from POFs are very thin and ultra lightweight. Prototypes have been made by weaving a PMMA POF as the weft with a natural or synthetic textile fibre as the warp (see Figure 7.15). The POF used for this purpose has a diameter of 0.5 mm; the fibres cannot be so thick that they resist bending in the weave structure and give a screen which is too rigid, but neither can they be so thin that the display fabric is too floppy.

Bending of the POF is a problem in woven or knitted textile structures since it can cause mechanical damage leading to a reduction in signal strength. A sateen woven structure, in which the weft yarns cross over five (or sometimes more) warp yarns is used (see Figure 7.16).

It is necessary to form light emitting 'pixels' in the textile screen, and this can be achieved by generating micro-perforations in the cladding that enable light to escape from the core (see Figure 7.17).

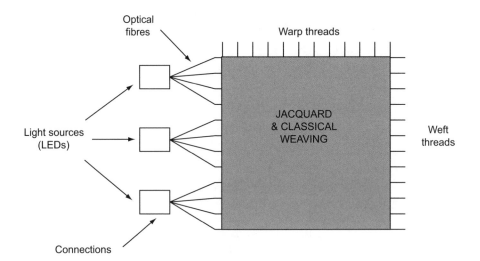

Figure 7.15 Woven fabric structure of a display screen. (Diagram courtesy of V Koncar, ENSAIT, France.)

Figure 7.16 PMMA optical fibres woven with silk. (Photograph courtesy of V Koncar, ENSAIT, France.)

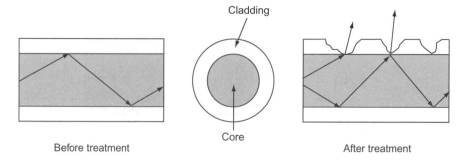

Figure 7.17 Micro-perforations to facilitate lateral light emission.

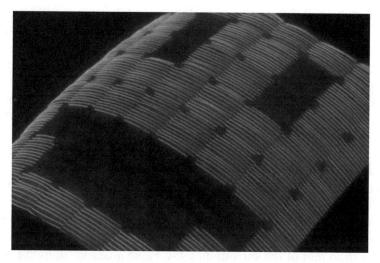

Figure 7.18   An optical fibre flexible display. (Photograph courtesy of V Koncar, ENSAIT, France)

The microperforations can be induced in the cladding by blasting micro particles on to the fibre cladding or by treatment with solvents.

The optical fibres can be connected to highly luminous LEDs that operate on small voltages. By using LEDs of different colours, varying the intensity or even having flashing lights, a variety of visual effects can be produced (see Figure 7.18).

The main application of flexible optical fibre displays is for exhibiting visual information such as simple texts or pictograms. The screens can display images which can be used in clothing in a design/fashion context, where animations could be downloaded from the internet for example then displayed on the garment. Fibres based on total internal reflection that have been treated to emit light out through the cladding layer in this way are also used as backlighting panels for medical and industrial applications.

7.3.5   Photonic Crystal Fibres (PCFs)

In the foregoing description of optical fibres, light is transmitted internally along the fibres by total internal reflection. Where it is required that light is emitted from the surface of the fibre, either the fibre has to be bent along its length so that when light strikes the cladding it does so at an angle which is greater than the critical angle, causing it to be refracted out through the cladding, or the cladding is etched to enable light scattering to take place. The process by which light is guided along photonic crystal fibres is quite different and is referred to as the 'photonic band gap effect'. Consequently, these fibres are often called photonic band gap (PBG) fibres and they are used in fibre-optic communications and as gas sensors, but because they can be tuned to emit visible light of certain wavelengths, they also have potential design applications in textiles.

PBG fibres are characterised by a periodic dielectric microstructure, which are regions of high and regions of low dielectric constant that repeat in a regular, periodic arrangement in their cross-section at a distance called the 'band gap'. These regions can be created by micron-sized air holes (the so-called 'holey' fibres), as illustrated in Figure 7.19a, or by forming concentric rings of polymers of different refractive index around a central core, which may be hollow or solid (Bragg fibres), as illustrated in Figure 7.19b. In each case the cross-sectional structures of these fibres are uniform along the length of the fibre. The refractive index of the core is lower than that of the surrounding layers, so guidance of light through the fibre cannot occur by total internal reflection. Instead, light is confined within a PBG fibre and guided along it by the photonic structure.

If an incident wave is in the band gap, it partially reflects off each layer of the dielectric slab. The reflected waves are in phase, reinforce each other and combine with the incident wave to produce a standing wave that does not travel through the fibre. In the case of an incident wave which is not in the band gap, the reflected waves are out of phase and cancel each other out, with the result that the light propagates through the material and is only slightly attenuated.

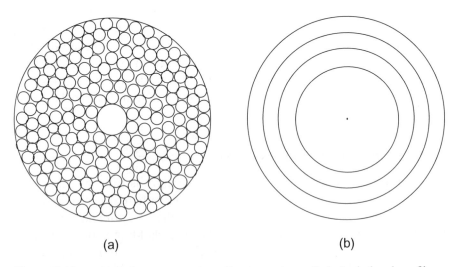

(a)   (b)

Figure 7.19   Schematic representation of cross-sections of photonic bandgap fibres: (a) holey fibre, (b) Bragg fibre.

The first PCFs were made by drawing a tube of silica glass, about 20 mm in diameter, to give a much thinner capillary of about 1 mm. Many of these capillaries were then stacked together, the stack being 1–2 cm in diameter. The stack was then fused and drawn to a very fine diameter (about 100 µm), preserving the cross-section of the preform along the entire fibre length. The presence of the periodic regions of high and low dielectric character allowed the transmission of electromagnetic light of some wavelengths, but not of others, so a photonic band gap existed, in a manner analogous to the existence of an electronic band gap where certain electron energies are disallowed.

The wavelengths of light that are allowed to propagate are confined to the air holes or to the central core of the Bragg fibre types. The forbidden wavelengths are diffracted out through the cladding layers and, depending upon their wavelengths, give a coloured effect to the fibres. Photonic crystal Bragg fibres made of chalcogenide glasses have been developed for the transmission of infrared radiation and as such they can be used as sensors. If a gas containing an analyte is introduced into the core, analysis by absorption of the infrared radiation can be made. Chalcogenide glasses absorb electromagnetic radiation in the visible region, however, and are therefore unsuitable for use in this region of the spectrum due to their strong orange colour. Alternatives for use in the visible region are silica glasses or polymer-based PBG fibres.

A hollow core PBG Bragg fibre has a complex structure. The air-filled hollow core is surrounded by a periodic sequence of alternating layers of high and low refractive index. In practice light can leak out through the reflector layers, and the fewer the layers, the greater is the leakage. Thus for the effective transmission of light, a large number of layers is required. Conversely, if it is preferred for light to escape, to produce optical effects for example, then only a small number of reflector layers are necessary. Another property of PBG fibres is that, depending on the refractive indices of the layers, a photonic band gap exists in the fibre and light with a wavelength greater than the band gap is effectively contained within the hollow core. All other wavelengths are irradiated out through the reflector layers during the first few cm of the fibre length. This latter situation provides opportunities for novel applications in textile fabrics. The colour of the irradiated light depends on the number and thicknesses of the reflector layers so by engineering their dimensions, the optical coloured effects can be tuned.

The fabrication of such fibre structures is very difficult and as a cheaper alternative, solid core PBG Bragg fibres, in which the refractive index of the core is very low, have been developed. The core of these fibres is made of polymethylmethacrylate (PMMA) with a diameter of about 1 cm. This core is surrounded by layers of a polystyrene

(PS)/PMMA blend (high refractive index) and a polycarbonate (PC)/ polyvinylene difluoride (PVDF) blend (low refractive index) and they are applied at small nano-sized layers round the core. The thicknesses of the layers can be varied by stretching (drawing) the fibres. The centre wavelength, λ_c, of the reflector band gap is given by the formula:

$$\frac{\lambda_c}{2} = d_h\sqrt{n_h^2 - n_c^2} + d_l\sqrt{n_l^2 - n_c^2} \qquad (7.7)$$

where n_c is the refractive index of the core, n_l and n_h are the refractive indices of the low and the high refractive index layers respectively, and d_l and d_h are their thicknesses. The periodicity of the high and low dielectric regions of the fibres has to be half the wavelength of the electromagnetic radiation to be guided, so for visible light (*ca.* 400–700 nm) the periodicity has to be 200–350 nm.

An additional property of these fibres is that they appear coloured under ambient lighting conditions, that is, when illuminated transversally with no illumination into the length of the fibre. Additional creative design opportunities are thus possible for such fibres by mixing the colour of the ambient reflected light with that of the irradiated guided light. It is likely that such fibres, as their production processes become more mature, will become cheaper and more commonly available for use by textile manufacturers. Potentially these fibres provide an alternative to dyeing and printing as a method for producing coloured effects.

7.4 LUMINESCENT FIBRES

7.4.1 Introduction

Luminescence is the emission of light by a material by a process other than heating, and it can result from a number of different processes, such as chemical reactions, mechanical action or the absorption of radiation. Of the various forms, only two types of luminescence will be covered in this section: photoluminescence and electroluminescence, since these are of most interest in textile fibre science.

7.4.2 Photoluminescent Fibres

Photoluminescence occurs when a material absorbs electromagnetic radiation then emits the absorbed energy as light, usually of a longer wavelength than that absorbed. The most common form of photoluminescence is fluorescence, which occurs when the absorbed energy causes a molecule to be raised from the ground electronic state (S_0) to one of the higher vibrational levels in the first excited singlet state (S_1). Through collisions with other molecules, the molecule quickly loses vibrational energy, falling to the lowest vibrational energy level in S_1, before returning from S_1 to the ground electronic state S_0. Since the transition from S_1 back to S_0 now involves less energy than that which was absorbed, the light emitted has a longer wavelength.

Fluorescence is a fast process and the emission ceases directly the illuminating source is removed. With some molecules, however, the excitation involves what is known as 'intersystem crossing' into the triplet state, in which the electrons have a higher spin multiplicity. The relaxation process occurs more slowly, because the transition is forbidden in quantum mechanical terms, with the result that the emission of light occurs for a considerable time, sometimes for a few hours, after the irradiating source is removed. This phenomenon is called 'phosphorescence'. Much of the interest in producing photoluminescent textile fibres is centred around phosphorescence, so that long-lasting afterglow characteristics can be imparted to them. These fibres have applications in clothing, especially safety clothing, carpets for theatres and aircraft interiors, as well as toys and novelty items.

There are various methods for imparting photoluminescence to fibres. The fibres can be coated with phosphorescent pigments, but these coatings can wear away through abrasion and during washing. Coatings do not give high brightness levels either and after-glow periods are short. An alternative method that can be used for synthetic fibres is to extrude the fibres from a melt containing the phosphorescent pigment, so the pigment is incorporated into the bulk of the fibre. The phosphorescent particles used are sulfide phosphors, mainly zinc sulfide. Again luminous intensities and after-glow times are low and large quantities (5–10% by weight of polymer) are required to produce a reasonable effect. Furthermore, such high loadings reduce the spinnability of the polymer during extrusion and are detrimental to the physical properties of the fibre. Pigment particle size is also of importance, with larger sizes (*e.g.* $\sim 40\,\mu m$) giving a much greater brightness than smaller ones (*e.g.* $\sim 2\,\mu m$), but larger particles can damage spinning equipment and take-up rollers and cause difficulties in yarn cutting. The pigments used are susceptible to moisture and as most textile fibres naturally absorb water (see Section 1.6), luminous performance gradually deteriorates.

Two methods to overcome the difficulties in producing phosphorescent fibres with high brightness and acceptable durability have been developed:

- Encapsulating the particles. One process that has been patented involves encapsulating pigment particles, composed of strontium oxide, aluminium oxide, dysprosium oxide and europium oxide, in a maleic anhydride-grafted polypropylene. The encapsulated pigments are then pelleted with the polymer (polyamide, polyester or polypropylene) to give a master batch. The master batch is then mixed with pure polymer and additives, such as softener and dispersing agent, then melt spun.
- Bicomponent spinning. In this process, one component is a photoluminescent polymer and the other component is a non-photoluminescent polymer, the polymer in each case being a polyolefin such as polypropylene. The particular bicomponent structure used is the sheath/core type in which the core is photoluminescent, containing alkaline earth metal aluminates activated by europium, and the sheath is non-photoluminescent. Polypropylene is very hydrophobic, so deterioration of the luminous intensity due to the influence of moisture on the photoluminescent particles is minimal.

7.4.3 Electroluminescent Fibres

Electroluminescence is a complex phenomenon and is of wide interest outside the scope of textile fibres in the fabrication of organic light-emitting diodes. An electroluminescent device has three layers (see Figure 7.20):

- A transparent indium tin oxide layer, which acts as the anode;
- An electroactive material, which is a fluorophore that is dissolved in a polymeric binder and coated from a solvent onto the anode; and
- A metal cathode, *e.g.* magnesium or aluminium.

A buffer layer on each side of the electroactive layer is usually incorporated, which enhances the transfer of electrons between the electroactive layer and the two electrodes.

The transparent substrate is usually glass or plastic, but there has been a growth in interest in recent times for using the devices for flexible colour displays. For this application, polyester and polyimides have been found useful as flexible substrates. For integration into clothing, a nylon monofilament woven fabric has also been used as a flexible

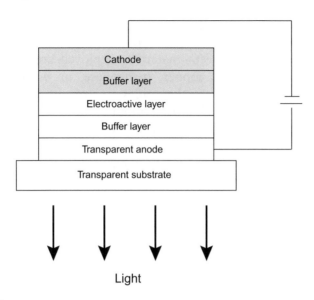

Figure 7.20   Basic structure of an electroluminescent device.

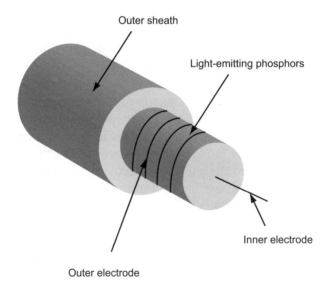

Figure 7.21   Basic structure of an electroluminescent cable.

substrate. This structural concept can be applied to the fabrication of electroluminescent cables (see Figure 7.21).

The outer wire is in contact with a very thin conducting film, thereby creating the outer electrode. When an alternating current is passed between the two electrodes the phosphors emit light. Depending on the phosphors and the voltage used, a wide range of colours can be produced. The electroluminescent materials employed are based on zinc sulfide doped with either copper, manganese, or both copper and manganese. The diameter of the cables is large in relation to conventional textile fibres, ranging from about 1–5 mm.

There is interest in electroluminescence for optical fibre sensor technology, especially for sensing electric field intensity. A sensor containing an electroluminescent material can be connected to an optical fibre (see Section 7.3.1). If the sensor is placed in an electric field, electroluminescence is then emitted which can be transmitted along the optical fibre to a remote photodiode detector.

7.5 BIODEGRADABLE FIBRES

7.5.1 Introduction

Fibres which are biodegradable are those which are susceptible to attack by enzymes, the action on the polymer chains of the fibres being either hydrolysis or oxidation. The attack can be limited to the ends of the polymer chain or may take place at random points along the polymer, the latter causing chain scission. When depolymerisation is complete, the fibre has effectively been broken down into the monomer units which were the building blocks of the polymer.

The characteristic features of biodegradable polymers are related to both their chemical and morphological structures. In chemical terms, fibres that are highly hydrophilic and contain oxygen or nitrogen in their repeating unit are highly biodegradable. The rate of attack by enzymes depends on the ability of the enzyme molecules to penetrate the polymer matrix, so fibres which are highly amorphous, with low degrees of crystallinity (molecular order) will be attacked more readily than highly ordered, crystalline fibres. Thus, fibres that are both hydrophilic, contain oxygen (or nitrogen), and are highly amorphous such as viscose, will biodegrade readily. At the opposite end of the scale, polyolefin fibres, which are totally hydrophobic, highly crystalline, and contain no oxygen (or nitrogen), are completely resistant to biodegradation.

There are many fibres that possess chemical and physical constitutions that are in conflict in terms of facilitating biodegradability. Flax is highly hydrophilic and is a very absorbent fibre (see Section 2.3.1), but it is quite resistant to biodegradation since it has a high degree of crystallinity and the crystalline regions are highly ordered. Indeed, flax has the greatest resistance to biodegradation of all the cellulosic fibres because of this feature. Generally, however, cellulosic fibres are bio-degradable, the only difference between the various types being the rate at which they degrade. A similar situation exists with synthetic fibres such as polyester and polyamide. These fibres contain oxygen (and also nitrogen in the case of polyamide), and have a small but not insignificant absorbance. As with flax they are highly crystalline, however, and so have the capacity to resist biodegradation.

7.5.2 Polylactic Acid (PLA)

7.5.2.1 Introduction. Polylactic acid is one of a group of biodegradable aliphatic polyesters with the potential to form fibres.

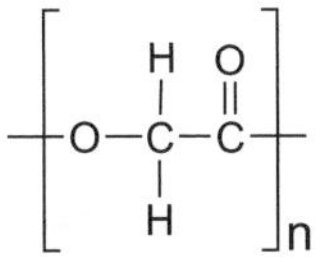

Polyglycolic acid (PGA)

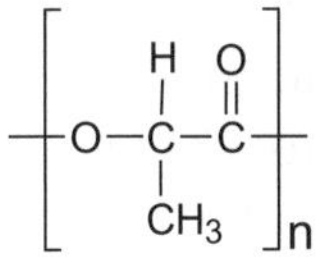

Polylactic acid (PLA)

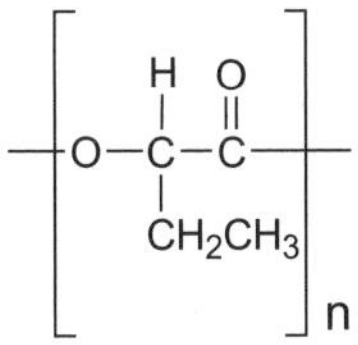

Polybutyric acid (PHB)

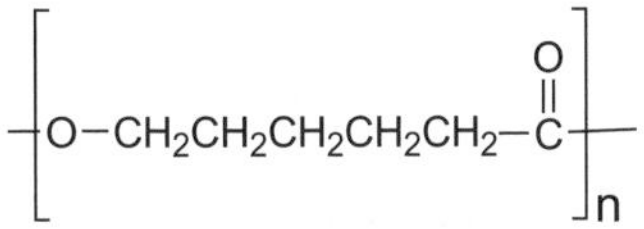

Polycaprolactone

Of these compounds, however, only polylactic acid has been successful commercially, but not to such an extent that it is a mainstream fibre for clothing manufacture. In this market sector, standard polyester remains dominant because of its superior technical performance (see Section 5.3). The main market for PLA fibres is now in the plastics sector, in applications such as packaging.

Although considerable interest in PLA has developed over the last ten years or so, because of its biodegradability, it is not a newly discovered polymer. Indeed, PLA was first synthesised by Wallace Carothers (the pioneer of nylon – see Section 5.2.1) in 1932. However, Carothers synthesised PLA by heating lactic acid in a vacuum and the product he obtained had only modest molar mass. Since the 1950s PLA has become established in the manufacture of sutures and medical implants, so it has importance due to its biocompatibility, as well as its biodegradability.

The attraction of PLA today is the fact that it has very good eco-friendly credentials. It is produced from renewable resources (corn starch in the USA, but also maize, sugar or wheat) and it is 100% biodegradable, so there is minimal requirement for landfill at the end of product life. Growth of the carbohydrate-rich plant involves the process of photosynthesis, in which water and carbon dioxide from the atmosphere are converted into starch.

7.5.2.2 Manufacture of PLA Fibres.   The process for manufacturing PLA fibre can be broadly split into two stages. The first stage involves the production of lactic acid monomer from the starch produced by photosynthesis within the cells of the plants, and the second stage concerns the conversion of the lactic acid monomer to polylactic acid. In the first stage, the corn is sent by the farmers to milling plants where it is cooked for about 30–40 hours at 50 °C, then ground and screened to isolate the starch.

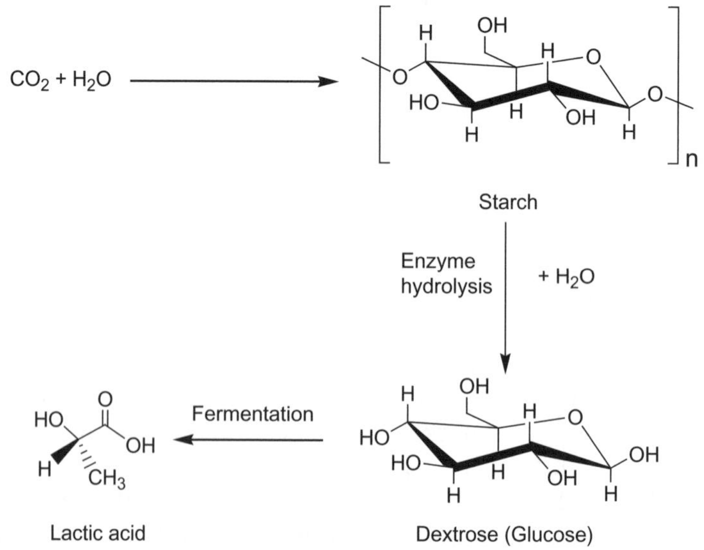

The starch is then converted to dextrose (glucose) by enzymatic hydrolysis and the dextrose then converted to lactic acid by fermentation, using a strain of *Lactobacillus*. Improvements in the fermentation process in recent years have considerably reduced the costs of making lactic acid. Lactic acid exists as two enantiomers: L- and D-lactic acid.

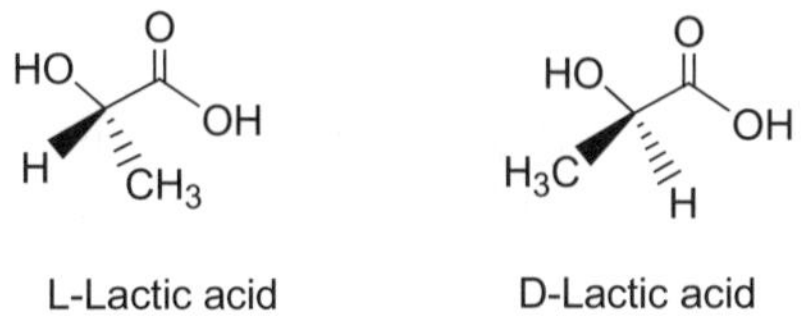

L-Lactic acid   D-Lactic acid

L-lactic acid rotates the plane of polarised light clockwise, whilst D-lactic acid rotates it anticlockwise. Whilst chemical synthesis yields a racemic mixture (50% L and 50% D) of lactic acid, the product by the fermentation is >99.5% of the L-isomer. The stereochemistry of the L- and D-isomers is such that polymers (and fibres) of different structural characteristics can be formed. At >99.5% L-isomer, the polymer formed is very crystalline, but as the proportion of the D-isomer is increased, the amorphous character increases.

The second stage involves the polymerisation of the lactic acid into polylactic acid and extrusion into fibre form. There are two methods by which the polymerisation can be carried out. The first method is the direct polycondensation of the lactic acid. In this process water is removed by the use of a solvent under conditions of high vacuum and high temperature. This is the method used by Carothers in 1932, but there are difficulties in removing the water which results in the production of a polymer of low molar mass (between 10 000–20 000). A polymer of higher molar mass can be obtained by azeotropic distillation using a high boiling point aromatic solvent. Versions of PLA fibres are manufactured in Japan by Mitsui Toatsu by a direct polymerisation from lactic acid.

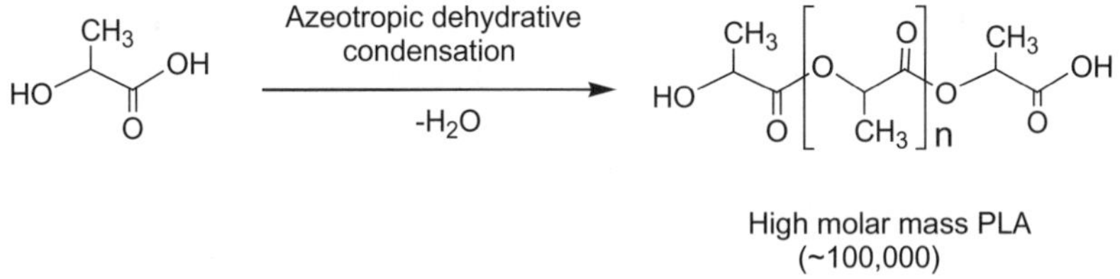

High molar mass PLA
(~100,000)

The second method is called ring-opening polymerisation. It is a solvent-free process, in which water is firstly removed under mild conditions, giving rise to lactic acid oligomers, which are effectively prepolymers of low molar mass (between 1000–5000). Through depolymerisation, the prepolymer is converted into a cyclic dimer, called a 'lactide'. Three forms of the cyclic dimer are possible, the LL-, the meso- (LD-) and the DD-lactides:

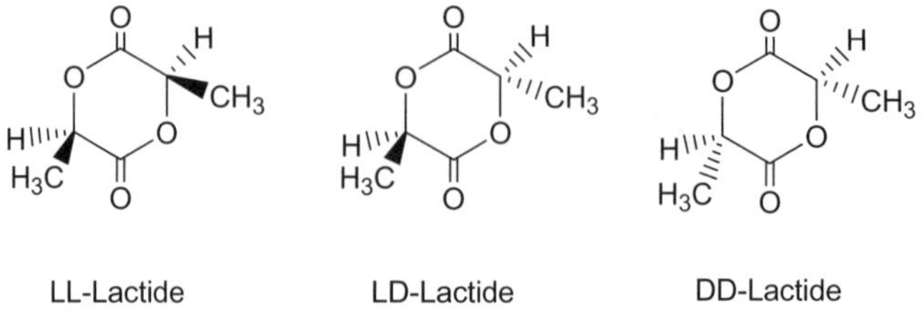

| LL-Lactide | LD-Lactide | DD-Lactide |

The lactide mixture is purified by vacuum distillation, then with the aid of a catalyst, typically stannous (II) *bis*-2-ethylhexanoic acid, and at a temperature between 180–210 °C, ring opening is brought about.

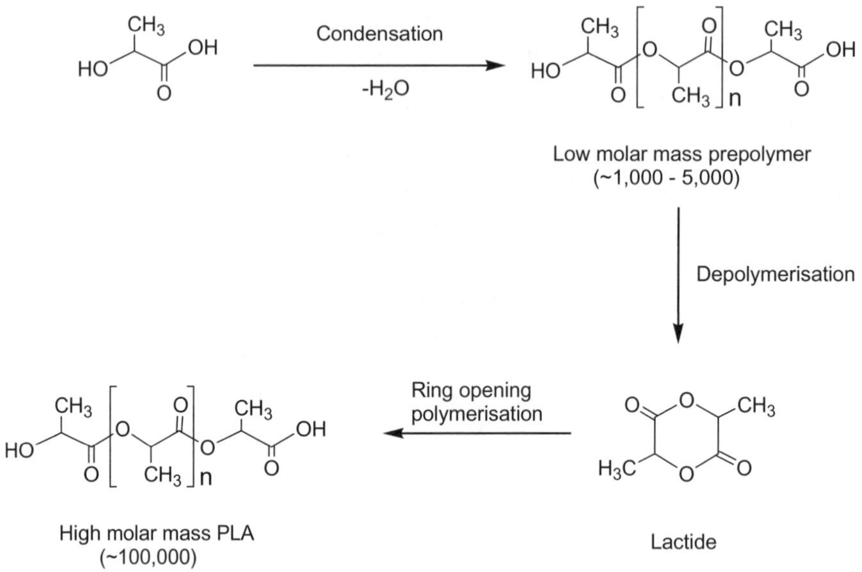

Low molar mass prepolymer
(~1,000 - 5,000)

Depolymerisation

Ring opening
polymerisation

High molar mass PLA
(~100,000)

Lactide

The reaction mechanism proposed occurs *via* a coordination insertion mechanism, yielding PLA of molar mass in excess of 50 000. This is the commercially favoured method for the production of PLA and has been developed and refined by one of the major producers of PLA, NatureWorks LLC, an independent company wholly owned by Cargill, in Nebraska, USA. The trade name it gives to its PLA fibre is Ingeo™.

Co-ordination insertion mechanism for ring-opening
and polymerisation of a lactide

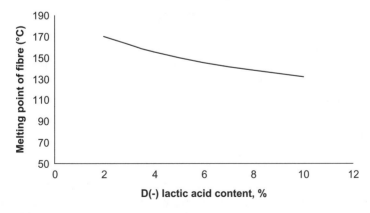

R = growing polymer chain

Polymers of a wide range of molar masses can be produced by varying the proportions, and the sequences of the L- and D-isomers. A polymer comprising only L-isomer is called P(L)LA and a polymer containing both isomers is called P(LD)LA. The greater the proportion of the L-isomer, the greater is the degree of crystallinity and the higher is the melting point of the resultant polymer (see Figure 7.22).

Fibres of PLA are usually formed by melt spinning, though both dry and wet spinning methods are possible.

Figure 7.22 Melting point of PLA fibres as a function of D-isomer content.

Table 7.3   Properties of PLA fibres.

Tenacity	32–36 cN tex^{-1}.
Elongation	30–40%.
Elastic recovery	93% from 5% stretch.
Specific gravity	1.25 g cm^{-3}.
Moisture regain	0.4–0.6%.
Melting point	120–175 °C.
Flammability	Hardly flammable, with good self-extinguishing properties. Low smoke generation.
Chemical resistance	Resistant to most solvents but dissolves in chloroform and methylene chloride. Very sensitive to alkali, but relatively stable in acid.

7.5.2.3   Properties of PLA Fibres.   PLA is a type of polyester, and in many of its properties it resembles standard PET fibres. Typical fibre properties are given in Table 7.3, though there is some variation in data because of the differences in the L- and D-isomer contents of the fibres produced by different manufacturers.

The main advantages claimed for PLA fibres are:

1. The only melt-processable fibre made from renewable resources. The use of solvents for extrusion is therefore avoided.
2. Similar in properties to standard PET in some ways and to PP and PE fibres in other ways.
3. Fully biodegradable under composting conditions. The degradation occurs firstly through hydrolysis to low molar mass fragments and then attack by microorganisms *via* lactic acid to carbon dioxide and water. The rate of degradation depends on temperature, humidity and the proportion of D-isomer in the polymer. Pure P(L)LA can take ten years to decompose, but P(LD)LA only takes a few months.
4. Low flammability.
5. Good elongation and recovery from stretch.
6. Low moisture absorption and excellent wicking properties.
7. Lower density than natural fibres, enabling light fabrics to be produced.
8. Excellent handle and touch.
9. Excellent UV resistance.
10. Biocompatible, for use in medical applications.

The fibres are especially good for their moisture management properties and are targeted at fabrics for sportswear, for example. The wicking characteristics are very effective, allowing the wearer to feel dry, even when undergoing high levels of physical activity. The fibre is also claimed to have lower odour retention than standard PET fibre.

There are some disadvantages to PLA fibres however. They have to be dyed using disperse dyes and such dyes are applied to standard PET fibres at 120–130 °C. However, to avoid degradation of the fibre, the maximum temperature which can be used is 110 °C, and dye uptake is only about 1/6th as much as on standard PET. Consequently to obtain the same depths of shade, considerably more dye has to be added to the dye bath, affecting light fastness and dry rubbing fastness properties.

The fibres have a high surface friction coefficient and low resistance to wear, so it is not always suitable for applications where heavy wear will be encountered. The low softening and melting temperatures also create problems, both for the garment manufacturer and the user. Garment manufacturers cut garment shapes simultaneously from several layers of fabric laid on top of each other. PLA fibres can fuse with each other due to the large shearing heat generated between the cutter and the fibre. For the user, considerable care has to be taken in laundering garments. Many detergents are formulated for washing cotton under slightly alkaline conditions, the very conditions to which PLA fibres are susceptible to degradation. When ironing PLA garments after washing, it is also important to ensure that the iron is set at only a very low temperature to avoid softening (or even melting) the fibres and distorting the fabric.

7.5.2.4 Uses of PLA Fibres. The uses of PLA fibres are widespread. In clothing, the main applications are in sportswear and shirts. The fibre can also be used in soft furnishings, for example for curtains and seat covers. There are many medical applications, such as sutures, tissue engineering, artificial implants and materials for drug release implants. It also has applications as technical textiles, where the biodegradability characteristic is important, such as in geotextiles for erosion protection and the reinforcement of embankments. Use is made of the susceptibility of the polymer to alkaline hydrolysis in the manufacture of microfibres and nanofibres by the 'islands-in-the-sea' method (see Section 7.1.2.3). A 'guest' polymer is extruded in a 'sea' of PLA and when the PLA is dissolved away, the fine filaments of the 'guest' polymer remain.

7.5.3 Polycaprolactone (PCL)

Polycaprolactone is another biodegradable aliphatic polyester. It is prepared by the ring-opening polymerisation of ε-caprolactone (2-oxepanone), a cyclic ester. As with the ring-opening polymerisation of the cyclic lactide to form polylactic acid, the reaction is catalysed by a tin salt, in this case stannous octanoate.

e-caprolactone   Polycaprolactone

The polymer has a very low melting point (around 60 °C), so it is limited to applications where it will not be subjected to temperatures much higher than room temperature. The main applications of PCL are those where biodegradability is a requirement, because the polymer is gradually degraded under alkaline conditions, though at a slower rate than that of PLA. It is used therefore for specialised medical products, such as long-term implantable devices, sutures and drug delivery devices. Encapsulating drugs in a shell of PCL enables a slow-release function and targeted delivery. Tissue engineering is a widely used technique in medicine, and another area where resorbable fibres are required, for the construction of scaffolds. PLA fibres are much used, especially for applications where long resorption times, for example in excess of one year, are required. PCL fibres have a higher compliance than PLA fibres and are therefore an attractive alternative. There is considerable interest in the electrospinning of PCL so it can be produced as webs of nanofibres, which are likely to have medical applications.

Another important application of PCL fibres is in the manufacture of sanitary items, such as diapers and sanitary napkins, disposable hand towels and wiping cloths. For these disposable items a self-digesting (microbial or biodegradable) fibre is required. In the case of sanitary towels, a biodegradable fibre that is also hydrophobic, has high strength and can be thermally bonded as a non-woven web to act as the surface sheeting is required. PCL fibres meet these requirements very well, though the conditions for the melt extrusion of the polymer have to be very carefully selected to produce fibres with the tenacity required.

7.5.4 Other Biodegradable Fibres

Other biodegradable fibres that have specialised applications, mostly for sutures, include poly-3-hydroxybutyrate and polyglycolide. Of the resorbable polymers, polyglycolide is the most commonly used. It is designed to maintain wound closure for fairly short periods, since the natural repair processes in skin and soft tissue take place relatively quickly – in just a few weeks. The polymer can be used individually, or as copolymers, examples being Vicryl which is a 90 : 10 poly(glycolide-L-lactide) co-polymer, poly(glycolide-ε-caprolactone) and poly(glycolide-trimethylene carbonate), all of which are used to make sutures. These polymers or copolymers degrade in the body by hydrolytic or enzymatic cleavage of the polymer chains. Their degradation products are not harmful to humans.

Polyglycolide is made by ring-opening polymerisation of glycolide, using a stannous octoate (tin (II) 2-ethylhexanoate) catalyst at approximately 200 °C for two hours.

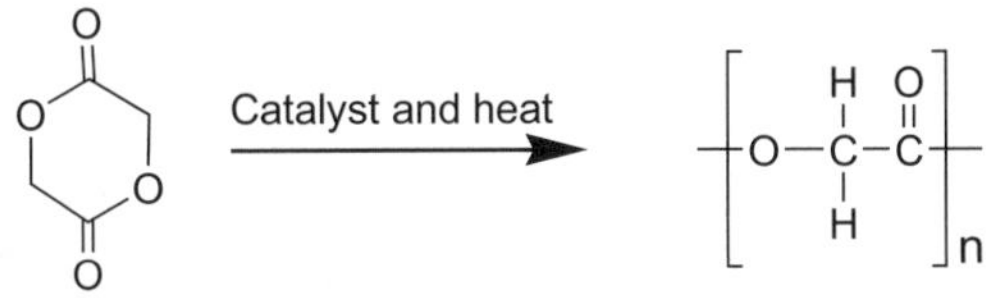

$$
\text{glycolide} \xrightarrow{\text{Catalyst and heat}} \left[O-\underset{H}{\overset{H}{C}}-\overset{O}{\overset{\|}{C}} \right]_n
$$

Poly-3-hydroxybutyrate is produced by microorganisms, but the cost of bulk scale production is uneconomic and efforts are being made to produce it from plants instead.

$$-[-O-CO-CH_2-CH_2-CH.CH_3-]-n$$

The polymer is difficult to process into fibre form by melt spinning, due to thermal degradation at temperatures above the melting point (175 °C) and its low melt viscosity. To some extent, the reduction in molar mass occurring through thermal degradation can be compensated for by starting with polymer of higher molar mass than that required, so although the final extruded polymer has a lower molar mass, it is still satisfactory for fibre-forming properties. On extrusion, the fibres tend to crystallise slowly, during which time large crystallites form, yielding a rather brittle fibre. A variety of techniques have been explored to overcome these difficulties, key amongst which is the need for drawing the extruded fibres quickly after they extrude from the spinneret. Gel spinning has also been explored as a way of overcoming the problems arising from thermal degradation.

SUGGESTED FURTHER READING

Nanofibres

1. P. J. Brown and K. Stevens, ed., *Nanofibers and Nanotechnology in Textiles*, Woodhead Publishing, Cambridge, UK, 2007.
2. D. Lukas, A. Sarkar, L. Martinova, K. Vodsedalkova, D. Lubasova, J. Chaloupek, P. Pokorny, P. Mikes, J. Chvojka and M. Komarek,

Physical principles of electrospinning (electrospinning as a nano-scale technology of the twenty-first century), *Textile Prog.*, 2009, **41**(2), 59–140.
3. S. Ramakrishna, K. Fujihara, W. E. Teo, T. C. Lim and Z. Ma, *An Introduction to Electrospinning and Nanofibres*, World Scientific Publishing, Singapore, 2005.
4. T. Subbiah, G. S. Bhat, R. W. Tock, S. Parameswaran and S. S. Ramkumar, Electrospinning of nanofibres, *J. Appl. Polym. Sci.*, 2005, **96**, 557–569.

Electrically Conducting Fibres

1. A. Dhawan, T. K. Ghosh, & A. Seyam, Fibre-based electrical and optical devices and systems, *Textile Prog.*, 2004, **36**(2/3), 1–84.
2. X. Tao, ed., *Wearable Electronics and Photonics*, Woodhead Publishing, Cambridge, UK, 2005.

Optical Fibres

1. R. Paschotta, *Encyclopedia of Laser Physics and Technology*, accessible at: www.rp-photonics.com
2. J. Zubia and J. Arrue, Plastic optical fibers: An introduction to their technical processes and applications, *Opt. Fibre Technol.*, 2001, **7**(2), 101–140.
3. F. Benabid, Hollow-core photonic bandgap fibre: New light guidance for new science and technology, *Philos. Trans. R. Soc. London, Ser. A*, 2006, **364**, 343–3462.

Biodegradable Fibres

1. R. S. Blackburn, ed., *Biodegradable and Sustainable Fibres*, Woodhead Publishing, Cambridge, UK, 2005.
2. B. Gupta, N. Revagade and J Hilborn, Poly(lactic acid) fiber: An overview, *Prog. Polym. Sci.*, 2007, **32**(4), 455–482.
3. B. Linnemann, M. Sri Harwoko and T. Gries, Polylactide acid fibers (PLA), *Chem. Fibres Int.*, 2003, **53**, 426–433.

CHAPTER 8

Enhancement of Fibre Performance by Surface Modification

8.1 INTRODUCTION

The textile fibre types discussed in previous chapters clearly have a variety of chemical and physical properties. In the case of man-made fibres, it is apparent that properties can be engineered by modifications to process conditions, such as those influencing the molar mass of the polymer, extrusion conditions, as well as treatments after extrusion (such as drawing, texturising, *etc.*). It is also possible to modify the performance characteristics of fibres after production by appropriate after-treatments, which may take the form of wet or dry processes.

Many of these after-treatments modify the surface features of fibres, which in turn influence the way they behave when in contact with liquids, other fibres or surfaces. Some treatments influence the bulk fibre rather than just the fibre surface, such as flame retardant and easy-care finishes. The main focus of this chapter is on treatments that influence the surface characteristics of fibres, indicating the nature of the treatments and the effects which they offer for improving the performance of fibres for particular applications.

8.2 WETTING AND WICKING PROPERTIES OF TEXTILES

8.2.1 Introduction

The way in which textile fibres respond to liquids is of paramount importance since it influences not only the ease of processing, but also has a

The Chemistry of Textile Fibres
By Robert R Mather and Roger H Wardman
© Robert R Mather and Roger H Wardman 2011
Published by the Royal Society of Chemistry, www.rsc.org

significant influence on the comfort of wear of garments and on the effectiveness of functional materials, such as medical wipes and hygiene products. During textile manufacture, fibres are treated with a variety of liquids, such as aqueous media (in scouring, desizing, bleaching, dyeing and printing operations) and organic liquids (in the application of spinning oils to reduce inter-fibre friction during weaving or knitting). The speed and effectiveness by which the treatment liquors or solvents wet the surface of the fibre govern the efficiency of the textile process concerned.

In terms of comfort of wear, there is a need for base-layer garments (garments worn next to the skin) to deal adequately with perspiration. Whilst the body is working at normal levels of activity, it is most comfortable to have an absorbent base-layer fabric, such as cellulosic materials like cotton, viscose or Lyocell. When the body is engaged in strenuous levels of activity, however, the amounts of perspiration produced are much larger and absorbent materials can become wet and uncomfortable. In this case base-layer fabrics which are able to wick the moisture away from the skin to outer garment layers are preferable. Thus for sportswear, for example, hydrophobic fibres such as polyester are preferred since they are non-absorbent, especially those which have grooves along their length (non-circular, cross-sectional shapes). The polyester fibres are given a hydrophilic coating to enable their surfaces to attract perspiration from the skin, however, after which the wicking action takes place. The efficiency of moisture transport by a garment also influences body temperature. If the rate is low, the humidity level within the clothing microclimate around the skin increases and the removal of sweat is suppressed. This in turn causes an increase in rectal and skin temperatures, leading to heat stress. In the case of garments worn in cold climates, the accumulation of moisture in the base-layer reduces the thermal insulation of clothing, so an unwanted loss in body heat can result. The efficient removal of perspiration from clothing is therefore essential to maintain comfort during periods of high activity, in both warm and cold climates.

Wetting and wicking in textile fabrics are complex processes. The liquid comes into direct contact with fibres, forming a solid–liquid interface (wetting), and is then transported in the capillaries between the fibres and yarns of a woven or knitted structure (wicking). Before wicking can occur, the fibres must first be wetted, so wetting is an essential precursor to wicking. The wetting and wicking characteristics of fabrics depend on the chemical nature of both the liquid and the fibre, and on the fabric structure itself.

8.2.2 Wetting

In the bulk of any liquid, the molecules experience attractive forces with neighbouring molecules which are equal in all directions. The molecules at the surface of the liquid, however, experience a net attraction towards the bulk of the liquid, giving rise to surface tension (see Figure 8.1).

The net attraction downwards experienced by the molecules at the surface of a liquid gives rise to free energy at the surface, the effect of which is to minimise the surface area and inhibit the spreading of the liquid over a fabric. If the liquid is to wet the fabric, the fabric must have a sufficient surface energy to overcome the free surface energy of the liquid.

During wetting, the fabric–air interface is displaced by a fabric–liquid interface. For a liquid drop at equilibrium with a fabric surface, the interfacial tension between solid and vapour (γ_{SV}) is balanced by the combined interfacial tensions between liquid and vapour (γ_{LV}) and between solid and liquid (γ_{SL}). This is illustrated in Figure 8.2, where there is partial wetting (a) and (b), and complete wetting (c). It is often considered, however, that wetting only occurs if $\theta < 90°$.

The equilibrium between the forces is expressed by the Young–Dupré equation:

$$\gamma_{SV} = \gamma_{SL} + \gamma_{LV}\cos\theta \tag{8.1}$$

where θ is the equilibrium contact angle, defined as the angle between the tangent to the liquid–vapour interface and the tangent to the solid–liquid interface. The contact angle decreases (and $\cos\theta$ increases) as

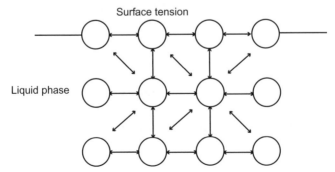

Figure 8.1 Surface tension due to unequal forces experienced by molecules at the surface of a liquid.

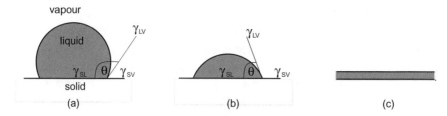

Figure 8.2 Forces in equilibrium in the case of (a) and (b) partial wetting, and (c) complete wetting of a liquid droplet on a solid surface.

wetting increases. The interfacial tensions, γ_{SV}, γ_{SL} and γ_{LV}, are also defined as surface energies between the respective interfaces, and the shape of a drop of liquid changes so that the overall surface energy is minimised. If $\theta < 90°$, the overall surface energy of a drop is lowest when it has a larger area covering the surface. Conversely if $\theta > 90°$, the energy is lowest when contact with the surface is low, that is when wetting does not occur. Strictly Equation 8.1 applies to a liquid drop on a hard, smooth, homogeneous and impermeable surface, such as glass, and of course textile surfaces do not possess these attributes. However, in practical terms the equation is difficult to apply because only the values of γ_{LV} and θ can be measured experimentally. Nevertheless, the equation provides a useful basis to characterise the wetting behaviour of liquids on fabrics.

The method generally accepted as the best for determining the contact angle on fibres is that devised by Wilhelmy. In this method a fibre is lowered into the liquid of interest (*e.g.* water), as illustrated in Figure 8.3. The liquid exerts an attractive force on the fibre, in addition to any buoyancy effects. The wetting force, F_W, is measured by weighing the fibre, before and after lowering it into the liquid, using a very sensitive microbalance. The cause of this attractive force is the surface tension of the liquid and Wilhelmy derived the equation:

$$F_W = \gamma_{LV} P \cos \theta \qquad (8.2)$$

where F_W = wetting force (mN), determined from the measured weight, mg (m = mass, g = acceleration due to gravity). P = perimeter of the fibre ($= 2\pi r$ for a fibre of circular cross-section and radius r, and assuming no diffusion of liquid into the fibre).

Equation 8.2 can be re-arranged to

$$\cos \theta = mg/2\pi r \gamma_{LV} \qquad (8.3)$$

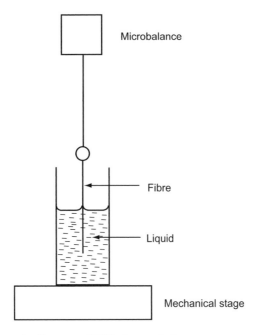

Figure 8.3 Schematic diagram of Wilhelmy technique.

In practice the fibre is usually held taut in the liquid by attaching a small weight to it. It is then necessary to take this weight, and the buoyancy effect it will itself experience in the liquid, into account.

Unfortunately it is often difficult to measure θ with a high degree of precision, and errors in its measurement are reflected in a non-uniform manner in the cos θ values obtained from it. As θ approaches zero, cos θ values hardly change, yet as θ approaches 90°, cos θ varies considerably, and in this range small errors in θ given rise to large errors in cos θ.

A measure that has been devised to indicate whether a liquid will wet a surface is the 'critical surface tension' of a solid, γ_C. The value of γ_C is determined by plotting a graph of the cos θ values of a range of different liquids against their respective surface tension values, γ_{LV} – the so-called 'Zisman plot'. Extrapolation of the graph to cos $\theta = 1$ gives γ_C. The method is not particularly robust in that it suffers from all the problems inherent in measuring θ, and there is an error inherent in the extrapolation of the plot to cos $\theta = 1$ (when $\theta = 0$). Nevertheless, it does give some useful indicative values of the attraction, or repellency, between liquids and surfaces. If the value of γ_{LV} of a liquid is higher than that of γ_C of the surface, the liquid will be repelled by the fibre. Conversely, if it is lower, attraction will occur and the liquid will wet the surface. These

Table 8.1 Critical surface tension values of some fibres and fibre finishes, and surface tension values of some liquids, at 20 °C.

Fibre	γ_C, mN m^{-1}	Liquid	γ_{LV}, mN m^{-1}
Fluorocarbon –CF$_3$	6	*n*-heptane	20
Silicone –CH$_3$	22	*n*-octane	22
Hydrocarbon –CH$_2$–	31	olive oil	32
Polypropylene	29	water	72.8
Polyester	43		
Bleached cotton	44		
Wool	45		
Polyamide 6,6	46		

data, some examples of which are shown in Table 8.1, are useful in the context of evaluating the effectiveness of water-repellent finishes for textiles.

Water, with its very high surface tension, is very effectively repelled by the three surface finishes (fluorocarbon, silicone and hydrocarbon), but the values show that only the fluorocarbon surface finish has a value of γ_C low enough to repel the alkanes and olive oil. The chemistry of water repellent finishes is discussed in Section 8.3.3.3.

Another useful parameter that can be derived from Equation (8.1) is the work of adhesion, W_A, which is the change in surface free energy that occurs when the liquid and solid are separated. In such a situation, the solid–liquid interface is replaced by liquid–vapour and solid–vapour interfaces, and the work of adhesion is the resultant difference in surface tension:

$$W_A = \gamma_{LV} + \gamma_{SV} - \gamma_{SL} \tag{8.4}$$

Combining Equations (8.1) and (8.4) gives:

$$W_A = \gamma_{LV} + \gamma_{LV} \cos \theta \tag{8.5}$$

or

$$W_A = \gamma_{LV}(1 + \cos \theta) \tag{8.6}$$

It is possible to measure values for γ_{LV} and cos θ experimentally, from which W_A can be computed using Equation (8.6). Some interesting re-sults can arise from the application of Equation 8.6. Table 8.2 contains data for contact angle, surface tension and the corresponding value of W_A, for some liquids on fibres.

The values of θ and W_A have to be interpreted individually because they can give apparently contradictory information about the

Table 8.2 Work of adhesion and contact angles.

	Surface tension γ_{LV}, mN m^{-1}	Nylon		Polypropylene	
		$\theta°$	W_A, mN m^{-1}	$\theta°$	W_A, mN m^{-1}
Ethanol (95%)	22.1	18	43.1	47	37.2
Toluene	28.4	57	43.9	–	–
Ethane-1,2-diol	47.7	57	73.7	74	60.8
Water	72.8	71	96.5	87	76.6

interaction between liquids and fibres, as illustrated by the following three examples:

1. The contact angles of toluene and ethane-1,2-diol on nylon are both 57°, yet the W_A values indicate that the ethane-1,2-diol is more strongly attracted to the nylon.
2. The contact angle of ethanol is much less than that of toluene on nylon, yet their W_A values are almost the identical, showing they are about equally attracted to the nylon.
3. Ethanol has a much lower contact angle (47°) on polypropylene than water (87°), indicating it is more readily attracted to the fibre surface, yet the W_A values show the attraction of ethanol to be much lower than for water.

So, contact angle is not the only useful indicator of wetting; the work of adhesion should be considered as well. Arguably W_A values give more information about the interaction between liquids and surfaces than do θ values. It must be remembered, however, that since W_A values are calculated from θ values (using Equation 8.6), any errors in measuring θ will be reflected in the W_A values obtained. Equations 8.1–8.6 strictly apply to perfectly flat surfaces, a condition that is not met by textile fabrics.

A value called the 'spreading coefficient', S, has been defined to measure the spreading behaviour of a liquid on a fibre as follows:

$$S = \gamma_{SV} - \gamma_{SL} - \gamma_{LV} \tag{8.7}$$

If $S > 0$ complete wetting of the fibre occurs, but if $S < 0$ spreading does not occur and the liquid forms a droplet on the fibre surface. In this latter case, wicking will not therefore take place either. For S to be positive, the value of γ_{SV} must be higher than γ_{LV}. For polyester, $\gamma_{SV} = 44.6$ mN m^{-1}, but for water, $\gamma_{LV} = 72.8$ mN m^{-1} (values reported at 20 °C) and S is negative, so wetting does not occur. This is the reason for coating polyester fibres with a hydrophilic agent, mentioned in Section 8.3.1.

In addition to wetting, which results from the interaction between a liquid and a solid, a related attribute is that of 'wettability'. Wettability is a measure of the potential of a surface to be wetted and is used to indicate not only the ease with which fabrics become wetted, for example after they have been subjected to a scouring process, but paradoxically to indicate the effectiveness of water repellency treatments.

The basis of the current British Standard method (BS 4554:1990) of determining the wettability of a fabric is the time taken for a drop of water or sugar solution to sink into the fabric, measured under standard conditions of 20 °C and 65% relative humidity. However, whilst this is a very simple concept, the method brings into play the complexity of textile fabrics, which do not behave as flat, smooth surfaces. Fabric structures, whether they are woven or knitted, contain small gaps between the component yarns, which form capillaries, and the rate at which a drop of liquid will sink into a fabric, depends not only on the wetting of the liquid on the fibre surfaces, but also on the sizes and shapes of the capillaries between them. Wettability measured in the way prescribed by the British Standard is of very practical use to the textile industry, but it does not yield information about the specific ability of a fibre surface to be wetted by a given liquid.

8.2.3 Wicking

Wicking is the term given to the movement of a liquid by capillary forces. Small capillaries exist in fibrous assemblies, such as between the individual fibres in yarns or in non-woven structures, or between the yarns in woven or knitted fabric structures. Wicking can only take place if the liquid is first capable of wetting the surfaces of the capillaries. The transport of a liquid through a fabric has significant implications for the effectiveness of many wet-processing treatments for textiles, especially dyeing operations, where it is essential for dye liquor to penetrate all the fibres in the yarns uniformly in order to produce even dyeings. Wicking is also an important requirement of base-layer garments for active sportswear, where the requirement is for perspiration to be transported away from the skin to outer clothing layers and thence to the external environment (as discussed in Section 8.2.2). By this mechanism, the wearer can be kept in a warm, but dry, state.

The mechanism of wicking involves the displacement of a solid–vapour interface by a solid–liquid interface in the confined volume of a capillary. A useful parameter to determine in this case is the work of penetration, W_P, which is the energy required for movement of liquid in a capillary, given by the difference between the solid–vapour and solid–liquid interfacial tensions:

$$W_P = \gamma_{SV} - \gamma_{SL} \tag{8.8}$$

The values of γ_{SV} and γ_{SL} are very difficult to measure experimentally, but all is not lost because from Equation (8.1) it will be seen that $\gamma_{SV} - \gamma_{SL}$ is equivalent to $\gamma_{LV}\cos\theta$, and γ_{LV} and θ can be measured. For capillary action to occur spontaneously, free energy has to be gained and W_P must be positive. This means that the value of γ_{SV} must be greater than γ_{SL}. In the simplest model to represent capillary action, that of a liquid rising up a capillary tube, a pressure, P, is caused by the surface tension of the liquid (see Figure 8.4).

The capillary pressure is given by the Young–Laplace equation, which in the case of a capillary tube of circular cross-section, radius, r, is:

$$\Delta P = \frac{2\gamma_{LV}\cos\theta}{r} \tag{8.9}$$

The equation indicates that for a given liquid and fibre type, ΔP varies inversely with radius. A necessary requirement for wicking, therefore, is that the pores in the fabric must be very small in comparison to the interfacial areas (fibre–liquid, fibre–vapour), just as they are in a glass capillary tube.

In comparison with glass capillary tubes, textile fabrics are complex structures because wicking can take place not only in the capillaries between the fibre filaments that constitute the yarns, but also in the capillaries between the yarns themselves. These capillaries are not uniform in size and their edges may be rough. In addition, most textile fibres are absorbent, so as well as the movement of liquid through a fabric by capillary action, there will be the competing process of diffusion of the liquid into the fibre. As liquid diffuses into the fibre, it will cause the fibres to swell, thereby reducing the size of the capillary pores. Consequently, fabric structures are not consistent and their wicking properties vary as the liquid moves through (and into) them. In practice when the wearer of a wicking garment is engaged in a sports activity for example, the garment flexes with body movement, this in turn causes dimensional changes in the pore structure of the fabric. In the application of Equation 8.9 to textile fabrics, an effective radius r_E for the capillary pores has to be used.

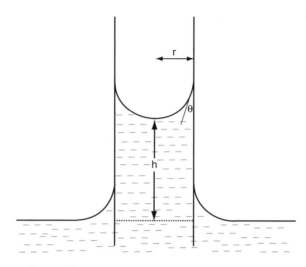

Figure 8.4   Capillary action.

It follows from the above, that the rate of wicking of a liquid in a system of capillaries in a textile fabric is a highly complex process to model. Most analyses of wicking begin with the classical Washburn–Lucas equation, which applies for a liquid moving up a single glass capillary tube against gravity:

$$\frac{dh}{dt} = \frac{\gamma_{LV} r \cos\theta}{4\eta h} - \frac{r^2 \rho_L g}{8\eta} \tag{8.10}$$

where r = diameter of the capillary, ρ_L = density of the liquid, g = acceleration due to gravity, η = viscosity of the liquid and h = height of the liquid rise in the capillary. Eventually the upward capillary force is countered by the downward gravitational force, so the liquid stops moving and an equilibrium height, h_E, is achieved, given by:

$$h_E = \frac{2\gamma_{LV}\cos\theta}{r\rho_L g} \tag{8.11}$$

In the application of Equation 8.10 to the early stages of wicking in real textile fabrics, when the height, h, achieved by a liquid wicking up a vertical fabric immersed in it is small, then:

$$\frac{\gamma_{LV} r \cos\theta}{4\eta h} \gg \frac{r^2 \rho_L g}{8\eta}$$

So,

$$\frac{dh}{dt} \rightarrow \frac{\gamma_{LV} r \cos\theta}{4\eta h}$$

$$\int_0^h h\,dh = \frac{\gamma_{LV} r \cos\theta}{4\eta} \int_0^t dt$$

$$\frac{h^2}{2} = \frac{\gamma_{LV} r \cos\theta t}{4\eta}$$

and this results in the following equation:

$$h_\tau = \left(\frac{r\gamma_{LV}\cos\theta}{2\eta} t \right)^{1/2} \tag{8.12}$$

This equation can be simplified to:

$$h_t = kt^{1/2} \tag{8.13}$$

Whilst implying that the liquid never stops rising, this square root relationship applies well to the early stages of wicking. The measurement of wicking is frequently made by the vertical wicking test (see Figure 8.5), in which one end of a length of a piece of fabric held vertically is immersed in a liquid (usually water coloured with a dye) and the rate of travel of the liquid front up the fabric is measured.

It is also useful to know how much water has been absorbed by the fabric when it has reached its maximum height, and this is usually expressed as the mass per mass of the length of the dry fabric.

In addition to vertical wicking, other wicking measurements of use are:

- Transverse wicking, which is the transport of water through the thickness of a fabric, knowledge of which is important in the context of wicking perspiration from the skin through a fabric to the external environment. This is not an easy measurement to make and various items of specialised equipment have been devised.
- Planar wicking, which is the spreading of a drop of liquid applied to a fabric surface, assessed by measuring the rate of increase in wetted area. This information is useful for measuring characteristics such as water repellency and stain resistance. It is also of value in printing, especially ink-jet printing where very tiny droplets (in the order of pico-litres in volume) are jetted onto the fabric surface. In order to produce clearly defined printed patterns, transverse wicking is preferred and the outward spreading of the droplets has to be minimised.

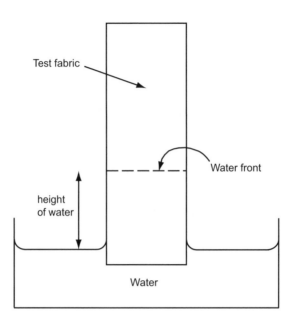

Figure 8.5 Vertical wicking test.

Wetting and wicking are important characteristics of textile fabrics which strongly influence their functional properties. Whilst textile fabrics of optimal wicking behaviour are essential for providing comfort of wear, especially for active sportswear, this is not the only application of textile fibres. Other applications, such as textiles for medical and hygienic applications where high wettability and absorbency are required, include highly functional products, for which the market is huge and extremely valuable.

8.3   SURFACE TREATMENTS

In the preceding chapters, especially those dealing with natural fibres, reference has been made to their complex morphology and especially to their surface characteristics. In order to modify the surface properties of textile fibres to confer beneficial performance attributes, different techniques can be used, such as:

- Plasma treatment;
- Enzyme treatment; and
- Chemical finishing.

The reasons for modifying fibre surfaces can be to reduce friction; improve wetting and wicking behaviour (see Section 8.2); or confer new properties, such as soil-release and water-repellency. Some of the techniques for achieving these properties will be discussed in this section.

8.3.1 Plasma Treatment

8.3.1.1 Introduction. The use of plasmas for the modification of surfaces is a well-established technology, going back some 40 years. Applications of the technology to textile substrates began in the 1980s and although there has been a considerable amount of research carried out into the modification of textile materials by plasma treatments there has been very little uptake by the textile industry. Also, until recently, plasma treatments on an industrial scale were not really achievable. The process has various attractions, because it is environmentally benign; does not involve the use of water or organic solvents; and is versatile in the modifications it can make to textile surfaces. Likely reasons for the reluctance by industry to invest in the technology are the high capital cost of the equipment and the fact that some of the effects, certainly those in which the surface reactivity is modified through the introduction of new functional groups, are not permanent. Indeed, some plasma activated effects can be lost after just one day.

Plasmas are gaseous mixtures of ions, radicals, electrons, excited neutral molecules and photons in the form of high energy UV radiation. They are therefore complex mixtures of reactive species capable of interacting with materials exposed to them. Plasmas can be created by high temperature, but the temperature required (around 1000 K) destroys textiles and most other materials. An alternative is to create the plasma at low temperature using an electric discharge. The electric discharge is passed through a process gas, but gases are insulators rather than conductors, so a high voltage is required. When the gas breaks down it conducts electricity. The various species formed in the plasma interact with the textile substrate that is present within the plasma, though the action is very much limited to the surface (<100 nm penetration), the bulk of the material remaining unaffected.

8.3.1.2 Methods for Generating Plasmas. Plasma treatments of textiles are carried out at ambient temperature, or slightly above it, and either at low pressure (<100 Pa), or at atmospheric pressure. The latter is more desirable, because for the continuous treatment of fabric lengths the equipment is less complex and therefore cheaper. Low pressure machines can run on a continuous basis, however, and can accommodate fabrics greater than 1 m in width. The process gas is continually fed into the plasma chamber and a power of up to 5 kW is required to maintain the plasma. An advantage of low-pressure plasma treatment is the highly uniform treatment it gives over the fabric. It also uses less process gas, an important consideration when using gases such as fluorocarbons.

There are three different types of atmospheric pressure plasmas:

1. Corona discharge, generated by applying a high voltage (up to 10 kV) between two electrodes positioned very close to each other (~1 mm apart) through which the fabric has to pass. One of the electrodes is flat. The other electrode is highly curved and the discharge sprays outward from it to the planar electrode, but becoming weaker as it does so. The process gas is air, but the density of the plasma produced is not really sufficient to give significant, long-lasting effects on textiles and not all fibres of thick textile fabrics will be exposed to the plasma. Although much research has been carried out on the corona treatment of textile fabrics, other methods are preferred and corona discharge is usually limited to treating polymer films.

2. Dielectric barrier discharge (DBD), which is also called silent discharge, was originally developed for the production of ozone, but has since been developed to produce a broad range of plasmas. A very high voltage, up to 20 kV, is passed between two electrodes separated by about 1 cm. One or both of the electrodes is covered by an insulator such as glass or a ceramic, which acts as a dielectric barrier preventing arcs. A large number of randomly distributed transient microdischarges called 'streamers' are formed, but their random nature creates a significant risk of uneven treatment. However, by optimising the process conditions (power applied, distance between electrodes, time of treatment, *etc.*) a more homogeneous discharge can be created so that the textile fabric is treated more uniformly.

3. Glow discharge, which is generated by applying a low voltage of around 200 V between two parallel plate electrodes, just a few millimetres apart. The discharge between the electrodes is highly uniform, producing a more uniform treatment of textile fabrics than the DBD system. The process gas is usually helium or argon.

8.3.1.3 Effects of Plasma Treatments.   Low temperature plasma technology can modify the surfaces of textiles in five ways: cleaning, etching, activation, deposition and grafting. The cleaning function is brought about using a non-polymerisable gas such as an inert gas (usually argon or helium) and removes contaminants such as oils or waxes by ablation from fibre surfaces. The process initially abstracts hydrogen from the contaminant, forming a free radical. Then, under attack by the various particles of the plasma, the contaminant undergoes molecular breakdown until remnants are left which are small enough to evaporate away in the vacuum. By a similar process but to a lesser extent, the polymer surface of fibres degrades a little, causing an etching effect which increases their roughness. In general, the effect of etching depends on the contact angle, θ, of a liquid on a flat surface of the material (see Section 8.2.2). If $\theta < 90°$, increasing the roughness will decease θ further and wetting will occur more easily. However if $\theta > 90°$ on a flat surface, increasing roughness will increase θ, making the surface more hydrophobic (if the liquid is water). Usually it is found that increasing surface roughness decreases the contact angle, θ, and increases the spreading of water over the fibre and some values that have been reported are shown in Table 8.3. It must be remembered though, that the effects of plasma treatment can be complex and that in addition to etching, the effect of activation may also exert an influence on wetting.

Activation of textile surfaces occurs when more reactive process gases, such as oxygen, ammonia and nitrous oxide are used. New functional groups are introduced, such as $>C=O$, $-COOH$, $-CHO$ and $-OH$ (from oxygen) and $-NH_2$, $-C\equiv N$ and $-C=NH$ (from ammonia or nitrogen). These groups alter the polarity of the surface but are also useful in that they can act as anchoring sites for other functional molecules. An interesting example of the etching and activation effects is shown in Figure 8.6 where the surface of polylactic acid fibre can be seen to be considerably rougher after plasma treatment (compare Figures 8.6a and b).

Table 8.3 Contact angles of water on textile fibres, before and after air plasma treatment.

Fibre type	Contact angle θ, degrees	
	Untreated	*Treated*
PP	87	22
PE	87	42
PA	63	17
PET	71	18
PTFE	92	53

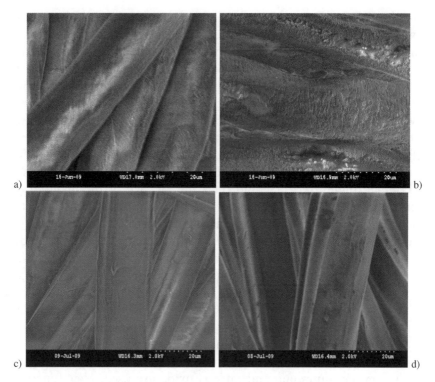

Figure 8.6   Effect of oxygen plasma treatment on PLA: (a) before treatment, (b) after treatment; and PET (c) before treatment, (d) after treatment. (Photographs courtesy of A. Abdrabbo, Heriot-Watt University).

In the case of the PLA fibres, the oxygen plasma treatment increases only surface roughness and does not alter the chemical nature of the surface. Treatment of the PET fibres, however, has little influence on surface roughness but introduces polar groups into the surface (see Figures 8.6c and d). The effect is to increase wettability of both fibre types, though the impact of the increased polarity on the wetting of PET is much greater than the increase of roughness on the wetting of the PLA.

The fourth effect that plasma treatment can confer is deposition. Some process gases can undergo fragmentation in the plasma then self-combine to form polymeric substances which become deposited on the textile surface. Examples of such gases are tetrafluoromethane and hexafluoroethane, the action of which is to deposit a fluoropolymer onto the surface, resulting in hydrophobic character. This process is known as plasma-enhanced chemical-vapour deposition and the effects induced are permanent.

Finally, low temperature plasmas can be used to graft monomers onto fibre surfaces. An inert process gas (*e.g.* argon) is used to form free radicals on the fibre surface. A monomer that is capable of reacting with the free radicals is introduced into the process chamber, when reaction with the fibre surface occurs and the monomer builds up on it as a polymer graft.

8.3.1.4 Characterisation of Plasma-Treated Textile Surfaces. Clearly a wide range of surface modifications can be produced by plasma treatment and whilst these modifications manifest themselves in changes to the performance characteristics of the fibres or fabrics, it is useful to assess the effects instrumentally. The techniques available split broadly into two categories, those which provide information on surface topology and those which indicate the chemical characteristics at the surface. The techniques available to examine surface topology are:

- Scanning electron microscopy (SEM). This is a long-established technique that is capable of providing images of fibres of excellent resolution at very high levels of magnification (up to ×10 000 or higher). It is necessary to coat the sample with a very thin metallic layer (usually gold) to provide conductivity for the electrons and the sample has to be mounted in an evacuated chamber. The images obtained do not readily provide any quantitative assessment but they enable effects such as etching to be observed. Examples of SEM images can be seen in Figure 8.6.
- Atomic force microscopy (AFM). An image of the surface topology of a sample is obtained by scanning a sharp tip made of silicon nitride, at the end of a cantilever across its surface. The forces generated as the tip undulates across the surface are recorded, so that a type of relief map of the surface is generated. The technique is relatively more straightforward than the SEM process, because no surface coating is required and measurements can be made at normal atmospheric pressure.

The chemical characteristics of surfaces can be assessed by:

- X-ray photoelectron spectroscopy (XPS). This is a spectroscopic technique that provides quantitative information about the surface chemistry of material. A focussed beam of X-rays at 1.5 kV is directed at the surface, and the number and kinetic energy of the photo-emitted electrons from the top 10–12 nm are measured. From their kinetic energies, the binding energy of the electrons is

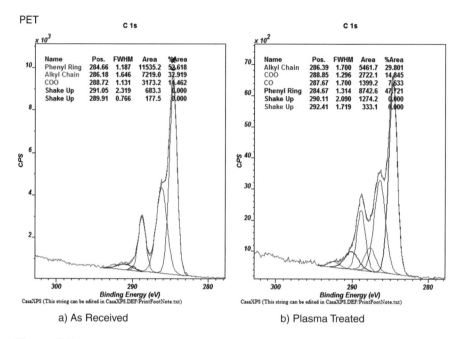

Figure 8.7 XPS spectra of (a) untreated and (b) oxygen plasma treated PET. (Photographs courtesy of A. Abdrabbo, Heriot-Watt University).

determined. Each element emits electrons from its core (inner shell) orbitals at characteristic binding energies, so hydrogen and helium, which only contain electrons in the outer valency shell, cannot be detected. The spectrum produced, a plot of the number of electrons at each binding energy, indicates the elements present and their relative amounts, in the area of the surface analysed. The sample has to be measured under ultra high vacuum conditions, but no surface treatment is required and the technique is non-destructive. An example of the XPS spectra of PET before and after oxygen plasma treatment is shown in Figures 8.7a and b. The spectra show the presence of new carbonyl (CO) groups at 287.87 eV in the treated samples.

- Fourier transform infrared spectroscopy (FTIR). Infrared spectroscopy is a technique which measures the absorption of electromagnetic radiation in the infrared region of the spectrum by a substance. The wavelengths of IR radiation absorbed depend on the vibrational frequencies of the various chemical bonds in a compound, so each compound has a characteristic plot of absorption (or transmission) against wavelength. FTIR is a refinement of this principle, though rather than scanning the sample with monochromatic radiation, a beam containing a wide range of wavelengths is used, and the absorption of that beam measured. The spectral composition of the beam is then modified and the absorption of that is measured. This process is continued and from all the measurements, the absorption at each wavelength is computed. FTIR is just a different numerical procedure for obtaining the spectrum but it is faster than a conventional monochromatic instrument. In order to obtain information about the chemical nature of the surface of a textile material, the sampling technique of total internal reflectance (ATR) is used. In this technique, the sample is mounted on a crystal, typically made of germanium, through which the infrared beam is directed. Each time the beam strikes the surface of the crystal in contact with the sample, it forms an evanescent wave that penetrates the top micrometre layer of the sample, before reflecting back into the crystal. In combination with FTIR, an absorption spectrum of the sample surface is obtained.

8.3.1.5 Examples of Fibre/Fabric Treatments. The effects that can be produced on textile materials depend on a range of variables, such as the nature and concentration of the process gas used, the power applied, and the time of treatment. Studies have been made on most fabric types in various research laboratories, sometimes with contradictory outcomes, though this is usually due to the different process conditions employed. Examples of some of the characteristics which can imparted by plasma treatment for selected natural and synthetic fibre types are given below.

8.3.1.5.1 Wool. The attraction of plasma treatments for wool is that they can modify its hydrophobic surface character, which exists due to the epicuticle (see Section 3.2.2) acting as a barrier to treatments such as dyeing and printing. In addition, the scale structure of wool fibres which gives rise to the directional frictional effect (DFE) is often modified to avoid unwanted shrinkage. The traditional processes for removing this layer involve chlorination, so plasma treatment is a more environmentally friendly alternative. Plasma treatment acts in two ways on wool fibres. Firstly the outer epicuticle layer is oxidised and often completely abraded, thereby reducing the hydrophobic character of the fibre surface. Secondly, the plasma treatment introduces hydrophilic groups such as $-OH$, $-C=O$ and $-COOH$ into the surface, so the hydrophilic character is increased. A further action of the plasma is attack on disulfide crosslinks, the residues being oxidised to sulfo-sulfonate and cysteic acid moieties:

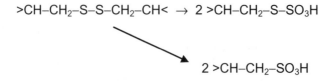

$$>CH–CH_2–S–S–CH_2–CH< \rightarrow 2 >CH–CH_2–S–SO_3H$$

$$2 >CH–CH_2–SO_3H$$

The resulting changes brought about by the plasma treatment significantly increase the wettability of wool and therefore its dyeability. The DFE is also reduced, but usually not to the extent that is achieved using the chlorine/Hercosett process (see Section 3.2.10), so additional coverage by resin is still required. The data in Table 8.4 illustrate the shrinkage due to the DFE, and the effectiveness of various treatments.

8.3.1.5.2 Silk. Fewer studies of plasma treatments of silk have been made than of wool. Plasma treatment using air or oxygen as the process gas has been shown to aid the removal of sericin from the fibre during subsequent degumming (see Section 3.4.1). The use of sulfur hexafluoride as process gas has been used to increase the hydrophobicity of silk fibres, the increase being due to the introduction of fluorine atoms into the fibre surface.

8.3.1.5.3 Cotton. Like wool, cotton fibres contain a protective outer layer, in this case the cuticle (see Section 2.3.1) which contains waxes, fats and pectins. Although some research work has been carried out on the removal of these substances by plasma treatment with a degree of success, it does not appear that there has been much success in their complete removal. The reason is probably that the coating of fats and waxes extends into the fibre surface further than can be efficiently removed by plasma treatment, and the traditional scouring process is still required.

The surface of cotton can be made to be either more hydrophobic or more hydrophilic, depending on the nature of the process gas used in the

Table 8.4 Shrinkage in area of knitted wool fabric after 50 simulated washing cycles.

Treatment	% shrinkage in area
Untreated	69
Plasma treated	21
Plasma/resin treated	1.3
Chlorine/Hercosett treated	1

plasma treatment. If the process gas is oxygen, then plasma treatment introduces hydrophilic groups such as $>C=O$, –COOH into the fibre surface. Even if an inert process gas such as argon is used, surface oxidation can occur in addition to the etching effect that would be expected by such a gas. The reason is that the plasma treatment generates free radicals in the fibre surface which then initiate oxidation when the treated sample is exposed to air afterwards.

It is also possible to impart hydrophobic character to cotton fibres. To achieve this, the process gases used contain fluorine compounds such as tetrafluoromethane (CF_4) or hexafluoropropene (C_3F_6). Sulfur hexafluoride (SF_6) has also been used. The result of plasma treatments using hexafluoropropene is to introduce polymeric fluorocarbon structures at the fibre surface and it is the presence of the fluorine that provides water repellency. Tetrafluoromethane is a non-polymerising gas but nevertheless still forms thin films on the fibre surface. It has been found that the contact angle for water on cotton increases from 30° on untreated cotton to between 90–150°, depending on the conditions of plasma treatment (such as gas pressure, time of exposure, *etc.*).

8.3.1.5.4 Nylon 6 and 6,6. Polyamides are already fairly hydrophobic fibres, so there is little interest in effecting plasma treatments using fluorine-containing compounds. Much of the research on the plasma treatment of polyamides has been carried out with a view to improving dyeability, wettability and surface properties. When oxygen or air is the process gas there is usually an increase in wettability, though somewhat unexpectedly the surface, rather than being roughened, has been found to become smoother. Nitrogen-containing plasmas, typically those using ammonia as the process gas, introduce functional groups such as $–NH_2$, $–CH=NH$ and $–C\equiv N$. The polarity of these functional groups reduces the hydrophobicity of the fibre surface, the effects of which are to increase wettability and printability.

8.3.1.5.5 Polyester. The plasma treatment of PET fibres has been extensively researched. The changes in the surface properties conferred depend on the process gas used and are consistent with the behaviours of other fibre types. Thus the use of fluorocarbons and sulfur hexafluoride causes an increase in hydrophobicity, together with some increase in roughness. The effects are not stable and subsequent oxidation when the fibres are exposed to air can take place, leading to the formation of carbonyl groups.

Increases in hydrophilicity are obtained when the process gas is air or oxygen, due to an increase in the oxygen content of the surface. If air is used, nitrogen can also be increased, through the formation of surface amide (–CO–NH–) groups. Air and oxygen plasma treatments create a rougher surface structure as well. The use of tetrachlorosilane ($SiCl_4$) also causes an increase in hydrophilicity when subsequent air oxidation has been found to convert the Si–Cl bonds to Si–OH.

A useful application of plasma treatment of PET is its potential for enabling the adhesion of metals, especially aluminium. The use of NH_3 process gas is very effective in promoting adhesion of PET film with Al, the increase being due to the creation of basic functional groups, such as –NH_2, –CH=NH and –C≡N. Al-coated PET films are used as dielectrics in capacitors, high-barrier performance packaging and as laminates for barrier foils.

8.3.1.5.6 Polypropylene. PP is a very hydrophobic fibre with an extremely low surface energy. The fibres are also quite smooth and whilst these characteristics are useful for certain applications, there are also occasions where an increase in wettability is required, for example to enable coatings to be applied. PP fibres can only be wetted with liquids of surface tension $<35\,\mathrm{mN\,m^{-1}}$ and since γ_{LV} for water is $72.8\,\mathrm{mN\,m^{-1}}$, no water can pass through non-woven filters, unless forced under high pressure. Plasma treatments using oxygen or argon process gases increase the hydrophilicity of PP, thought to be due to the formation of hydroxyl, carbonyl, ester and ether groups. Nitrogen-containing plasmas introduce the same functional groups as those in nylon. The use of fluorocarbons in plasma processing acts to increase hydrophobicity and also makes PP surfaces more oleophobic (oil re-sistant) which is useful for some filtration requirements.

8.3.1.6 Uses of Plasma Treatments. Plasma treatments of textile fibres are carried out to modify their surface characteristics with the aim of improving performance or processibility. Examples are:

- Improvement in dyeability. In general, plasma treatment improves the dyeability of most textile fibres, though there can be some ad-verse effects. Positive improvement seems to be most attained with wool fibres, where the surface modification considerably facilitates penetration of dye. Also the use of nitrogen-containing process gases, which introduce basic groups into the keratin structure, provides greater uptake of acid dyes. The use of oxygen plasma for pre-treating cotton fibres increases the uptake of direct dyes. Again the dyes are able to penetrate the eroded fibres more easily, but in addition the creation of carbonyl and carboxylic acid groups also have an influence. In the case of synthetic fibres, whilst an increase in surface roughness can promote dye absorption, the surface erosion may occur preferentially in the non-crystalline regions, so that overall the treated fibre surface has a more highly crystalline character, which is less accessible to dye molecules.

- Increase in either hydrophobic or hydrophilic character. Increase in hydrophobic character is generally brought about by the introduction of atomic fluorine or fluoropolymers through the use of fluorine-containing process gases such as CF_4 or C_3F_6. Conversely, if an increase in hydrophilicity is required, a process gas containing oxygen (either oxygen itself or air) is used, where polar hydrophilic groups (such as $>C=O$, $-COOH$ or $-OH$) are created in the surface.
- Increase in wettability. Generally, as the hydrophilic character of a surface is increased, so too is the wettability. Also, increasing the surface roughness can bring about an increase in wettability, and whilst many process gases have an etching effect, it is most pronounced with the inert gases, such as argon or helium.
- Creation of functional groups to act as anchor points for other agents. The introduction of polar functional groups into a fibre surface enables the reaction of the surface with other chemicals. This is especially useful in providing reaction sites to molecules of biological activity, and to increasing the biocompatibility of textile fibres. A good example is the introduction of chitosan to PET to confer antibacterial properties. Acrylic acid is firstly grafted to the PET. Chitosan and quaternised chitosan are then attached to the bound polyacrylic acid through amide links. Another useful example is the creation of basic functional groups in PET fibre surfaces by plasma treatment that provide sites capable of attracting metal atoms, such as aluminium (see Section 8.3.1.5.5).

It is likely that plasma treatment will become more attractive to the industry in the future, driven by the need to process textiles in more environmentally friendly ways. Certainly further development work is necessary in some processes to ensure uniformity, consistency and durability of treatment. However, there are ecological and economic benefits to be gained once the processes become mature and established.

8.3.2 Enzyme Treatments

Enzymes are proteins of high molar mass produced by all living organisms and are very efficient catalysts in chemical reactions. Unlike most other types of catalysts that are able to catalyse a wide range of

Table 8.5   Some enzymes for textile treatments.

Enzyme	Use in textile processing
Cellulase	Break down cellulose. Used for bio-finishing and bio-polishing.
Protease	Break down proteins. Used in detergents for removal of protein-based soils. Also used for degumming of silk and bio-antifelting of wool.
Lipase	Split fats (lipids) into glycerol and fatty acids. Used in detergents.
Amylase	Breaks down starch into simple sugars. Used in desizing processes.
Pectinase	Break down pectins. Used in preparation (cleaning) of cotton fabrics.

reactions, enzymes are highly specific and they are categorised according to the substrates on which they act. They are named by adding 'ase' to the name (or part name) of the substrate for which they are specific and some examples are given in Table 8.5.

The conditions of pH and temperature under which enzymes function can be quite specific. This is because the catalytic activity of enzymes very much depends on the shapes of the protein chains, which are determined by the intra- and intermolecular bonds that hold them in their secondary and tertiary structures. These bonds are sensitive to changes in pH or temperature. Enzymes function most efficiently at their characteristic optimum temperature and as the temperature rises above that level then the enzyme structure begins to break down.

Enzymes are used widely in the textile industry, both during manufacturing processes and in after-care treatments, such as washing. During manufacturing, they are used in desizing processes. Sizes are chemical formulations which are put onto textile yarns to reduce friction in spinning and weaving operations. They act as lubricants and are usually starch-based, but starch is difficult to remove by conventional chemical treatments. Amylases are very efficient enzymes for removing the starch and so are frequently used for desizing operations. In after-care treatments they are used in washing powder formulations where they are very effective in removing soil and stains from fabrics at low temperatures. They therefore reduce the need for bleaching, and smaller quantities of detergent and their associated phosphates are required.

Another major use of enzymes in the textile industry is in bio-finishing. The purpose of this process is to modify the appearance of fabrics, particularly cotton fabrics, by removing surface hairs, which might cause pilling. Smoother, softer fabrics are obtained through bio-finishing processes. A fashion driver in the jeans industry is the 'worn' look. Traditionally this has been achieved by subjecting denim fabric to stonewashing, using pumice stones. Such treatment could, however, actually damage a garment. Enzymes can be used to give the desired 'worn' appearance without incurring damage.

8.3.3 Chemical Finishing of Textiles

8.3.3.1 Introduction. In order to enable textile fabrics to meet the functional performances required of them, it may be necessary to give them a finishing treatment of some type. Finishing treatments are broadly classified into two main areas – chemical finishing and mechanical finishing. Mechanical finishing usually involves modification to appearance or dimensional stability, usually brought about by processes involving heat, or modification to the 'handle', that is the extent to which they feel soft, smooth and fluffy. Such processes involve raising the fibres on the surface by teezles, then cropping, which is cutting the raised fibres to leave a pile of fixed height on the fabric surface. Mechanical finishing is sometimes referred to as 'dry' finishing, in contrast to chemical finishing which is sometimes referred to as 'wet' finishing.

In chemical finishing processes, additives are applied to fabrics to impart specific functional properties, either to meet performance requirements for certain end-uses, or to meet legislative requirements, the most important example of which is the application of flame retardants to fabrics to be used in the manufacture of soft furnishings. The most common chemical finishes applied to textiles are:

- Chemical softeners
- Water repellents
- Soil-release/anti-soiling agents
- Antistatic agents
- Flame retardants
- Shrink-resist treatment for wool
- Easy-care finishes for cotton

It is often the case that for a specific end-use of a fabric, it is necessary to apply more than one type of chemical finish to meet the performance standards required. For example, there may be a need to apply a flame retardant as well as a softening agent. In situations like this it is important that the chemical finishes applied are compatible and the effectiveness of one finish is not compromised by the presence of another finish, or *vice versa*.

The chemistry of shrink-resist treatments for wool has already been covered in Section 3.2.10 and that of easy-care finishes for cotton in

Section 2.2.5.5, so only the first five types of finishes will be covered in this chapter.

8.3.3.2 Chemical Softeners.   An important selling attribute of textile fabrics for clothing or items such as towelling is their 'handle'. To improve handle, chemical softening agents are often applied to fabrics during manufacture. Not all treatments are permanent and after a few washes the effect can diminish. As a result, drying clothes after laundering, especially those made of cotton, can leave them feeling quite harsh and not particularly supple to the touch. To overcome this problem fabric conditioners, which are formulations of softeners and other ingredients such as fragrances and colorants, are sold direct to consumers to add after the washing cycle is completed.

A wide variety of softeners are used, their structural features being similar to those of detergents. By far the most commonly used softeners in fabric conditioners are cationic, mainly because they are the type most suited to use on cotton fabrics, but also anionic, amphoteric and non-ionic softeners are used for other fibre types. The improvement in handle imparted by them arises from their interaction at the surface of the fibres, though the mechanism by which the handle is improved depends on the nature of the substrate and the softener.

Cationic Softeners.   Cationic softeners are often quaternary ammonium compounds containing long alkyl chains, in which the active ion is cationic, such as *N,N*-distearyl-*N,N*-dimethyl ammonium chloride and di-(stearylcarboxyethyl)-hydroxyethylmethyl ammonium methylsulfate.

$$C_{17}H_{35}-\overset{\overset{\displaystyle CH_3}{|}}{\underset{\underset{\displaystyle CH_3}{|}}{N^{+}}}-C_{17}H_{35} \qquad Cl^{-}$$

N,N-distearyl-N,N-dimethyl ammonium chloride

$$R_1-\overset{\overset{\displaystyle CH_3}{|}}{\underset{\underset{\displaystyle \underset{\displaystyle \underset{\displaystyle OH}{|}}{CH_2}}{|}}{\underset{\displaystyle CH_2}{\overset{+}{N}}}}-R_2 \qquad X^{-}$$

$$R_1, R_2 = -(CH_2)_2O\overset{\overset{\displaystyle O}{\|}}{C}-C_{17}H_{35}$$

$$X^{-} = {}^{-}OSO_2(OCH_3)$$

Di-(stearylcarboxyethyl)-hydroxyethylmethyl ammonium methyl sulfate

The surfaces of cotton fibres in aqueous environments acquire a slight negative charge, probably through dissociation of the hydroxyl groups of

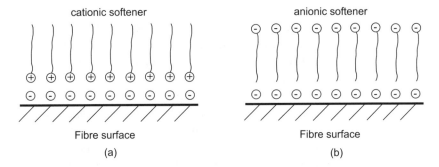

cationic softener

anionic softener

Fibre surface

(a)

Fibre surface

(b)

Figure 8.8 Bonding of (a) cationic and (b) anionic softening agents to cellulose fibre surfaces.

the glucosidic rings. The positive end of the cation is attracted to the negative fibre surface, with the long hydrophobic chains pointing outwards from the surface (see Figure 8.8a). These hydrophobic groups enhance the softness and smoothness, but because they are hydrocarbon chains they have affinity for soil and they can adversely influence absorbency.

Anionic Softeners. Anionic softeners are usually straight chain alkyl sulfates, of general formula $R{-}O{-}SO_3^- Na^+$, where $R =$ a long alkyl chain. On a surface with a slight negative charge, such as cotton, the negative head of the anion is oriented outwards (see Figure 8.8b). The softener molecules are attracted to the fibre by hydrogen bonding and van der Waals forces. These types of softeners are used more for functional textiles, such as those for medical applications, rather than in domestic laundering products.

Amphoteric Softeners. These types of softener contain both positive and negative charges, such as alkyldimethylamine oxides and betaines.

$$\begin{array}{c} CH_3 \\ | \\ R{-}\overset{+}{N}{-}O^- \\ | \\ CH_3 \end{array}$$

an alkyldimethylamine oxide softener

$$R = C_nH_{2n+1} \ (n = 11 \text{ to } 17)$$

$$\begin{array}{c} CH_3 \quad\quad O \\ | \quad\quad\quad || \\ H_3C{-}\overset{+}{N}{-}CH{-}C{-}O^- \\ | \quad | \\ CH_3 \; R \end{array}$$

a betaine softener

Non-Ionic Softeners. There are a wide variety of non-ionic structures, typical ones being ethoxylated fatty acids, amides or amines, yielding polyglycol ethers. In addition to acting as softeners, they confer antistatic properties to textiles, through their hydrophilic character. The polyglycol ethers are not especially good as softeners, but silicone non-ionic compounds have much better efficacy. The silicone products are based on polydimethylsiloxane and give very high levels of softness, especially their functional derivatives. They also confer increased hydrophobic character as well, however, which is not always desirable if only a softening function is required. Polydimethylsiloxanes are used also to impart water-repellency and this function is discussed in Section 8.3.3.3.

The alignment of non-ionic molecules at a fibre surface depends on the hydrophobic–hydrophilic character of the surface. If the fibre surface is predominantly hydrophilic, the ethoxylate groups will be attracted preferentially. Conversely, on a hydrophobic fibre surface the alkyl groups will be attracted.

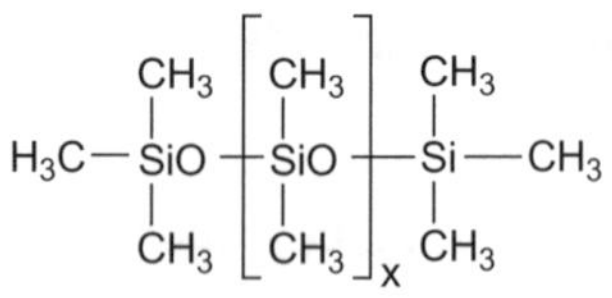

Polydimethylsiloxane

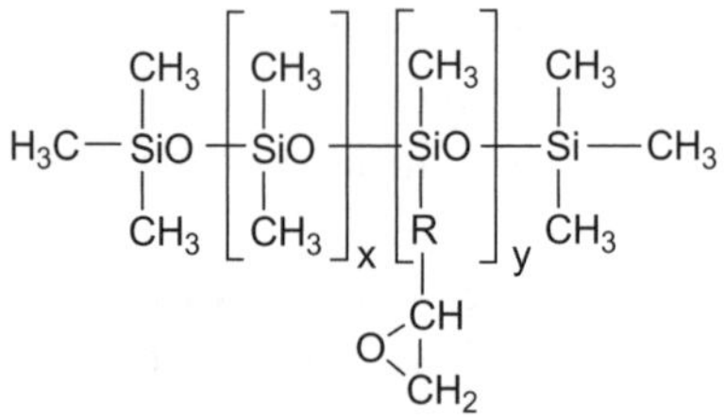

Epoxy functional silicone softener

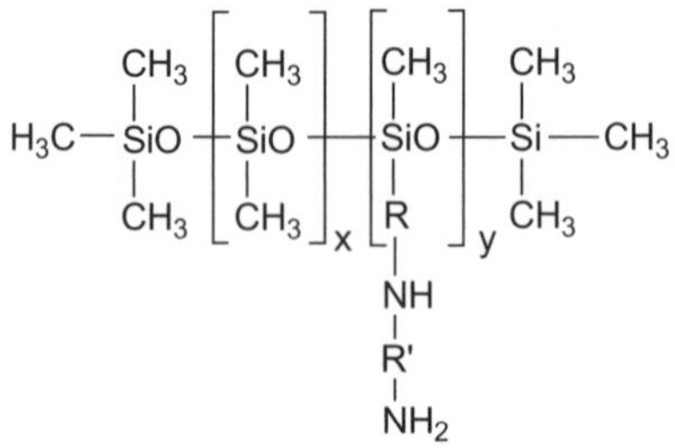

Amino functional silicone softener

Care has to be taken in the application of softeners if other finishing chemicals are applied as well. For example, fabrics for base-layer garments for active sportswear can be treated with agents to aid wicking. The functioning of these agents can be negated if the garments are treated with cationic softening agents after laundering, and garment labels on such items often advise users not to use fabric conditioners.

8.3.3.3 Water Repellents. A large number of terms are used to indicate the water-repellent qualities of garments, such as shower-proof, rain-proof, shower-resistant, rain-resistant, shower-repellent, rain-repellent and so on. In general, water-repellency is the relative resistance of a fabric to surface wetting, water penetration or water absorption. A fabric which is waterproof is totally resistant to the absorption or penetration of water, and is generally made by laminating the fabric with a polymer coating that is entirely impervious to water. Typical of such coatings are polyurethanes, PVC and neoprene. Such coatings also make the fabric stiff and impervious to air, however, so they are more suitable for technical products such as awnings or tarpaulins, than for apparel products.

Wax repellents were amongst the first agents used, typically paraffin wax alone or in combination with other waxes such as beeswax or carnauba wax (wax from the leaves of palm trees). The wax is applied by padding an aqueous wax emulsion onto the fabric, followed by melting the wax to distribute it evenly over the surface, by 'hot calendering'. The treatments are cheap and effective, but they are not durable to laundering and confer low breathability.

Another group which was used successfully for many years was based on melamine, the compounds being formed by its reaction with stearic acid and methanal (formaldehyde). The result was a very large molecule, with considerable hydrophobic character due to the stearic acid groups.

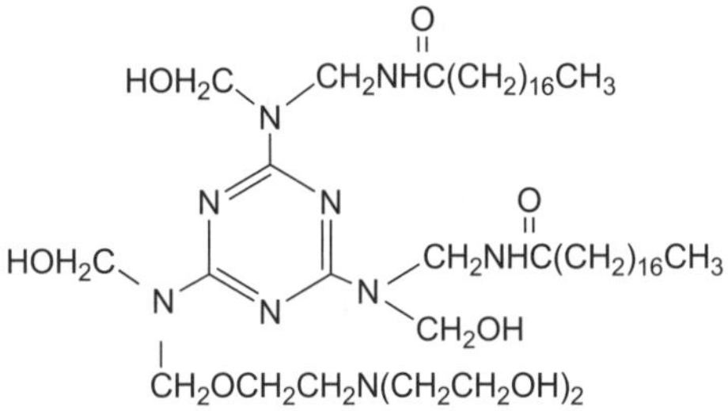

Stearic acid-melamine agent

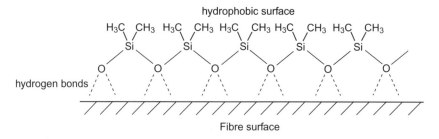

Figure 8.9 Schematic representation of PDMS on hydrophilic fibre surface.

This type of agent has been largely superseded by the silicone and fluorocarbon agents.

Between the 1970s and 1990s, silicone-based water repellents became very popular because they are very effective, even at low concentrations on the fibre, and they also impart a softening effect and give fabrics a silky handle. They are based upon the polydimethylsiloxane structure (shown in Section 8.3.3.2). The polymer chains are able to orientate on the fibre surface in such a way that the oxygen atoms form hydrogen bonds with polar groups in the fibre surface, whilst the hydrophobic methyl groups face outwards from the fibre conferring water-repellency (see Figure 8.9).

The structure of the final product formed on the fibres is rather more complex than the scheme in Figure 8.9 suggests, because in practice a formulation is used that contains a silane, a silanol and tin(II) octoate as catalyst.

$$HO-\underset{\underset{CH_3}{|}}{\overset{\overset{CH_3}{|}}{Si}}-\left[O\underset{\underset{CH_3}{|}}{\overset{\overset{CH_3}{|}}{Si}}\right]_n\overset{\overset{CH_3}{|}}{\underset{\underset{CH_3}{|}}{OSi}}-OH \qquad \text{Silanol}$$

$$H_3C-\underset{\underset{CH_3}{|}}{\overset{\overset{CH_3}{|}}{Si}}-\left[O\underset{\underset{CH_3}{|}}{\overset{\overset{CH_3}{|}}{Si}}\right]_n\left[O\underset{\underset{H}{|}}{\overset{\overset{CH_3}{|}}{Si}}\right]_m\overset{\overset{CH_3}{|}}{\underset{\underset{CH_3}{|}}{OSi}}-CH_3 \qquad \text{Silane}$$

$$\left[CH_3\,(CH_2)_6CO\right]^{-2}Sn^{++} \qquad \text{Tin octoate}$$

After the formulation is padded onto the fabric, the fabric is dried when a catalysed reaction between the silane and silanol occurs to form a three-dimensional cross-linked structure on the fabric surface. The methyl groups of the polymer are oriented outward from the fabric, presenting a hydrophobic exterior surface. During the polymerisation reaction, the silanol attacks the hydrogen atom linked directly to the Si atom in the silane to form a –Si–O–Si– crosslink. Rather than hydrogen being eliminated, as simple stoichiometry would suggest, however, it is likely that the –Si–H group is first hydrolysed by water in the formulation to –Si–OH and then a condensation reaction with the silanol takes place. Overall, the process can be represented by:

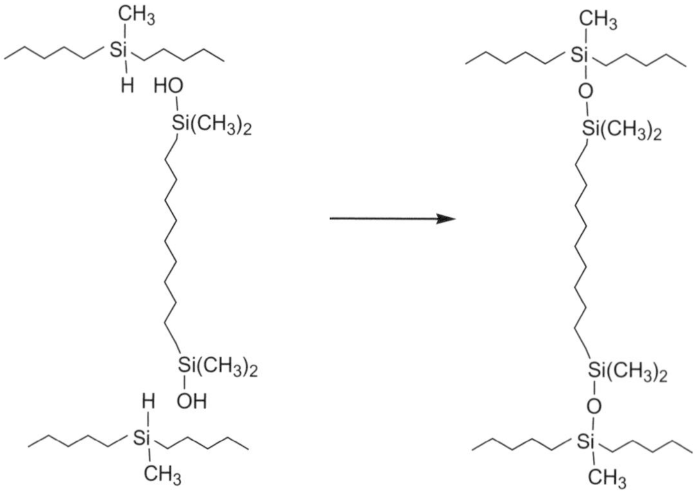

Although silicone finishes are highly effective as water-repellents, they do not repel oil or soils and in fact can be attractive to dirt.

Since the early 1990s a group of compounds called fluorocarbons have become the preferred water-repellents, mainly because they can be formulated to provide a wide range of fabric properties. All fluorocarbons contain perfluoroalkyl residues in which all the hydrogen atoms have been substituted by fluorine atoms. The basis of their water-repellent character is the formation on the fabric surface of an outer layer that is rich in highly hydrophobic –CF$_3$ groups. Fluorocarbons create surfaces of such low surface energy that in addition to providing water-repellency they are also oil and soil repellent (see Table 8.1, Section 8.2.2).

Fluorocarbon formulations contain perfluorinated acrylate polymer as the active component. Rather than using the pure fluoroacrylate homopolymer, however, they often contain long-chain fatty alcohol acrylates such as lauryl or stearyl acrylate as comonomers as well, which are found to enhance water repellency. Other comonomers used in the formulations include vinyl chloride, methyl methacrylate and acrylonitrile, if additional

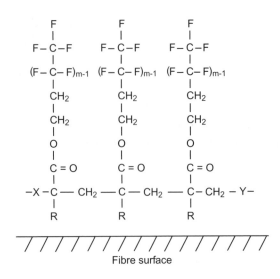

Figure 8.10   Perfluorinated acrylate polymer on a fibre surface. m = 8–10. X and Y are
                   stearylacrylate co-monomers.   R = H or CH$_3$ (polyacrylic or poly-
                   methacrylic esters).

effects such as soil repellency or solvent resistance are required. The
overall structure is depicted in Figure 8.10 It has been found that to
obtain the orientation of projecting side chains which offers optimum
performance, the perfluorinated side chain must contain at least seven
carbon atoms. Laundering can spoil this orientation, though it can be
reinstated by ironing. Some newer formulations avoid this requirement
and automatically reform in the required orientation on air drying (the
so-called 'laundry-air-dry' or LAD fluorocarbons).

Fluorocarbon finishes have very high resistance to washing and dry
cleaning and are very durable. They are expensive, however, so their
benefits have to be weighed against their performance attributes.
Nevertheless, they are applied very widely to textiles for use in applications from industrial textiles, to protective clothing and workwear,
home textiles, carpets, sports and leisurewear.

8.3.3.4   Soil-Release/Anti-Soiling Agents.   There are many types of
soils and they are often complex mixtures of substances such as mud,
clay, silica and oily substances. The effective removal of soils can be a
major challenge, especially for items such as industrial workwear, uniforms and active sportswear. Soil-release agents act to facilitate the
removal of soils during laundering, either at a domestic level, in industrial cleaning operations or in dry-cleaning processes.

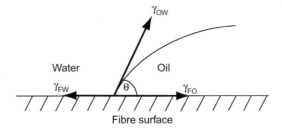

Figure 8.11 Oil, water and fabric interfaces of soiled fabric.

The removal of soil involves the action of a detergent at the fibre–soil interface and solubilising or emulsifying the soil by the micellar structure of the detergent. The process of soil removal also involves mechanical action to dislodge soil from fibres and from between fibres, so it is carried away by hydrodynamic fluid flow in the washing machine.

The effectiveness of the removal of fatty soils depends on how much work is required, which in turn depends on the interfacial tensions. In the case of an oily fabric in a detergent solution (see Figure 8.11), the forces acting are the interfacial tensions γ between:

- The fibre (F) and the water (W), γ_{FW}.
- The oil (O) and the water, γ_{OW}.
- The fibre and the oil, γ_{FO}.

The work required to remove the soil, W_{FS}, is given by Equation (8.14):

$$W_{FS} = \gamma_{FW} + \gamma_{OW} - \gamma_{FO} \qquad (8.14)$$

When a detergent is adsorbed at the liquid–fibre and liquid–soil interfaces, the surface tensions at these interfaces, γ_{FW} and γ_{OW} are reduced, and it follows from Equation (8.14), that W_{FS} is also reduced, so the soil is removed. Oily soils are removed by what is termed a 'roll-up' mechanism, where the contact angle, θ, gradually increases as the shape of the oil layer becomes more spherical, and then detaches from the fibre. The force responsible for this action, R, is the resultant of the interfacial tensions, given by Equation (8.15):

$$R = \gamma_{FO} - \gamma_{FW} + \gamma_{OW} \cos \theta_d \qquad (8.15)$$

where θ_d is the dynamic contact angle.

When the value of $(\gamma_{FO} - \gamma_{FW}) < \gamma_{OW}$ the 'roll-up' action occurs, until $R = 0$ and θ_d approaches $180°$ ($\cos\theta_d \rightarrow -1$). This analysis indicates that removal of oil from a surface will be facilitated if the value of γ_{LF} is low and the value of γ_{SF} is high, which is achieved by a hydrophilic/oleophobic finish on the fabric surface.

The chemicals used to provide soil-release properties are polymers which are amphiphilic in character, that is, they contain both hydrophilic and lipophilic groups. The ratio of these two components in the polymer structure significantly influences the effectiveness of the agent. There are four main chemical classes of polymers used.

Carboxyl-Based Polymers. The polymers are formed from ethyl acrylate and either acrylic acid or methacrylic acid, so typically have the structure:

$$-[CH_2-CH]_x-[CH_2-CH]_y-$$
$$\qquad\quad |\qquad\qquad\qquad |$$
$$\qquad\quad COOH\qquad\quad COOC_2H_5$$

The content of the acrylic acid in the polymer is important: if it is too low (below 15%) it has insufficient hydrophilic character, but if it is too high (above 25%) durability to washing is lost. The acrylic acid component dissociates to form the carboxylate anion under the alkaline conditions of domestic detergents, giving the fabric surface a negative charge which repels soil particles that have themselves become negatively charged by adsorption of detergent molecules. It is also thought that the polymer provides soil release by a swelling mechanism. These soil-release products were developed for application with easy-care finishes for cotton and cotton/polyester blends, and they are normally applied by padding with a cross-linking agent such as DMDHEU (see Section 2.2.5.5).

Hydroxy-Based Polymers. These are water-soluble polymers containing hydroxyl groups, typically starch, ethyl cellulose, carboxymethylcellulose and other similar products. They form a protective coating over the fabric but are washed off with any soil during laundering. Their effect is therefore only short lived.

Ethoxy-Based Polymers. These polymers contain oxyethylene groups and are formulated for synthetic fibres. Typical of this group is a product for application to polyester, formed from the copolymerisation of terephthalic acid with ethylene glycol and polyethylene glycol:

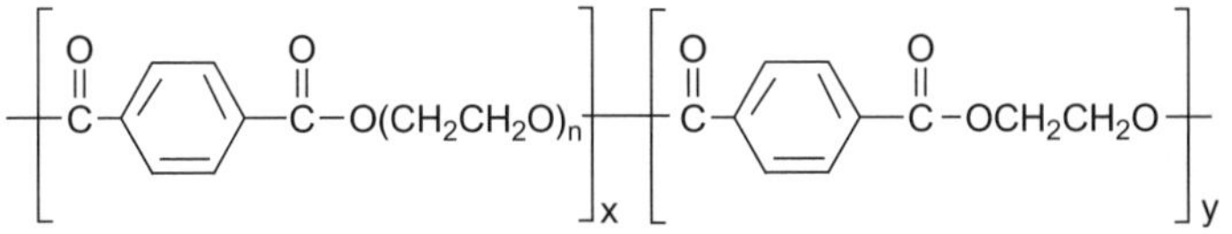

The resulting polymer contains the hydrophilic regions required for soil-release, but also hydrophobic regions which confer affinity for the polyester substrate, making them very durable.

Hybrid-Fluorocarbon Polymers. The use of fluorocarbons for imparting water-repellency has already been mentioned and it is reasonable to question how such agents could function in a soil-release capacity where hydrophilic character is required. The potential use of fluorocarbons lies in their ability to considerably lower the surface tension of a surface, and hence reduce the spreading of soil over a fabric surface. The required hydrophilic character can be incorporated by constructing a block copolymer structure, one block comprising hydrophilic polyoxyethylene segments and the other block comprising the lipophilic fluorocarbon. An example of such a polymer is:

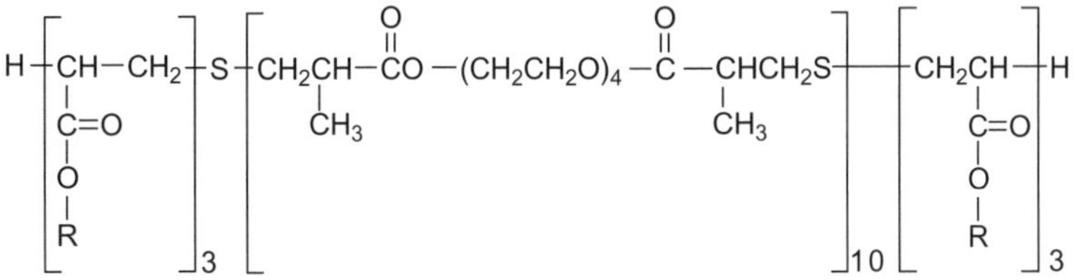

$$R = -CH_2CH_2(CF_2)_7CF_3$$

These types of polymer finishes are quite remarkable in that they are able to function in two ways. In the dry state, the polyoxyethylene segments are coiled and the fluorocarbon segments are exposed at the fabric surface, making it hydrophobic (water-repellent) and oleophobic (soil-resistant). In water, during a washing process, the polyoxyethylene segments swell so their hydrophilic character dominates the surface, which is the condition required for soil release. This is why these polymers are called 'hybrid' fluorocarbons. This is clever chemistry and these products are a good example of the versatility of the fluorocarbon group of compounds. They are not cheap, but they are effective and only small quantities are required on a fabric surface.

8.3.3.5 Antistatic Finishes. Reference was made in Section 1.6.3 to the build up of static electricity in textile fibres. Textile fibres are not conductors of electricity and indeed they can be considered to be insulators (see Figure 7.9). When surfaces of two different materials come into contact, electrons will flow between them and insulating materials, such as textile fibres, will retain the charge. Frictional charging, brought about by rubbing two surfaces together, increases the flow of electrons. When the surfaces are separated the transfer of charge that has taken place manifests itself as an electrostatic build-up. When the charge difference between the surfaces exceeds the electrical discharge potential of air, it is released by an electrical discharge, seen as a spark or heard by the crackling sound, for example when a wool sweater is pulled over a polyester T-shirt. The term given to the generation of charge by friction is 'triboelectrification' and it is possible to put the various fibre types into a series, the 'triboelectric series', where a fibre becomes positively charged if it is rubbed with another fibre type below it in the series. An example of the order of fibres is given in Table 8.6.

Whilst static electricity causes discomfort in wearing garments, it can have serious consequences in certain environments, especially for workers in areas where there are flammable gases or vapours, or in areas where there are sensitive electronic components. Water is a good conductor of electrical charge, so in general terms, fibres with high moisture regain values (see Appendix) have higher conductivity and are less prone to the build-up of static electricity. The moisture content of fibres depends very much on the relative humidity of the air, however, so in the summer, when humidity levels tend to be low, moisture content in fibres

Table 8.6 Triboelectric series of fibres.

Glass	
Human hair	positive
Nylon	
Wool	
Cotton	↑
Acetate	
Polyester	
Acrylic and modacrylic	
Polyurethane	
Polyvinyl chloride	↓
Polyethylene and polypropylene	
	negative
Polytetrafluorethylene	

is low and static charges build up more easily. This is especially a problem with fibre types that have only low moisture regain values anyway, which is most of the synthetic fibre types. Static electricity generated by friction causes considerable difficulty in textile manufacturing, especially directly after extrusion, when yarns are fed at high speed over guides and spindles to take-up cones. It is necessary to apply 'spin-finishes', which comprise lubricating oils and anti-static agents to aid processability, but these finishes are usually removed during fabric preparation for dyeing. It is therefore necessary to apply an antistatic agent to the final processed fabric.

The mode of action of an anti-static agent is to increase the conductivity of a fibre at its surface and simultaneously to reduce triboelectrification by reducing the friction coefficient. The agents used to reduce static electricity have hydrophilic character, and are mainly either polyglycols (or their derivatives) or ionic compounds. Typical of the former are:

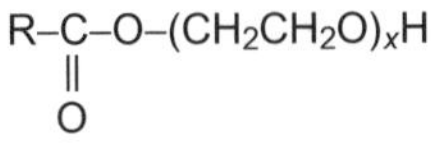

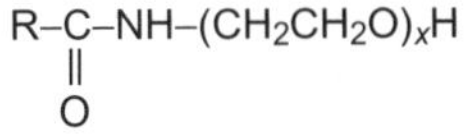

$$R-\underset{\underset{O}{\|}}{C}-O-(CH_2CH_2O)_xH$$

$$R-\underset{\underset{O}{\|}}{C}-NH-(CH_2CH_2O)_xH$$

Typical of the ionic antistatic agents are quaternary ammonium compounds, *e.g.*

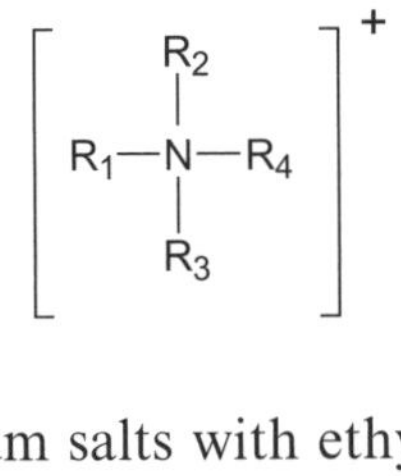

$$\left[\begin{array}{c} R_2 \\ | \\ R_1-N-R_4 \\ | \\ R_3 \end{array}\right]^{+} \quad X^{-}$$

and quaternary ammonium salts with ethylene oxide moieties, *e.g.*

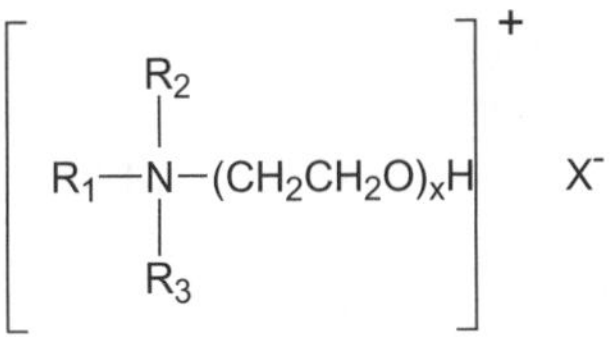

$$\left[\begin{array}{c} R_2 \\ | \\ R_1-N-(CH_2CH_2O)_xH \\ | \\ R_3 \end{array}\right]^{+} \quad X^{-}$$

Other important ionic types are phosphoric esters:

$$R-O-(CH_2CH_2O)_x-PO_3^{2-}-2M^+ \quad \text{where } M = NH_4 \text{ or } Na$$

It is very difficult to achieve an antistatic treatment that is permanent to repeated laundering. Some agents have been developed that are cross-linked hydrophilic polymers but these can interfere with other surface treatments such as water-repellency and soil-release so their use is not widespread.

8.3.3.6   Flame Retardant Agents.   All textile fibres commonly used for apparel and interior textiles will burn, some more readily than others, but nevertheless their flammability is crucially important in terms of safety, to the extent that for certain products flame-resistance to specified standards is a legal requirement. For firefighters, highly specialised, inherently flame-resistant fibres have been developed (such as those marked with * in Table 6.1) and these are described in Sections 6.2 and 6.4. For those domestic products which have to be treated with flame-retardant agents, however, there is a wide selection of agents available. The more important agents and their mode of action will be discussed in later sections.

The reaction of fibres to heat is an extremely complex process. In general, as their temperature increases fibres start to undergo pyrolysis, producing both non-combustible gases (such as carbon dioxide, water and oxides of nitrogen), and combustible gases (such as hydrogen and oxidisable organic molecules), together with carbonaceous chars. The effect of heat on the various fibre types has been mentioned in the preceding chapters. As the temperature rises further, the flammable gases reach their ignition point and combustion proceeds. During combustion the gases react with oxygen in a series of reactions involving the formation of free radicals, which are continually reformed. The heat of the combustion maintains the temperature for further combustion, so the reactions are self-sustaining. Oxygen is an essential requirement for combustion and a useful measure of the ease with which fibres will burn is their Limiting Oxygen Index (LOI). The LOI value is the minimum concentration of oxygen in the atmosphere required to sustain burning of the fibre, and the lower the value the more easily the fibre will burn. LOI values for typical fibres, and some high-performance fibres, are shown in Table 6.1. Given that air comprises about 20.8% oxygen, it is clear that the cellulosic fibres burn very readily. In general, fibres with LOI values of more than 26% are regarded as flame retardant.

For the treatment of those fibres which are flammable, flame-retardant agents have been developed to act in one of five ways:

1. Agents which act as a heat sink and remove heat from the burning material so the combustion is not maintained. This is achieved by using compounds which thermally decompose but in a strongly endothermic reaction. Examples of such agents are organic phosphorus compounds, aluminium hydroxide or 'alumina trihydrate'.
2. Agents which inhibit the formation of flammable gases and promote the formation of residual char. This is the most commonly used technique, especially since cellulosic fibres are so common and the agents typically contain compounds of phosphorus.
3. Flame dilution and decreasing the availability of oxygen to the flame. The agents used are usually halogen-containing compounds.
4. Increase the ignition temperature. Again halogen-containing compounds, usually in conjunction with antimony compounds are used.
5. The use of intumescents. These are substances that generate an expanded, foamed char, which provides an insulating layer on a fabric surface, which is resistant to radiant heat and flame. Further ignition is prevented or retarded and thermoplastic material is prevented from softening. Intumescent coatings comprise a charformer, catalyst and gas-former in a binder and are very effective.

Many of the agents used have been established for many years, with few new developments over the last 20 years or so. Many of the treatments are highly effective, but the reluctance to introduce new agents is partly due to environmental and legislative concerns (especially REACH legislation). The development of intumescent systems, which has taken place in recent years, is a notable exception.

The type of agent used for a particular application depends on the fibre type and the durability required, since not all flame-retardant treatments will withstand repeated laundering. However, for some textile applications, *e.g.* curtains, this is not a problem. The detailed chemistry of the modes of action of the treatments is beyond the scope of this book so only outline structures of agents are shown for selected fibres.

8.3.3.6.1 Cellulosics. Lewis acids are effective in promoting char formation and a mixture of borax ($Na_2B_4O_7 \cdot 10H_2O$) and boric acid (H_3BO_4) have been much used. However they begin to release acid at fairly low temperatures (around $130\,°C$) which can be normal ironing temperatures for cotton. Good alternatives are ammonium polyphosphates, used alone or in conjunction with urea. High loadings (of

up to 15%) are required on the fibre, however, which can significantly
affect the handle of the fabric.

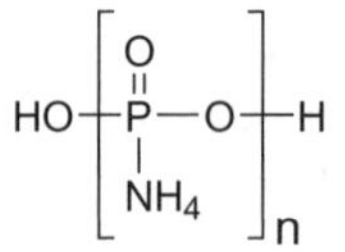

$$\left[HO - \overset{\overset{O}{\|}}{\underset{NH_4}{P}} - O \right]_n H$$

The most successful, durable agents are based on phosphorus- and
nitrogen-containing compounds such as methylolated phosphonamides
and tetrakis(hydroxymethylol) phosphonium chloride (THPC). The
former is applied by padding to the cellulosic fabric, together with a
methylolated melamine. After padding, the fabric is cured in an oven,
where condensation reactions take place in which the phosphonamide
becomes chemically bonded to the cellulose *via* the melamine, which acts
as a bridge. The final product has the following structure:

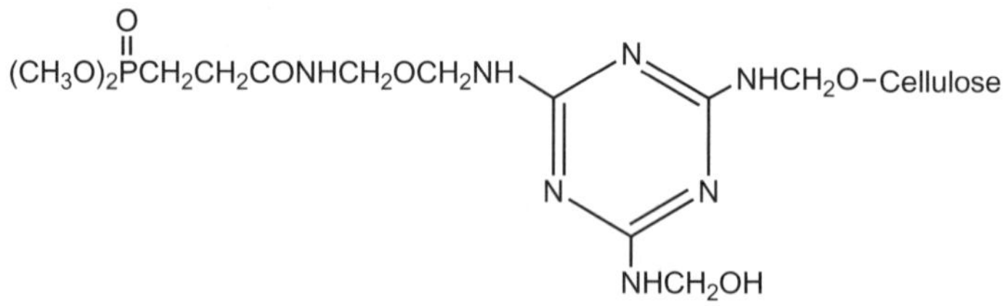

$$(CH_3O)_2\overset{\overset{O}{\|}}{P}CH_2CH_2CONHCH_2OCH_2NH \underset{}{\overbrace{}} NHCH_2O-\text{Cellulose}$$

NHCH₂OH

The THPC system is a little different in that it is applied with urea, again
by padding, but the curing takes place in an ammonia atmosphere,
when a cross-linked poly(phosphine) polymer is formed within the fibre
matrix. This polymer is stabilised by a final treatment of the fabric in
dilute hydrogen peroxide:

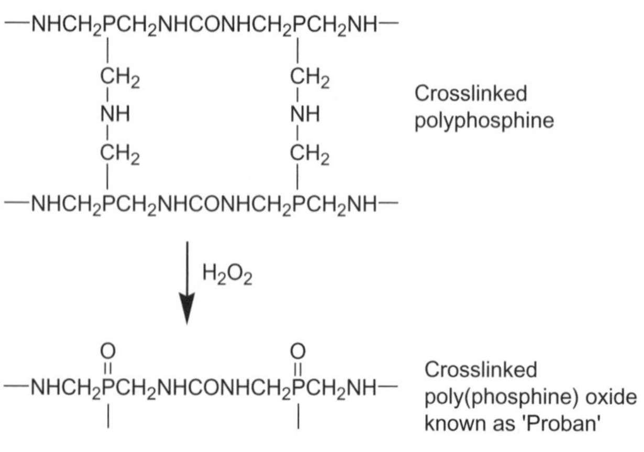

$$\begin{array}{l}
-NHCH_2PCH_2NHCONHCH_2PCH_2NH- \\
\qquad | \qquad\qquad\qquad\qquad\quad | \\
\qquad CH_2 \qquad\qquad\qquad\quad CH_2 \\
\qquad | \qquad\qquad\qquad\qquad\quad | \\
\qquad NH \qquad\qquad\qquad\quad NH \\
\qquad | \qquad\qquad\qquad\qquad\quad | \\
\qquad CH_2 \qquad\qquad\qquad\quad CH_2 \\
\qquad | \qquad\qquad\qquad\qquad\quad | \\
-NHCH_2PCH_2NHCONHCH_2PCH_2NH-
\end{array}$$

Crosslinked
polyphosphine

$\downarrow$ H₂O₂

$$-NHCH_2\overset{\overset{O}{\|}}{P}CH_2NHCONHCH_2\overset{\overset{O}{\|}}{P}CH_2NH-$$

Crosslinked
poly(phosphine) oxide
known as 'Proban'

8.3.3.6.2 Wool. Of all the 'conventional' fibres wool has the lowest inherent flammability and can self-extinguish small sparks. It has a natural tendency to char and the char acts as an effective thermal barrier which serves to inhibit further burning. For some applications such as curtains, upholstery fabrics or protective clothing, however, treatment with a flame retardant is necessary. In these types of applications wool is not laundered regularly and durability is not a main requirement. Often only fastness to dry cleaning is required.

Effective flame retardants for wool are those which promote char formation and so ammonium phosphates or polyphosphates, which function as Lewis acids, are often used. The most commonly used durable treatment is the Zirpo process developed by the International Wool Secretariat. In this process, potassium hexafluorozirconate (K_2ZrF_6), used either alone or mixed with potassium hexafluorotitanate (K_2TiF_6), is applied to the wool from an acid solution. The ions form a complex with the wool, $[wool - NH_3^+]_2[ZrF_6]^{2-}$. The zirconium ions are thought to promote char formation. The application can be carried out in a dye bath during a dyeing process and the treatment is compatible with other finishes such as shrink-resist and insect-resist finishes.

8.3.3.6.3 Synthetic Fibres. An option that is available for synthetic fibres is to incorporate a flame-retardant finish during manufacture, by adding it to a polymer system just prior to extrusion. In this way it can be assured that the agent is well distributed through the final fibre matrix. Agents for polyester and polyamide are usually cyclic phosphonates or organohalogen compounds, such as hexabromocyclododecane.

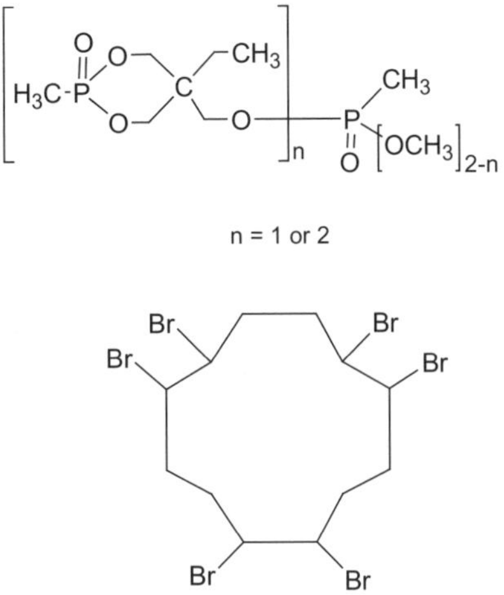

n = 1 or 2

There is not a suitable flame retardant available for acrylic fibres, but modacrylics made using halogen-containing comonomers have excellent inherent flame-resistant properties. A durable, effective treatment for polypropylene fibres has not been found. They can be treated with bromine- and phosphorus-containing agents but the large amounts necessary adversely affect their handle. Some intumescent flame retardants, such as ammonium polyphosphate and polytriazinyl-piperazine, have also been evaluated for polypropylene fibres. For polypropylene carpets, halogen-containing compounds with antimony trioxide are used in a back-coating. A similar system is used for nylon carpets.

SUGGESTED FURTHER READING

Wetting and Wicking

1. A. Patnaik, R. S. Rengasamy, V. K. Kothari and A. Ghosh, Wetting and wicking in fibrous materials, *Textile Prog.*, 2006, **38**(1), 1–105.
2. N. Pan and W. Zhong, Fluid transport phenomena in fibrous materials, *Textile Prog.*, 2006, **38**(2), 1–93.
3. B. Das, A. Das, V. K. Kothari, R. Fanguiero and M. de Araujo, Moisture transmission through textiles, Part I: Processes involved in moisture transmission and the factors at play, *Autex Res. J.*, 2007, **7**, 100–110, accessible at http://www.autexrj.org
4. B. Das, A. Das, V. K. Kothari, R. Fanguiero and M. de Araujo, Moisture transmission through textiles, Part II: Evaluation methods and mathematical modelling, *Autex Res. J.*, 2007, **7**, 194–216, accessible at http://www.autexrj.org

Surface Treatments

1. R. Shishoo, *Plasma Technologies for Textiles*, Woodhead Publishing, Cambridge, UK, 2007.
2. C. M Pastore and P. Kiekens, *Surface Characteristics of Fibres and Textiles*, Marcel Dekker, Inc., New York, 2001.
3. R. Morent, N. De Geyter and C. Leys, DBD treatment of textiles, *Int. Rev. Phys.*, 2007, **1**, 272–279.

Chemical Finishes

1. W. D. Schindler and P. J. Hauser, *Chemical Finishing of Textiles*, Woodhead Publishing, Cambridge, UK, 2004.

2. D. Heywood, *Textile Finishing*, The Society of Dyers and Colourists, Bradford, UK, 2003.
3. M. Lewin and S. B. Sello, *Handbook of Fiber Science and Technology, Vol II, Chemical Processing of Fibers and Fabrics, Functional Finishes, Part A and B*, Marcel Dekker, New York, 1983 and 1984.

CHAPTER 9

Fibre Blends in Textile Fabrics

9.1 INTRODUCTION

For some applications, the use of fabrics made of just one type of fibre is essential to meet specific performance requirements, but there is also considerable advantage to be gained from producing fabrics from mixtures of fibre types. It is not always the case that any one individual fibre can meet all of the attributes desirable of a fabric under all conditions in a certain end-use. By blending fibres of different types, a weakness or disadvantage in some aspect of one fibre can be compensated for by the presence of another fibre with strength or advantage in that aspect. In this way the fibres can complement each other in a blend so that optimum performance is obtained.

Some of the specific reasons for blending fibres are as follows:

- Enhancement of physical properties. The physical characteristics of fibres which are important in considering their suitability for a particular use include: tenacity, initial modulus, moisture regain (absorbency), extensibility and elastic recovery and specific gravity. It is quite common to blend natural fibres, with their relatively high absorbency, with synthetic fibres which have lower absorbency, but perhaps confer greater resilience or abrasion resistance qualities.
- Improvement of resilience and durability. The presence of a strong durable fibre in a blend may compensate for a fibre that might be weaker but have a softer handle.

The Chemistry of Textile Fibres
By Robert R Mather and Roger H Wardman
© Robert R Mather and Roger H Wardman 2011
Published by the Royal Society of Chemistry, www.rsc.org

- Enhancement of appearance. Different properties such as colour and lustre of fibre types can be combined to produce novel visual effects.
- Optimisation of price. A wider penetration of the market place can be obtained by blending expensive 'luxury' fibres with fibres that are cheaper, but do not detract too much from the inherent qualities of the more expensive component in the final blend. An example is to blend expensive cashmere with cheaper wool or silk. However cost considerations are not normally the driving force in blending; it is more the aim to maximise fabric performance.

9.2 MAKING BLENDED MATERIALS

There are different methods for producing blended fabrics. At the first level, a fabric blend can be produced either from two or more yarns, each of 100% of a different single fibre type, or from yarns which are themselves made by intimately mixing two or more different fibre types.

9.2.1 Blended Yarns

There are three ways of blending different fibre types in forming yarns.

9.2.1.1 Intimate Yarns. In this method the staple yarns of the different fibre types are mixed together, usually during the carding stage, so that when the yarn is spun an intimate mixture, in which the two fibre types are evenly distributed, is obtained. This is the most usual method for producing fabrics of fibre blends. There are many factors that have to be considered in producing a blend, however, since the operation is not necessarily a straightforward one. Of importance are the physical properties of the fibre types to be blended, such as their staple length, fibre diameter, initial modulus and extensibility. If these properties are substantially different, the fibres can be difficult to process by spinning machines and also the fabrics made from them may not benefit from the optimum properties of the individual component fibres. To this end, not only is careful selection of the individual components important, but also the blend ratios in which they are mixed. It is possible to compensate to a certain extent for a difference in properties, for example by using a higher tex (larger diameter) of a weak fibre in admixture with a lower tex (smaller diameter) of a stronger fibre.

The benefits to be gained from blending different fibre types, in terms of tenacity and extensibility, are complex and depend not only on the

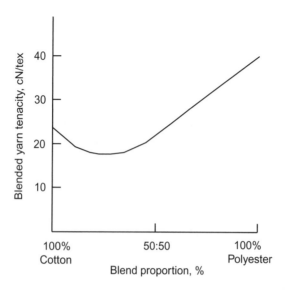

Figure 9.1 Variation of tenacity with blend proportion for cotton/polyester blends.

blend proportion, but on the actual fibre types being blended. It might be expected that the tenacity of a blended yarn would be equal to the sum of the tenacities of individual component fibres, weighted by their respective proportions in the blend. This is not the case, however, and it is possible for a blend of two fibre types to have a lower breaking strength than either of the two component fibres individually. This situation arises in blends of polyester and cotton, where the cotton, which has a lower extension at break and slightly lower initial modulus than polyester, takes most of the strain, and breaks first. The polyester has a higher breaking strength and is also more extensible. A blended yarn containing 75% of cotton and 25% polyester is weaker than that of either fibre type individually, and a blend containing more than 50% of polyester is required to produce a higher strength (see Figure 9.1). In the case of blending wool with polyester (or nylon) the situation is different, because wool fibres have a much lower initial modulus than polyester or nylon, so blending wool with small quantities of either fibre will give an improvement in strength.

Soon after the launch of polyester in the early 1950s, its value in blending with cotton to enhance wrinkle resistance was recognised. The presence of polyester was also found to improve the abrasion resistance. Polyester is more suited than nylon to blending with cotton because its initial modulus is closer to that of cotton than of nylon. A wide range of ratios of cotton and polyester are used, but to improve wrinkle resistance effectively, around 50% or greater of polyester in the blend is required. With the application of durable press finishes, however, lower quantities of polyester are possible and the contribution of the cotton in the blend, of providing comfort through its absorbency and handle, is not overshadowed. Thus 60/40 cotton/polyester blends are now very common for shirting fabric.

Other commonly used blends are those of polyester or nylon with wool, where the synthetic fibres again provide the strength and abrasion resistance. In this case the contribution of the wool is to provide warmth and soft handle character. The high durability of polyester and nylon contribute to the abrasion resistance in blends with wool, even in small quantities, unlike their effect on tenacity, where an excess of the synthetic fibres in the blend is required. Blending polyester with wool has been shown to improve the crease recovery of wool by more than that predicted from the blend ratio (see Table 9.1), indicating a synergistic effect.

Two good examples where wool is blended with nylon are: in carpets with 80/20 wool/nylon, where the nylon enhances the wear properties and the wool provides the warmth and the resilience to the carpet pile, and in socks, where the nylon adds strength, abrasion resistance and durability. The presence of the nylon detracts from the softness, however, so there is a trade-off in terms of comfort of wear. Another interesting development is in socks for trekkers, where not only is abrasion resistance and durability important, but also the effective removal of perspiration to give comfort. Whilst blends of wool/nylon have been traditionally used, base-layer 'liner' socks comprising a blend of nylon (for hard wearing and durability), polyester and polypropylene (to provide wicking properties) and elastane (to provide stretch) have recently come on to the market. These socks are claimed to reduce blisters and allergies, and improve hygiene. Wool/nylon blends are also widely used for trousers, where again the nylon improves the abrasion resistance and durability.

Table 9.1 Crease recovery values of wool, polyester and wool/polyester blends.

Fabric	Crease recovery, %, after dyeing at 105 °C
100% wool	55
55:45 polyester/wool (calculated from blend proportion)	69
55:45 polyester/wool (actual)	74
100% polyester	80

One of the problems of blending synthetic fibres with wool is that of 'pilling'. Pilling is the formation of small fluffy balls of fibres (pills) on the surface of a knitted or woven garment, these pills being composed mainly of short fibres. Pills are formed by protruding fibres on the fabric surface during rubbing or abrasion as the garment is worn. The result is a visually unattractive garment and also a garment that has a poorer handle. With 100% wool garments, many of the pills fall off, but they tend to be held more firmly if a stronger fibre, such as nylon or polyester is intimately mixed in with them. Pilling is an important issue for manufacturers and possibly making it worse by blending wool with a strong synthetic fibre is of much concern to them.

When microfibre variants of synthetic fibres, notably polyester and nylon, came on to the market in the 1980s, there was immediate interest to use them to make blends with other fibres. Microfibres are those whose filaments are less than 1 dtex, though some manufacturers produce microfibres of up to 1.3 dtex. More recently supermicrofibres of <0.3 dtex have been introduced. These fibres are very fine indeed; the supermicrofibres are ten times finer than the standard polyester grades and are characterised by their soft handle and silky appearance. The intimate blending of micro- and supermicrofibres with natural fibres does present processing problems, however, especially in spinning and dyeing yarns, and such blends are not common. Polyester microfibre has been successfully blended with wool to produce fine, lightweight fabrics of soft handle and excellent draping qualities, but blends with cotton are much more difficult to produce due to separation of the fibres during spinning. Polyester microfibre is successfully blended with viscose in a 50/50 ratio for the production of casual shirts. Nylon microfibre is successfully blended with cotton and with wool, often with an elastane fibre as well, to produce stretchable fabrics for a wide range of garment types. One of the problems with producing microfibre-blended fabrics is dyeing them. About twice as much dye is required to produce the same visual depth of shade on polyester microfibres as on standard polyester. Consequently, producing solid shades on these fabrics is both more difficult and expensive. Nevertheless, with careful selection of dyes and dye application conditions these difficulties can be overcome.

9.2.1.2 Non-Intimate (Core Spun) Yarns. In yarns of this type the fibres of one fibre type form the core of the yarn, with the fibres of the second type spun around it, giving a sheath. These yarns are claimed to be softer and more flexible than intimately blended yarns. The properties of the sheath fibre, being on the outside of the yarn, however, have a considerable influence on the surface characteristics of the blended yarn and the fabric made from it, such as abrasion resistance.

9.2.1.3 Plied Yarns. These are yarns formed by wrapping (usually) two yarns of different fibre types around each other. This is not a common method of blending fibres, though the technique is common where the two yarns are of the same type. The purpose here is not to blend but to produce fuller yarns that give better handle and stretch to fabrics made of them.

9.2.2 Union Fabrics

An example of a 'union fabric' is one which is woven using a warp yarn of one fibre type (such as cotton), and a weft yarn of a different fibre type (such as polyester). Union fabrics provide a useful structure for obtaining novel colour effects such as:

- Reserve effect, where the warp is dyed but the weft remains undyed, or *vice versa*;
- Shadow effect, where the warp is dyed with the same hue but to a deeper depth of shade than the weft, or *vice versa*; or
- Contrast effect, where the warp is dyed to a colour of one hue and the weft to a colour of a different hue, usually both to the same depth of shade.

To achieve these selective coloration effects, it is necessary that the fibres for the warp and weft, and/or the application class of dyes selected to dye them, are chosen carefully to achieve the desired effect. For example, if extensive cross-staining of the weft with the dye intended for the warp occurs, then the impact on the reserve and the contrast effects can be considerably diminished.

Most often, fabrics are produced from yarns which themselves are made by blending two or more different fibre types together. There are various ways in which blended yarns can be produced, all of which apply to yarns made from staple fibres. It is not always the case that man-made fibres are blended with natural fibres; there are many instances of blended yarns of two or more different types of only natural fibres, or two or more different fibre types of only man-made fibres. It is not common practice for continuous filament yarns to be blended, an exception to this being the formation of core spun yarns of an elastomeric fibre, which is used as a continuous filament, with another yarn spun round it (usually cotton).

9.2.3   Double Face Knitted Fabrics

Clothing intended for a specialised application, such as sportswear, must be capable of providing thermophysiological comfort during periods of high activity when large amounts of heat and perspiration are produced by the body. If an absorbent fabric is worn next to the skin then perspiration accumulates within the garment during exercise, and the body feels uncomfortable. Clothing for sportswear must allow the perspiration to pass through, and if this happens effectively the body temperature will also reduce. To achieve this in functional sportswear it is usual for combinations of different fibre types to be used. In some cases intimately blended yarns are appropriate, but an alternative often used is to construct garments using knitted double face fabrics. The mode of action of the bilayer structure in sportswear is totally different from that of intimately blended fibres.

The type of structure used for sportswear is a simple knitted double face construction, in which the inner face (back side of the fabric) is made from a hydrophobic yarn and the outer face is made from a hydrophilic yarn. The inner hydrophobic yarn is usually a synthetic fibre such as polypropylene (which has a zero moisture absorbency – see Table 1.5) that has an effective capillary action to wick moisture away from the skin. The outer fabric is likely to be a cellulosic fibre (though nylon is also sometimes used) which absorbs the wicked moisture and allows it to evaporate. The effectiveness of these double face constructions for comfortable sportswear depends on the knitted structure as well as the fibres from which the two faces are made. Furthermore, it is not simply the chemical type of the fibres that influences hydrophobic/hydrophilic character, but also the fineness of the fibre filaments and their cross sectional shapes. Thus microfibres are found to be highly effective in promoting wicking of moisture. Fibres with irregular cross sectional shapes, rather than round shapes, are also found to be better. In addition, it is necessary to balance the two layers in terms of their respective functions. Wicking can occur more quickly than evaporation, especially if the external climate is cold and humid, in which case perspiration can transfer back to the skin, creating discomfort.

Microfibre bilayered structures for socks have been found to be very effective for wearer comfort. Inner layers of micro polypropylene or micro polyester fibre, with an outer layer of micro modal (a cellulosic fibre) have been found to have higher moisture vapour transport rates than socks made with a traditional single layer construction. The capillary action takes place through the very small spaces between the fibres – and the smaller the spaces, the greater the wicking. Microfibres pack very tightly to form narrow capillaries, so fabrics made from them transmit perspiration more effectively than those made from conventional fibres.

Another interesting development for the sportswear market is that of Sportswool Pro™, in which the inner face is a superfine merino wool and the outer face is polyester. The Sportswool Pro™ fabric is treated with a hydrophilic agent that increases the surface energy of the merino slightly, but increases it greatly on the polyester. The huge difference in surface energy and in the diameters of the two fibre types causes strong wicking from the inside to the outside of the bilayer fabric. Thus perspiration picked up from the skin by the merino is rapidly transmitted by the strongly positive wicking gradient to the outer layer, where the coarser polyester fibres, with their high surface area, promote rapid evaporation to the atmosphere.

9.3 USES OF BLENDED FIBRES

It is very difficult to obtain reliable information on the extent to which fabrics of blended yarns are produced. By far the most common fibre blends are polyester/cotton and polyester/viscose. Polyester/wool is also an important blend. The use of blends is extensive and covers a very wide range of products where textiles are used. In apparel, blends are used in all types, from intimate wear to garments for rugged outdoor use and sportswear. They are used in home furnishings, automobiles and public buildings. Probably the only uses for which blends do not have a role are those for which specialised performance characteristics are required, such as flameproof garments or fabrics that require resistance to attack from aggressive agencies. Protective clothing (*e.g.* chemical, radioactive, nuclear) or materials frequently in contact with sea water or sunlight, require construction using fibres specifically designed for that purpose.

SUGGESTED FURTHER READING

1. R. W. Moncrieff, *Man-Made Fibres*, John Wiley & Sons, London, 1969.
2. J. Shore, *Blends Dyeing*, Society of Dyers and Colourists, Bradford, 1998.
3. R. Alagirusamey and A. Das, in *Polyesters and Polyamides*, ed. B. L. Deopura, R. Alagirusamey and B. Gupta, Woodhead Publishing, Cambridge, 2008, ch. 8.

Appendices

In these two appendices, some of the terms that have been used in the text are explained. A useful and authoritative publication which readers may wish to consult for a full range of textile terms is: *Textile Terms and Definitions*, ed. J. E. McIntyre and P. N. Daniels, The Textile Institute, Manchester, UK, 10th edn, 1995.

APPENDIX I

Units of Measurement used to Characterise Textile Fibres

Note: Most textile fibres are able to absorb water, the extent to which varies with relative humidity and temperature of their environment. The amount of water absorbed considerably influences their mechanical properties. Therefore it is usual practice to measure the mechanical properties in specially conditioned rooms under the standard conditions of 20 °C and 65% relative humidity.

        The mechanical properties of fibres, particularly tenacity, are usually quite different when they are wet, so it is common for tenacity values to be quoted when dry (*i.e.* measured under the standard conditions) and when wet (measured when saturated with water).

Linear Density (Fineness of Fibres)

The unit used to measure the linear density of fibres is tex, which is the mass in grams of one kilometre of the fibre. Since most fibres are very

The Chemistry of Textile Fibres
By Robert R Mather and Roger H Wardman
© Robert R Mather and Roger H Wardman 2011
Published by the Royal Society of Chemistry, www.rsc.org

fine, decitex (dtex) is often used instead, which is the mass in grams of ten kilometres (10 000 metres) of the fibre.

For example, a fibre for which 10 000 metres has a mass of 5 grams is said to have a linear density of 5 dtex. If 10 000 metres of a second fibre type has a mass of just 2.5 grams (2.5 tex), then it is two times finer than the first.

Some companies use the word *titer* instead of linear density.

A unit that has been used extensively in the past (and unfortunately still is by some authors) is *denier*, which is the mass in grams of 9000 metres of fibre. However, interconversion between dtex and denier is relatively easy: they differ by a factor of approximately 1:1, so a fibre of fineness 5 denier will be approximately 5.5 dtex.

Typically, commercial fibres have a fineness of 1.5–5 dtex. Fibres that are classed as microfibres are <1 dtex.

Tensile strength

This is the tensile force at the point when a fibre ruptures, and is measured in Newtons (N).

Tenacity

Thicker, coarser fibres will be stronger than thinner, finer ones, so in order to compare the strengths of fibres more accurately, fineness needs to be taken into account. Tenacity is the measure by which the strengths of fibres can be compared. It is the tensile force per unit linear density and corresponds with the maximum force (in Newtons) a fibre can withstand before rupture or plastic flow. The unit usually used for fibres is cN tex^{-1}.

Fibres with tenacities of <15 cN tex^{-1} can be regarded as fairly weak. For fibres to form useful textile materials, say for apparel or home textile useage, tenacities in the range 15–40 cN tex^{-1} are required. Higher strength fibres (such as some polyesters and nylons used for industrial purposes) have tenacities in the range 40–80 cN tex^{-1}, and the ultra high tenacity fibres developed in recent years have tenacities of well over 100 cN tex^{-1}, with some even in excess of 200 cN tex^{-1}. What are described as 'superfibres' (such as Zylon) have tenacities of over 400 cN tex^{-1}.

A unit formerly used for tenacity was grams denier^{-1} (g d^{-1}), and as with linear density, it is still used by some authors. The conversion to SI units is straightforward, however: the value in g d^{-1} is multiplied by 8.85, so for example a tenacity of 5 g d^{-1} is equivalent to 44.2 cN tex^{-1}.

Initial modulus

This is the ratio of stress to corresponding strain in the initial region of a stress–strain graph, where the extension follows Hooke's Law. The unit of *initial modulus* (sometimes simply referred to as 'modulus') is the same as that of tenacity, cN tex^{-1}. The higher the value, generally the stiffer and more brittle the fibre is.

Extension at Break

This is the extent to which a fibre is capable of being stretched. It is expressed as a percentage and is calculated according to the formula:

$$\frac{L_B - L_O}{L_O} \times 100$$

where L_B = length of fibre at breaking point; L_O = original length of fibre.

Elasticity

When fibres are stretched, it is desirable that they will return to their original length when the stress is removed. The elastic recovery, expressed as a percentage, is given by the formula:

$$\frac{L_S - L_R}{L_S} \times 100$$

where L_S = increase in length under applied stress; L_R = remaining increase in length after stress is removed.

Moisture Regain

This is the ratio of the mass of moisture in a material measured under standard conditions, to the oven-dry mass, usually expressed as a percentage. It is calculated using the formula:

$$\frac{M_{Cond} - M_{dry}}{M_{dry}} \times 100$$

where M_{cond} = mass of material under standard conditions (65% R.H. and 20°C); M_{dry} = mass of oven – dry material.

APPENDIX II

Some Terms and Definitions used in Textile Manufacture

Auxiliary

An auxiliary is a chemical (or a manufacturer's formulated product) which is added to a process to improve its effectiveness. Auxiliaries are most often used in preparatory processes, and in dyeing, printing and finishing processes. Typically they are acids, alkalis, salts, wetting agents, softeners, dispersing agents, *etc.* The types of auxiliaries used depend on the particular process.

Conditioning

In conditioning, a textile material is allowed to reach equilibrium with the surrounding environment after manufacture or undergoing a wet process, such as dyeing or finishing. During conditioning, the textile absorbs moisture from the atmosphere to an extent which is a characteristic of the textile material. Textile properties such as strength and elasticity depend on the amount of moisture present in the material, so in order to measure such properties, the textile is always allowed to condition in a standard atmosphere of 20 °C and 65% relative humidity.

Curing

Some textile processes which involve the application of a chemical to a material require the chemical to undergo a change in order to be effective. Curing is the method by which the chemical reaction is brought about, and may involve heat treatment or exposure to UV radiation.

Delustrant

When man-made fibres are extruded, either by melt, wet or dry spinning, they are translucent. In order to make them opaque and appear white with a low lustre, a delustrant (usually the anastase form of titanium dioxide) is added to the spinning dope or molten polymer just prior to extrusion.

Dyeing

Dyeing is the application and fixation of a dye in a textile fibre to produce a solid colour effect. During the dyeing process dye molecules, which are soluble, or made soluble, in the medium from which they are applied (most often water), diffuse monomolecularly through the polymer matrix to produce a uniform distribution. The dyes selected for a particular fibre type are those with high substantivity for the fibre, that is, they will readily transfer from the dye bath to the fibre during dyeing and show good resistance to washing out afterwards. The dyes used have different 'application classes', common ones being:

- Acid dyes. The dyes are sodium salts of complex anions which are responsible for the colour. They are coloured soluble in water. The coloured anion is taken up at positive sites in the fibre. Typically they are applied to wool, silk and polyamide fibres.
- Direct dyes. Like acid dyes, direct dyes are the sodium salts of complex anions, but are designed to have affinity for cellulosic fibres such as cotton, viscose and Lyocell fibres.
- Basic dyes. The coloured part of the molecule of basic dyes is cationic, so the dyes are taken up at negative sites in the fibre. Typically they are used to dye acrylic fibres.
- Vat dyes. These dyes are non-ionic and insoluble in water. They can be applied as dyes, however, by firstly reducing them with alkaline sodium hydrosulphite, giving the 'leuco' form in which the coloured ion is anionic. In this form they behave similarly to direct dyes, and so can be applied to cellulosic fibres as well. After dyeing, an oxidising agent is added, when the dyes revert to their insoluble 'parent' form. Indigo, for the colouration of denim, is an example of a vat dye.
- Disperse dyes. These dyes are non-ionic and are largely insoluble in water. They possess polar groups which confer some affinity for water, and during the dyeing process they dissolve and diffuse into the polymer matrix. There is therefore no requirement for the complex chemical operation required in the application of vat dyes. Disperse dyes are applied from a dispersion stabilised with dispersing agents. They are typically used for fibres which do not contain charged groups, such as polyester.
- Reactive dyes. These dyes are similar in structure to direct dyes in that the coloured part of the molecule is anionic, but they contain one or more chemical groups that are capable of reacting with a fibre to form a stable covalent bond between the dye and the fibre. During the dyeing process, the dye molecules therefore become chemically bound to the fibre and the final dyed product has very high fastness to washing processes. Reactive dyes have been developed mainly for application to cellulosic and protein fibres.

Finishing

Finishing is a process applied to a textile fabric to confer the final performance properties required of it. Finishing operations are broadly classed as 'wet' or 'dry'. The 'wet' finishing processes generally involve the application of a chemical, such as a crease-resist or water-repellent agent. In 'dry' finishing, the process usually involves subjecting fabric to a mechanical process that alters, for example, its handle or lustre.

Fluidity

Fluidity is the ease with which a liquid flows, and is the reciprocal of viscosity, with the unit of $Pa^{-1} s^{-1}$. It is a measure often used in textile evaluation because fluidity is inversely proportional to the molar mass of a polymer. If a fibre has been subjected to damage, especially chemical damage, the molar mass of the polymer will be reduced, and the extent of degradation can be assessed by measurement of the fluidity of the fibre dissolved in a suitable solvent.

Gel Dyeing

In addition to applying dyes from a dye bath, it is also possible to apply them to man-made fibres directly after wet spinning, when they are still in their wet and swollen state. In this state they are highly receptive to dye molecules and colouration at this stage avoids the need for a separate dyeing process.

Geotextile

A geotextile is a permeable textile material used in ground engineering, for functions such as reinforcement, stabilisation and drainage.

Handle

Handle is a term used to describe how a textile material feels to the touch. It involves the sensations of roughness, harshness and thickness, *etc.*

Mass Colouration

As with gel dyeing, mass colouration avoids the need for a separate dyeing process. In mass colouration, dye or pigment is added to the spinning dope (or molten polymer) prior to extrusion. Since the dye or pigment becomes thoroughly mixed with the polymer, perfectly uniformly coloured fibres are obtained, and especially if pigments are used, with excellent fastness properties.

Non-Woven Fabric

Whilst most textile fibres are spun into yarns and then produced as woven or knitted fabrics, another form in which they are used is in the non-woven form. In this form, the structures are made of fibres instead of yarns and are formed by laying the fibres on each other as webs. The fibres are held in the structures to give stability by methods such as thermal bonding, adhesive bonding, or stitch bonding. The use of textile materials in non-woven form is extensive, with applications in areas such as carpets, noise reduction matting, filtration, geotextiles and medical textiles.

Padding

Padding is a process for the impregnation of a fabric with a chemical. It involves feeding the fabric from a roll through a trough containing a solution of the chemical and out through a pair of 'nip' rollers which squeeze out any excess liquid. The pressure on the two nip rollers can be adjusted so that a specific quantity of the solution is left on the fabric. After padding the fabric is usually fed into a unit to fix the chemical, such as a steamer, dry-heat or UV curing machine. Padding is used for continuous dyeing operations or for the application of chemical finishes to fabrics.

Partially Oriented Yarn (POY)

When filaments of synthetic fibres are extruded, their molecular chains can be made to possess a certain amount of orientation. In this state, the 'partially orientated yarn', the full strength has not been developed and so the yarns are subjected to the process of drawing, which stretches them and induces further orientation. The result is then 'highly oriented yarn' (HOY). Some fibre filaments when first extruded have very little orientation and are referred to as 'low oriented yarn' (LOY). Again, by drawing, HOY yarns can be produced.

Pigment

A pigment is a coloured substance which is insoluble in the medium in which it is applied. Pigments are insoluble in water for example. They generally possess strong, bright colours and have excellent fastness to light.

Preparatory Process

Natural fibres contain impurities which serve to act as a barrier to the uptake of dye during a dyeing process and influence the colour achieved. In preparation for dyeing, it is therefore necessary to remove such impurities and the processes to achieve this are called 'preparatory processes'. They typically involve the processes of scouring to remove oils, fats and waxes, and bleaching to improve the base colour of the textile.

Printing

Printing is a process by which a coloured pattern is produced on a fabric. The process involves the application of a dye or pigment as a paste onto the surface of the fabric, followed by a fixation process, usually steaming. If a dye is used, it diffuses into the fabric during the steaming process, but if a pigment is used, it remains on the fabric surface and is held there by a chemical also added to the pigment print paste, called a 'binder'. The 'binder' forms a clear polymeric film over the pigment particles, adhering them strongly to the fibre surface.

REACH

REACH is a new European Union regulation, the acronym standing for the Registration, Evaluation, Authorisation and restriction of Chemicals, which came into force in June 2007. Its aim is to provide protection of human health and the environment from the use of chemicals. A major part of REACH is the requirement for manufacturers or importers of substances to register them with a central European Chemicals Agency (ECHA). A registration package will be supported by a standard set of data on that substance. If a manufacturer or importer does not register the chemicals they deal with, data on them will not be available and as a result, they will not be able to manufacture or supply them legally.

In order to place on the market or use substances with properties that are deemed to be of 'very high concern', industry must apply for an authorisation. A company wishing to market or use such a substance must submit an application to the ECHA for authorisation. Decisions on authorisation are made by the European Commission, taking advice from the ECHA and member states. Applicants have to demonstrate that risks associated with uses of these substances are adequately controlled or that the socio-economic benefits of their use outweigh the risks. Applicants must also analyse whether there are safer suitable alternatives or technologies. If there are, then they must prepare substitution plans, and if not, then they should provide information on research and development activities if appropriate. This regulation therefore has implications for the textile industry, not just in the manufacture of fibres, but in their processing, such as scouring, bleaching, dyeing and finishing operations.

Size

A size is a substance applied to the warp yarns prior to weaving to protect them from abrasion through friction when in contact with the moving parts of the loom, such as the healds and reeds. The size can be a complex formulation, though most are based on carbohydrates such as starch, or on gelatine. Polyvinyl alcohol is also used. The size formulation may also contain oils and fats which act as lubricants. After weaving the size has to be removed in a 'de-sizing' process.

Suture

A suture is a surgical sewing thread used to close a wound. It is made of either natural or synthetic polymers and can be designed to degrade biologically at particular rates.

Web

A web is a sheet of fibres produced by a carding machine. The fibres usually lie in a parallel orientation. The term is also used in non-woven fabric production, though in this case the fibres do not necessarily lie parallel.

Subject Index